CAMBRIDGE LIBRARY COLLECTION

Books of enduring scholarly value

Travel, Middle East and Asia Minor

This collection of travel narratives, primarily from the nineteenth century, describing the topography, antiquities and inhabitants of the Middle East, from Turkey, Kurdistan and Persia to Mesopotamia, Syria, Jerusalem, Sinai, Egypt and Arabia. While some travellers came to study Christian sites and manuscripts, others were fascinated by Islamic culture and still others by the remains of ancient civilizations. Among the authors are several daring female explorers.

A Voyage into the Levant

Joseph Pitton de Tournefort (1656–1708) was originally destined for the church, but his interest in botany led him to become professor of botany at the Jardin des plantes in Paris, and to travel all over Europe and beyond in search of interesting specimens. He was chiefly interested in the classification of plants, but is now best remembered for the accounts he wrote of voyages undertaken for the purpose of scientific discovery. This illustrated two-volume work, published posthumously in French in 1717 and translated into English the following year, recounts a journey begun in 1700, around the eastern Mediterranean and the Black Sea, visiting Crete and other Greek islands, Istanbul, Armenia and Georgia. Tournefort notes not only plants, but geographical features, antiquities, the people he encounters, and their way of life, agriculture and industry. Volume 2 continues the journey around the Black Sea and describes the voyage home.

Cambridge University Press has long been a pioneer in the reissuing of out-of-print titles from its own backlist, producing digital reprints of books that are still sought after by scholars and students but could not be reprinted economically using traditional technology. The Cambridge Library Collection extends this activity to a wider range of books which are still of importance to researchers and professionals, either for the source material they contain, or as landmarks in the history of their academic discipline.

Drawing from the world-renowned collections in the Cambridge University Library and other partner libraries, and guided by the advice of experts in each subject area, Cambridge University Press is using state-of-the-art scanning machines in its own Printing House to capture the content of each book selected for inclusion. The files are processed to give a consistently clear, crisp image, and the books finished to the high quality standard for which the Press is recognised around the world. The latest print-on-demand technology ensures that the books will remain available indefinitely, and that orders for single or multiple copies can quickly be supplied.

The Cambridge Library Collection brings back to life books of enduring scholarly value (including out-of-copyright works originally issued by other publishers) across a wide range of disciplines in the humanities and social sciences and in science and technology.

A Voyage into the Levant

Perform'd by Command of the Late French King

VOLUME 2

JOSEPH PITTON DE TOURNEFORT
TRANSLATED BY JOHN OZELL

CAMBRIDGE
UNIVERSITY PRESS

CAMBRIDGE
UNIVERSITY PRESS

University Printing House, Cambridge, CB2 8BS, United Kingdom

Cambridge University Press is part of the University of Cambridge.

It furthers the University's mission by disseminating knowledge in the pursuit of
education, learning and research at the highest international levels of excellence.

www.cambridge.org
Information on this title: www.cambridge.org/9781108075237

© in this compilation Cambridge University Press 2014

This edition first published 1718
This digitally printed version 2014

ISBN 978-1-108-07523-7 Paperback

This book reproduces the text of the original edition. The content and language reflect
the beliefs, practices and terminology of their time, and have not been updated.

Cambridge University Press wishes to make clear that the book, unless originally published
by Cambridge, is not being republished by, in association or collaboration with,
or with the endorsement or approval of, the original publisher or its successors in title.

A

VOYAGE

INTO THE

LEVANT:

Perform'd by Command of the Late *French* King.

CONTAINING

The Antient and Modern STATE of the Iflands of the *Archipelago* ; as alfo of *Conftantinople*, the Coafts of the *Black* Sea, *Armenia*, *Georgia*, the Frontiers of *Perfia*, and *Afia* Minor.

WITH

PLANS of the principal Towns and Places of Note ; an Account of the Genius, Manners, Trade, and Religion of the refpective People inhabiting thofe Parts : And an Explanation of Variety of Medals and Antique Monuments.

Illuftrated with Full Defcriptions and Curious Copper-Plates of great Numbers of Uncommon Plants, Animals, *&c.* And feveral Obfervations in Natural Hiftory.

By M. *TOURNEFORT*, of the Royal Academy of Sciences, Chief Botanift to the late *French* King, *&c.*

To which is Prefix'd,

The Author's LIFE, in a Letter to M. *Begon :* As alfo his Elogium, pronounc'd by M. *Fontenelle*, before a publick Affembly of the Academy of Sciences.

Adorn'd with an Accurate MAP of the Author's Travels, not in the *French* Edition : Done by Mr. *Senex.*

VOL. II.

LONDON,

Printed for D. BROWNE, A. BELL, J. DARBY, A. BETTESWORTH, J. PEMBERTON, C. RIVINGTON, J. HOOKE, R. CRUTTENDEN and T. COX, J. BATTLEY, E. SYMON. M. DCC. XVIII.

The Contents of the Letters of the Second Volume.

LETTER I.

A 2   LET.

The Contents.

ERRATA *in* VOL. II.

PAg. 31. l. 6. read *Chiefs, whom they present with a Sum of Money for such Consent.* P. 37. l. 30. for 818819 r. 99999. P. 38. l. 17. for *a Croud* r. *Plattoons.* P. 39. l. 13, &c. r. *The Advantage he gain'd at Sea near the Islands of Spalmadori over the* Venetians, *won him the Island of* Scio, &c. P. 40. l. 23, &c. read thus: *consists of* 200 *Rowers, and Tallow for Careening. If the Captains are rich enough to substitute their own Slaves in the room of those Rowers, they make a considerable,* &c. l. 26. r. thus: *advantage also of their Slaves Day-labour, forcing 'em to work on shore as much as possible during,* &c. l. 28. for *press* r. *hire.* P. 43. l. antepen. r. *except as to.* P. 44. l. 3. r. *to bring them nearer to Reason.* P. 45. l. 3. r. *Hegira.* P. 53. l. 4. read thus: *and for those* whom they look upon as *Saints.* lin. 5. after *Dead,* insert *for the Sick.* lin. 30. after *victorious,* insert, *who turnest the Hearts and Thoughts of Men.* P. 55. l. 14. r. *Zoulcudé.* P. 55. l. 22. r. *the Day of the first Fast.* P. 57. l. 26. for *with* r. *to.* P. 59. l. ult. for *support* r. *cover.* P. 66. l. 20. for *temperate* r. *well heated.* P. 69. l. 8. read thus: *empty, at least not overcram'd with Apparel and Jewels.* P. 71. l. 3. r. *pink'd Waistcoat.* l. 17. r. *Isis.* P. 72. l. 20. r. *Greek Papas.* l. 22. for *Spout* r. *Common-Sewer.* l. 25. l. *The She Jew Slaves.* P. 76. l. 17. r. *kiss you, holding your Beard.* P. 78. l. antepen. for *curdled* r. *raw.* P. 84. l. 2. for *upon* r. *under.* P. 85. l. 4. for *red* r. *green.*

A
VOYAGE
INTO THE
LEVANT:

By the KING's Exprefs Command.

LETTER I.

To Monfeigneur the Count de Pontchartrain, Secretary of State, &c.

MY LORD,

IF you had not taken a Refolution to make thefe Papers, I fend you, publick, I fhould not prefume to entertain you with a world of things, which you know much better than my felf: But as you have obliged me to communicate to the Public an Account of the State of the *Levant*, I believe you will not be difpleas'd that I infert in the Letters I have the Honour to write to you, feveral things which are not generally known, or which have received

Of the Government and Polity of the Turks.

Vol. II.   B   fome

fome Change fince the laft Relations: I fhall alfo endeavour to explain the true Caufes of thofe Alterations. But it will be neceffary firft, to lay open, as I may fay, the Foundations of the Empire of the *Turks,* and difcover the Principles upon which their Government is eftablifh'd.

THOSE who do not reflect on the Original of this Empire, difcern at firft fight, that the *Turkifh* Government is extremely fevere, and almoft tyrannical: But if we confider that it began in War, and that the firft *Ottomans* were, from Father to Son, the moft formidable Conquerors of their Age, we fhall not be furprized, that they fet no other Limits to their Power, than merely their Will.

COULD it be expected that Princes, who ow'd their Greatnefs folely to their Arms, fhould diveft themfelves of their Right of Conqueft, in favour of their Slaves? It is natural for an Empire, which is founded in a time of Peace, and the People of which make choice of a Chief to govern them, to be mild and gentle ; and the Authority of it may, in a manner, be divided and fhared. But the firft Sultans owing their Promotion purely to their own Valour, and being full of Maxims of War, affected to have a blind Obedience, to punifh with Severity, and to keep their Subjects under an Inability to revolt ; and, in a word, to be ferv'd only by Perfons who ftood indebted to them for their Fortune, and whom they could advance without Jealoufy, and crufh without Injuftice.

THESE Maxims, which have continued among them for four Centuries, render the Sultan abfolute Mafter of his Empire. In poffeffing the whole Revenues of it himfelf, he does but enjoy the Inheritance of his Anceftors, and if he has an abfolute Power of Life and Death over his People, he regards them only as the Iffue of his Forefathers Slaves. His Subjects alfo are fo intirely perfuaded of the fame Opinion, that they make no refiftance, but fubmit to the firft Order which is fent to take away their Life or their Goods ; and by a refin'd piece of Policy, it is infus'd into them in their very Cradle, that this Excefs of Obedience is rather a Duty of Religion, than a Maxim of State. Under the Force of this Prejudice, the Prime Officers of the Empire themfelves conclude it to be the higheft Good-fortune and Glory to end their Days by the Hand or Order of their Lord. But the Savages of *Canada* are

more

more eafy and compos'd under this Circumftance than the *Turks* ; for without reading *Epictetus*, or the *Stoicks*, they naturally account Death a great Good, and deride us, who lament thofe who are appointed to die : they fing alfo in the middle of the Flames ; and the quickeft Pain affects them very little, becaufe they are fill'd with Hope of entring upon a happier Life.

T H E Grand Signior is ador'd by his Subjects, and wins them to him by the flighteft Favours ; for they have no Poffeffions, but what they hold of him. His Empire extends from the *Black Sea* to the *Red Sea* ; he has the better part of *Africa*, is Mafter of all *Greece*, and even to the Frontiers of *Hungary* and *Poland* ; and, in fhort, can boaft that his Predeceffors, or their Grand Vifiers, have befieg'd the Capital of the Weftern Empire, and have left only the Gulph of *Venice* between their Dominions and *Italy*. After this, would any Man believe there have been Sultans who have liv'd only on the Income of the Royal Gardens belonging to the Empire, tho even at prefent thefe Revenues amount but to an indifferent Sum ? and that feveral have liv'd by the Labour of their own Hands ? and that at *Adrianople* are fhewn the Tools Sultan *Morat* us'd in making Arrows, which he fold for his own Profit in the Seraglio ? The Courtiers, it is likely, paid dear enough for their Emperor's Work. One is very far from feeing the fame Frugality now-a-days in a Prince's Palace.

F O R fear of being furpriz'd in an unguarded Pofture, the Sultans have provided a Bulwark for themfelves and their Succeffors, by inftituting a formidable Militia, which is kept on foot as well in time of Peace as of War. The Janizaries and the Spahis balance the Power of the Prince, in fuch a manner, as abfolute as it is, that they have fometimes had the Infolence to demand his Head : and they depofe Emperors, and cre-ate new ones more eafily than the *Roman* Soldiers did of old. This is a Curb upon the Sultans, and reftrains their Tyranny.

T H E Revenues of the Emperor are partly fix'd, and partly cafual. The fix'd are the Cuftoms ; the Capitation impos'd upon the Jews and the Chriftians ; the Excife upon all the Produce of the Soil ; and the an-nual Tributes which the Cham of *Tartary* the lefs, the Princes of *Moldavia* and *Wallachia*, the Republick of *Ragufa*, and one part of *Mengrelia* and

Ruſſia pay in Gold. To which muſt be added five Millions of Livres return'd from *Egypt* ; for of twelve Millions furniſh'd by that ſpacious Kingdom, in Sequins coin'd upon the ſpot, the Pay of the Soldiers and the Officers confumes four ; and three more the Grand Signior ſends to *Mecha*, for an uſual Preſent, to maintain the Expences of the Religious Worſhip, and of filling the Ciſterns of *Arabia* with Water, which are on the Road where the Pilgrims paſs.

THE Treaſurers of the Provinces receive the Duties of their ſeveral Diſtricts, and defray all the Charges by Aſſignments from the Port. Theſe return the Money which is in their hands every three Months to the Treaſurers of the Empire, who are accountable to the Grand Viſier for what they receive from the Provinces.

THE caſual Revenues of the Grand Signior confiſt in Inheritances: For, according to the Laws of the Empire, the Prince is Heir both to great and ſmall, to whom he hath given Penſions during their Life; and in like manner, to the Soldiers, if they die without Children. If only Daughters are left, he receives two Thirds of the Eſtate, not out of the Fiefs, for they belong naturally to the Prince; but out of the Lands independent of the Fiefs, as of the Gardens, the Farms, the Caſh, the Moveables, and of the Slaves, the Clothes, Horſes, *&c.* The Relations dare not offer to alienate any part of the Eſtate, for there are Officers eſtabliſh'd to look after it; and if they ſhould attempt it, the whole would be forfeited to the Sultan.

THE Spoils of the great Men of the Port, and the Baſſa's riſe to an immenſe Sum, and make it impoſſible to know the Amount of the Grand Signior's Revenues. Very often he does not ſtay for their dying a natural Death, or give them time to conceal their Treaſure ; but their Gold, and Silver, and Jewels, and their Heads, are carry'd at once to the Seraglio. Nor is the Removal of the Baſſa's only of advantage to the Grand Signior; but he who ſucceeds a diſplac'd Baſſa, pays for his Preferment a conſiderable Sum. All whom the Sultan gratifies alſo with a Viceroyſhip, or any Poſt of Conſequence, are indiſpenſably oblig'd to make him Preſents, not according to their Riches; for the Perſons advanc'd are frequently taken out of the Seraglio, where they had no opportunity of laying up any thing; but the Preſents muſt be anſwerable

to

to the Favour they receive.   The Prefent of the Baffa of *Cairo* is com-
puted at fifteen hundred thoufand Livres, without reckoning feven or
eight hundred thoufand he muft diftribute to them who procur'd him the
Government, and who have Intereft enough to preferve him in it: thefe
are the chief Sultaneffes, the Mufti, the Grand Vifier, the Boftangi-
Bafhi, *&c.*

THESE Sums are not fuffer'd to lie in the hands of the Treafurers,
who might wafte them or ufe them to their own Profit; but they are
brought to the Seraglio into the Treafury-Royal, which is near the Hall
of the Divan.   It is divided into four Chambers; the firft two of which
are taken up with different Arms, and great Coffers of Vefts and Furrs,
Cufhions embroider'd, and fet with Pearls, with pieces of the fineft Cloth
of *England*, *Holland*, and *France*, and with Velvets, Brocades of Gold
and Silver, and with Bridles and Saddles cover'd with precious Stones.

IN the third Chamber are kept the Jewels of the Crown, which are
of an ineftimable Price: the Staff which bears the Plume of Feathers is
adorn'd with the richeft Stones, and is in the Form of a Tulip; this is
faften'd to the Grand Signior's Turbant, who wears it there.   If the Sultan defires to fee any of his Jewels, the Chief Treafurer, accompany'd
with 60 Pages belonging to that Chamber, gives notice to the Key-
Keeper to attend at the Treafury-Door; and firft the Treafurer examines
whether the Seal he plac'd the laft time upon the Lock be entire; after
which, he orders the Key-keeper to break it, and open the Door; and
acquaints him which of the Jewels it is the Grand Signior demands; and
receiving it, goes away to deliver it to him immediately.   In this Chamber are lodg'd alfo the nobleft Harneffes, and the richeft Arms in the
World: the Sabres, and Swords, and Poinards glitter with Diamonds,
Rubies, Emeralds, Turquoifes, and Pearls.   Thefe feldom lie long here,
but are generally circulated; for in proportion as the Emperor has given
any of them to the Baffa's, he receives others from them, when they
die, or are remov'd.

THE fourth Chamber is properly the Publick Treafury: It is full
of ftrong Coffers, arm'd with Bands of Iron, and fecur'd every one with
two Locks; in thefe are put all the pieces of Gold and Silver.   The Door
of the Chamber is feal'd with the Grand Signior's Signet, who keeps

one of the Keys, and the other is in the hands of the Grand Vifier. Before they proceed to take off the Seal, it is certified very ftrictly that it has fuffer'd no Alteration, and this is commonly done upon Council-Days; at which time they lock up the new Receipts in the Coffers, and take out Sums appointed for the Payment of the Troops, and other Services; after which, the Grand Vifier applies the Emperor's Signet again.

AS to the Gold, that paffes into the Grand Signior's Privy-Treafury, which is a fubterraneous Vault, in which no one enters befide the Prince, attended by fome Pages of the Treafury: The Gold is put into Bags of Leather, containing fifteen thoufand Sequins apiece, and the Bags are depofited in ftrong Chefts. When it appears there is Gold enough in the fourth Chamber to fill two hundred Bags, the Grand Vifier fignifies it to his Highnefs, who repairs thither to fee them remov'd into his Privy-Treafury, and to feal them up himfelf. At that time he ordinarily makes his Largeffes, both to the Pages who wait on him in the Privy-Treafury, and to the great Men who follow him to the Door, and ftay behind in the fourth Chamber with the Grand Vifier.

IF the Wars exhauft thefe Sums, or the State is in a preffing Neceffity, the Treafures of the Mofques, which are kept in the Caftle of the Seven Towers, are ftill a noble Supply to the Emperor.

THE Mofques are rich, efpecially that which is call'd the *Royal*: after the Officers are paid out of thefe Religious Revenues, the Remainder of the Money is put into that Treafury, of which the Grand Signior is the principal Guardian. This facred Treafure, it is true, cannot be made ufe of, unlefs for the Defence of their Religion; but does not fuch an Occafion offer it felf at every turn in the Wars with their Neighbours, who are either Chriftians or Schifmatical Mahometans? And the Mufti knows not how to difapprove the applying of this Money to fuch a War.

THERE is no Prince who is ferv'd with more Refpect than the Sultan. Such a Veneration for him is infpir'd into thofe who are educated in the Seraglio; and their Condition requires from them fo much Fidelity and Devotion to his Perfon, that he is not only regarded as the Lord of the World, but even as the Sovereign Arbiter of every Man's

Good

Good and Evil in particular: the Palace therefore is fill'd only with a
Train of Creatures entirely confecrated to him.   They may be di-
vided into five Claffes; the *Eunuchs,* the *Ichoglans,* the *Azamoglans,* the
Women, and the *Mutes*; to whom may be added, the *Dwarfs* and the
Buffoons, who deferve not to be accounted a diftinct Clafs by them-
felves.

T H E *Eunuchs* have the Charge of the whole Palace, and are in the
higheft Confidence; being incapable of pleafing the Fair Sex, and difen-
gag'd from Intrigues of Love, they refign themfelves wholly to Ambition, and the Care of enriching their Fortune.   They are eafily diftinguifh'd by their Colour; for fome are Black, and others are White.   The
White are employ'd in ferving the Perfon of the Prince, and overfeeing
the Education of the Children of the Seraglio.   The Black are the more
unhappy, for they are always fhut up in the Apartments of the Women.
They are forc'd to ufe a Pipe in making Water, being depriv'd of the
natural Conveyance in their Infancy : for the Sultans were jealous of them,
while the Operation was perform'd in any other manner; and to cure
this extravagant Imagination, they are cut fmooth clofe to the Belly.
The Operation is not without danger, and cofts many of them their
Lives: But the Eaftern People and the *Africans* facrifice every thing to
their Jealoufy.   Yet after this barbarous Precaution, they fcarcely fuffer
the poor Wretches to caft their Eyes upon their Women, and commonly permit them only to ftand Centinel at the outer Door of the
Chamber.

T H E Chief of the White *Eunuchs,* who has been handled in his Youth *The Chief of*
as feverely as the reft, is the great Mafter of the Seraglio ; he has the *the White* Eu-
Infpection of all the Pages of the Palace, and all Petitions, which are nuchs.
to be prefented to the Prince, are deliver'd to him: he is in the Secret
of the Cabinet, and commands all the *Eunuchs* of his own Complexion.
The principal of thefe Eunuchs are, 1. The Great Chamberlain, who
is firft of the Officers of the Chamber. 2. The Deputy-Supervifor of
the Pages Apartments, and other Buildings of the Palace: He never ftirs
out of *Conftantinople,* and gives his Orders to others who follow the Grand
Signior abroad.   3. The Privy-Treafurer, who keeps the Jewels of the
Crown, and one of the Keys of the Secret Treafury, and commands all
the

the Pages of the Treafury.   4. The Grand Expenditor of the Seraglio, who is alfo Great Mafter of the Wardrobe : it is his Charge to look to the Sultan's Sweet-meats and Drinks, the Syrups and Sherbet, and the Counter-poifons or Antidotes, as the Treacle and Bezoar, and other Drugs : he takes care alfo of the Grand Signior's *Porcelain* and *China* Ware. The other White *Eunuchs* are Preceptors to the Pages, the firft Prieft of the Palace-Mofque, and Overfeer of the Infirmatories.

The Chief of the Black Eunuchs.   T H E Chief of the Black *Eunuchs*, who may be call'd, *The Eunuch*, by way of eminence, has the abfolute Command of the Women's Apartment ; and all the Black *Eunuchs*, who are plac'd there for a Guard, obey him blindly.   He has the Super-intendence of the Royal Mofques of the Empire, and difpofes of all the Offices which belong to them. The principal Black *Eunuchs* are, the *Eunuch* of the Queen-Mother ; the Intendant or Governour of the Princes of the Blood ; the Comptroller of the Queen-Mother's Treafury, the Steward of her Perfumes, Sweet-meats and Liquors ; the two Chiefs of the Great and Little Chamber of the Women ; the Head-Janitor of their Apartment ; and the two Priefts of the Royal Mofque, whither the Women refort to Prayers.

Ichoglans and Azamoglans.   T H E *Ichoglans* are young Men, bred up in the Seraglio, not only to ferve about the Prince, but to fill, in time, the firft Pofts of the Empire.   The *Azamoglans* are train'd up there for inferior Employments.

T H A T Honours may not become hereditary or fucceffive, or any Family be advanc'd which may be able to form a confiderable Party ; the Children of the Vifiers and Baffa's are fo far from fucceeding their Fathers, that it is ordain'd they fhall not rife above the Degree of Captain of a Gally ; and if there are Inftances of the contrary, they are very rare.   It is not long fince the Emperors employ'd fuch only as had neither Relations nor Friends in the Seraglio : And out of the diftant Provinces were continually fent thither Numbers of Chriftian Children taken in the War, or levy'd by way of Tribute in *Europe*, for thofe of *Afia* were exempted ; the moft beautiful and well-made were chofen, and fuch as appear'd to have the greateft Spirit and Senfe.   Their Names, Age, and Country were regifter'd ; and the unhappy Infants foon forgetting Father and Mother, Brothers and Sifters, and their Country it felf,

felf, become wholly devoted to the Perfon of the Sultan. At prefent this Tribute of Children is difcontinu'd; not out of favour to the *Greeks*, but becaufe the *Turks* themfelves give Money to the Officers of the Se-raglio to have their own Children admitted there, in profpect of their arriving to the higheft Places in the Empire. According to the beft of their Capacity, thefe Children think of nothing but how to pleafe thofe who have the Care of their Education, in order to merit the Favour of the Court. The Emperor frequently makes his Choice of them, according as they are prefented, or appoints them to be review'd by the Heads of the White *Eunuchs*, who are good Phyfiognomifts: the greater part of them are kept at *Conftantinople*; but fome, I have been inform'd, are fent to *Adria-nople* and *Prufa* in *Afia*: the moft Graceful continue among the *Ichoglans*, and the others are diftributed among the *Azamoglans*.

IN the firft place they are requir'd to make a Profeffion of Faith, and are circumcis'd; during which Operation they repeat, *There is no God but God, and* Mahomet *is the Meffenger of God*. They are bred with an ex-emplary Modefty, and are no lefs fubmiffive and obedient, than the No-vices among our Religious: they are chaftis'd feverely for the fmalleft Faults by the *Eunuchs* who overlook their Behaviour, and are ftrictly held for fourteen Years under thefe Preceptors Eyes. Inftead of whipping, they receive the Baftinado upon the Soles of their Feet; which is fo fevere-ly inflicted for fome Tranfgreffions, that they expire under the Blows. The *Eunuchs* are very cruel, and being vex'd at their own miferable Condi-tion, difcharge their Anger upon thofe who have not fuffer'd in the fame kind. Thefe unhappy Youths therefore are forc'd to bear all their capri-cious Humours, and never leave the Seraglio till their time is finifh'd, unlefs they are willing to quit the Society; and then they lofe their For-tune, and receive but a trifling Acknowledgment at their Departure. The Seraglio is perfectly a Republick, the Members of which have Laws and Cuftoms peculiar to themfelves: Both thofe who command there, and they who obey, have no Notion of Liberty, and have no Commerce with the Inhabitants of the City; and the *Eunuchs* never ftir out thither, but to execute their Orders. The Sultan himfelf is in a manner a Slave to the Pleafures of his Palace: He alone, and fome of his Miftreffes, are heartily merry, the reft are dull and fad.

THE *Ichoglans* are divided into four Chambers, which are beyond the Hall of the Divan, on the left fide of the third Court. The firft, which is call'd the Little Chamber, contains ordinarily 400 Pages, who are all fubfifted at the Grand Signior's Charge, and receive every one four or five Afpers a day for their Pay. But the Education which is given them, is beyond any Price: Nothing is inculcated to them, but Civility, Modefty, Politenefs, Accuracy, and Honefty; above all, they are taught to keep filence, to hold down their Eyes, and fold their Hands acrofs their Breaft. Befide Mafters to teach them to read and write, there are fome whofe Care it is to inftruct them in their Religion, and efpecially to fhew them to fay their Prayers at the ftated Hours.

AFTER fix Years Practice, they pafs to the fecond Chamber with the fame Pay and the fame Habit, which is of common Cloth; they continue here alfo the fame Exercifes, but apply themfelves more particularly to Languages, and whatever may improve and brighten their Wit. The Languages are the *Turkifh*, the *Arabian*, and the *Perfian*. As their Strength comes on, they put them to draw the Bow, to fhoot, to throw the Dart, to handle the Pike or the Lance, to mount on Horfeback, and every thing belonging to the Art of Riding; as to dart on Horfeback, to dif-charge their Arrows before or behind, on the right hand and on the left. The Grand Signior takes a pleafure in feeing them fight on Horfeback, and rewards thofe who fhew the greateft Skill. The Pages continue four Years in this Chamber before they remove to the third.

IN that they learn to few, embroider, and make Arrows; and here they alfo fpend four Years, in order to become the better qualify'd to wait on the Sultan. To this end, befide Mufick, they practife Shaving, paring the Nails, folding Vefts and Turbants, attending in the Baths, wafh-ing the Grand Signior's Linen, and keeping Dogs and Birds.

DURING thefe fourteen Years of Noviciate, they never fpeak to one another but at certain Hours, and their Difcourfes are modeft and grave: If they go to fee one another at any time, it is under the Eyes of the Eunuchs, who follow them continually. In the Night, not only their Chambers are illuminated, but the Eyes of thofe *Argus's*, who are in-ceffantly walking the Round, difcover all that paffes. Between every fix Beds lies an Eunuch, who erects his Ears at the leaft Noife.

OUT

OUT of this Chamber are taken the Pages of the Treafury, and thofe
who ferve in the Laboratory, where they prepare the Treacle, the Cordials,
and fine Liquors of the Emperor: and it is not till after an Examination
of their Abilities and Senfe, that they are permitted to attend his Perfon.
Thofe who feem not to have fufficient Capacity, are fent back with a
flight Gratuity, and are generally entr*e*d among the Cavalry, which is
the Fortune of fuch alfo who do not hold out thro' the whole Probation;
for the infinite Conftraint, and the Blows of the Battoon often caufe them
to renounce their Station. This third Chamber is reduc'd to about two
hundred Pages, whereas the firft has four hundred.

IN the fourth there are but forty in Number, who are well-made,
polite and modeft, and thorowly prov'd in the three preceding Claffes:
their Pay is double, and amounts to near nine or ten Afpers a day. They
are drefs'd in Satin and Brocade, or Cloth of Gold, and are properly Gentlemen of the Chamber. They make their court with the utmoft Application, and have a liberty of vifiting all the Officers of the Palace:
but the Prince is their Idol; for they are of a proper Age for Ambition
after Employments and Honours. There are fome of them who never
leave the Sultan, but when he goes into the Apartment of the Women,
namely, they who bear his Sabre, his Cloak, his Veffel of Water to drink,
and to make the Ablutions, and he who carries the Sherbet, and holds
the Stirrup when he mounts on Horfeback or alights. The other Officers of the Chamber, who are lefs about the Prince's Perfon, are, the
Mafter of the Wardrobe, the Chief Mafter of the Palace, the Chief Bar-
ber, he who pares his Nails, and he who takes care of his Turbant, the
Secretary of his Orders, the Comptroller-General of the Houfhold, and
the Chief Supervifor of the Dogs. All thefe Officers expect to rife
to the firft Pofts, and with reafon, for it is natural to recompenfe thofe
whom we fee every moment.

NO Method feems better fitted to form skilful and great Men, than
the Education which is given to the Pages of the Scraglio; who pafs, as
one may fay, thro a courfe of all the Virtues: neverthelefs, in fpite of all
their Pains, when they are advanc'd to great Stations, they appear to be
indeed mere Scholars, who want to be taught how to command, after they
have learn'd how to obey. And tho the *Turks* imagine God gives Pru-

dence,

dence, and the other neceſſary Talents, to thoſe whom the Sultan raiſes to high Employments, Experience often teſtifies the contrary. What Capacity can Pages have, who are train'd up among Eunuchs, who treat them with the Baſtinado for ſo long a time? Wou'd it not be better to promote Youth by degrees, in an Empire where no regard is had to Birth? Beſides, theſe Officers paſs, at a ſtep, from a ſtate of the utmoſt Uneaſi-neſs and Conſtraint, to ſuch an extraordinary Liberty, that it is impoſſible they ſhould not let looſe their Paſſions; and yet they are intruſted with the Government of the moſt important Provinces. As they have neither Abilities nor Experience to perform the Duties of their Charge, they truſt to their Deputies, who are commonly great Robbers, or Spies of the Grand Viſier, to ſend him an account of their Conduct. Theſe New Governours are forc'd alſo to paſs thro the hands of the *Jews*; for as they have nothing when they come from the Seraglio, they have recourſe to thoſe Uſurers, who lead them to all manner of Rapine and Extortion. Beſide the Preſents a new Baſſa muſt make to the Grand Signior, the Sul-taneſſes, and the principal Men of the Port, he is alſo to provide for his own Living. The *Jews* alone are able to advance him the Money; and theſe honeſt Pilferers will not furniſh a Piece, but at *Cent. per Cent.* This Evil would not be ſo extreme, if they would be content to receive it a-gain by little and little: but as they are afraid every moment the Baſſa ſhould be ſtrangled or remov'd, they never let the Debt grow old, and the People muſt be ſqueez'd to repay them.

YET, if the Baſſa is ſuffer'd to remain there ſeveral Years, it is no Ad-vantage to the Province: for if he is a Man of Underſtanding, he labours not only to diſcharge the Debt he contracted at his receiving the Govern-ment, but to raiſe a Fund ſufficient for his Expences; and eſpecially to oblige his Protectors at Court, without whom, inſtead of being advanc'd he wou'd infallibly be recall'd, let him behave himſelf as he will. More-over, the *Jew*, or the *Chifou*, as the *Turks* call them, manages his Game all the while; and all the Money of the Baſſa's Houſe, not to ſay of the whole Province, goes thro his hands. The Avarice of Sultan *Morat* was truly the Source of all theſe Diſorders: for it was he who introduc'd the Cuſtom of receiving Preſents from the Great Men whom he promoted; and theſe, to make themſelves whole again, practis'd the ſame towards their

Infe-

Inferiours: fince which time, every thing is open to the higheft Bidder. Sultan *Solyman* alfo, who had a wonderful Affection to his Sifters and his Daughters, marry'd them to the Chief Officers of the Port, contrary to the Ufage of his Predeceffors, who beftow'd them on the Governours of very diftant Provinces. The Husbands of thefe Sultaneffes, under their Ladies Protection, made it their Bufinefs to get what they could from every one, to fupply the Expences of their Conforts. Thefe Diforders, it is vifible, are able to ruin the Empire; but the Evil is beyond a Cure: for the Emperor himfelf, the Sultaneffes, the Favourites, and the Great Ones of the Port, inrich themfelves wholly by this fort of means; and the Inferiours fucceed in no Suit, but by fubmitting to their Extortions. It is not furprizing therefore, that this great Empire fhould at prefent be in a kind of Declenfion.

FROM the *Ichoglans* we muft pafs to the *Azamoglans*, for thefe laft are *The Azamoglans.* only the Refufe of the former: In thefe the Qualities of the Body are regarded more than thofe of the Mind. If they happen to want Perfons for this Service, they purchafe them from the *Tartars* of *Tartary the Lefs*, who are continually making Inroads upon their Neighbours to carry off Children. Thefe Children are bred under the Difcipline of the White Eunuchs, as well as the *Ichoglans*. After the Circumcifion, and the Profeffion of Faith, they inftruct them in Matters of their Religion, and efpecially in their Prayer, which is the only Language, as the *Turks* fay, with which Men fpeak to the Lord; and thofe who are inclin'd, are taught to read and write. Their Habit is Cloth of *Salonica*, blue and very coarfe; and their Caps are yellow Felt, and fhap'd like a Sugar-Loaf. Their firft Exercifes are Running, Wreftling, Leaping, or Pitching the Bar: after this, they are appointed in the Seraglio to be Porters, Gardiners, Cooks, Butchers, Grooms, Waiters in the Infirmitory, Wood-Cleavers, Centinels, Footmen, Archers of the Guard, and Rowers of the Grand Signior's Gally: and many of them are employ'd to clean his Arms; others, under the Direction of the *Arabs*, to take care of his Tents; and fome look after the Baggage and the Chariots. But whatever be their Employment, their Pay is but from two Afpers a day, to feven and a half; out of which they are oblig'd to fubfift themfelves, for the Sultan allows them only Cloth and Linen. They live with a furprizing Oeconomy in their Chambers.

bers. The Janizary-Aga reviews them from time to time, and enters thofe whom he likes among the Janizaries of the Port. Some of them become Spahis; but neither thefe nor the others are lifted, till after their Bodies are throughly harden'd to Labour, and are able to endure all the Fatigues of War, by being accuftom'd to bear Cold and Heat, to cleave Wood, carry Burdens, and cultivate the Ground; and, in a word, to execute the loweft and moft painful Drudgeries: A great many are fent into *Afia*, among the Peafants, to learn Agriculture.

THOSE who remain in the Seraglio, are lodg'd by the Sea-fide, under Sheds: the principal of them are the *Boftangi's* or Gardiners, the Chief of whom is chofen out of thefe, and is call'd the *Boftangi-Bachi*; he is one of the moft powerful Officers of the Port, tho his Place, at firft view, feems not of the higheft Honour: but as he has the Prince's Ear, and waits upon him often in his Gardens, it is in his power to do good Offices or ill; and on that account he is courted by the firft Men in the Empire. Befide his Apartment by the Sea, the *Boftangi-Bachi* has a fine Kiofc upon the *Bofphorus*: he is Super-intendant of the Grand Signior's Gardens and Fountains, and Governour of all the Villages along the Channel of the *Black Sea*: he commands above ten thoufand *Boftangis* or Gardiners, who are in the Seraglio, or in the Royal Houfes about *Conftantinople*: he has the Charge of that Quarter of the *Bofphorus*, where the *Franks* inhabit; and punifhes feverely the Muffulmans and the Chriftians who are drunk, or caught in the Company of Women: but the moft honourable part of his Function is, to hold the Helm of the Sultan's Barge, when he diverts himfelf upon the Water, and to ferve him with his Back, inftead of a Footftool, as he mounts his Horfe, or alights, when he rides a Hunting, or to take the Air.

EVERY Friday the Head-Gardiners give an account to the *Boftangi-Bachi*, of the Money arifen by the Fruits of the Grand-Signior's Kitchen-Gardens; this Money is properly the Prince's Patrimony, for it is appointed for his Table. The Sultan often takes a pleafure in feeing the Gardiners work: but this is when he is alone; for if he is accompanied with any of the Sultaneffes, thofe poor Drudges vanifh in an inftant, or lie as clofe to the Ground as they are able: it would be a Crime beyond Remiffion in them, to be feen at fuch a time; and the wretched *Boftangi*

thus

thus taken, would be put to death upon the spot. The Honour of appearing in the Presence of the Women, is granted to none but the Black Eunuchs, who are capable of giving neither Temptation nor Jealousy.

I T is said at *Constantinople*, that Renunculus's are the chief Ornament of the Flower-Gardens of the Seraglio; but there are very few of these Flower-Gardens, in comparison of the Number of Kitchen-Gardens and Orchards, in which almost all the sloping and low Ground of the Palace is laid out. The Orchards are over-run with Cypress-Trees, and Pines, and Brambles; but it is natural in the *Turks* to neglect their Gardens, or at least to take care only of their Melons and Cucumbers. There are whole Families who live upon nothing but Cucumbers above half the Year: they eat them raw, without peeling, like Apples; or else they cut them out in thick slices, not to dress them in a Salad, but to throw them into a Bason of very sour Milk; and after they have eat plentifully of it, they drink a great Pot-full of fresh Water. These Fruits are admirable, and never occasion the Gripes. The Pages of the Seraglio dare not enter into the Places where these are set, ever since *Mahomet* II. caus'd even seven to be ript up, to discover who had eat one of his Cucumbers.

B E S I D E the Officers already mention'd, the Sultans have also in their Palace two sorts of People, who serve to divert them, namely, the *Mutes* and the *Dwarfs*. The Mutes of the Seraglio are a Species of *The Mutes.* rational Creatures by themselves: For, not to disturb the Prince's Repose, they have invented a Language among themselves, the Characters of which are express'd by Signs alone; and these Signs are understood by Night as well as by Day, by touching certain Parts of their Body. This Language is so much in fashion in the Seraglio, that they who would please there, and are oblig'd to be in the Prince's Presence, learn it very carefully: for it would be a want of the deep respect they owe him, to whisper one another in the Ear before him.

T H E Dwarfs are perfect Apes, and make a thousand Grimaces among *The Dwarfs.* themselves, or else with the Mutes, to set the Sultan a laughing, who sometimes does them the honour to give them several Kicks with his Foot. Whenever they meet with a Dwarf who is born deaf, and consequently dumb, they esteem him as a very Phenix of the Palace, and admire

mire him beyond the moft graceful Man in the World, efpecially if the Baboon is an Eunuch alfo. And thefe three Defeats, which ought to render a Man contemptible in the laft degree, make him the moft compleat of all Creatures, in the Eyes and Judgment of the *Turks.*

The Women of the Seraglio. I OUGHT now to fpeak of the Women of the Seraglio, but in that I muft be excus'd; for they fall no more under the Knowledge of the Senfes, than fo many pure Spirits. Thefe Beauties are entirely referv'd to entertain the Sultan, and vex the miferable Eunuchs. The Governours of the Provinces make prefents to the Grand Signior of the lovelieft Girls in the Empire, not only to ingratiate themfelves with him, but to plant fome Creatures of their own alfo in the Palace, who may be able to procure them an Advancement. After the Sultan's Death, the Women whom he honour'd with his Embraces, and their eldeft Daughters, are remov'd into the old Seraglio of *Conftantinople*; the younger are fometimes left for the new Emperor, or are marry'd to the Baffa's. However, fince it is a Crime to fee thofe who remain in the Palace, very little regard can be given to what is written about them: for tho Means might be found to get into the Seraglio; yet, who would be willing to die for a Glance of his Eyes fo unhappily employ'd? Whether thefe Ladies alfo enter the Sultans Bed at the Feet, as fome would have us believe, or at the Side, I fhall not determine; but content my felf with accounting them the leaft unfortunate Slaves in the World: Liberty is always preferable to fo flender and trifling a Happinefs.

WHAT can one fay concerning a Place, where even the Prince's chief Phyfician is admitted to vifit the Women who are fick, with the greateft difficulty? The Phyfician alfo can neither fee them, nor be feen by them; nor is he fuffer'd to feel their Pulfe, but thro a piece of Gaufe or Crape; and very often he cannot diftinguifh whether it is an Artery or a Vein which beats. The Women alfo who look after the fick, dare not acquaint him with what paffes; for they fly the Room in all hafte, and no one ftays about the Bed but the Eunuch, to prevent the Phyfician from feeing his Patient, and to lift up juft the Edge of the Curtain, as far as they fhall think neceffary for the fick Creature to put out her Arm. If the Phyfician fhould require to view fo much as the tip of her Tongue, or to touch any part, he would be ftabb'd upon the fpot. *Hippocra-*

tes

tes, with all his Knowledge, would have been ſtrangely embarraſs'd, if Letter I.
there had been Muſſulmans in his time: For my ſelf, who have been
bred up in his School, and according to his Maxims, I was extremely at
a loſs how to behave towards the Great Men, when I was call'd in, and
viſited the Apartments of their Wives: theſe Apartments are juſt like the
Dormitories of our Religious, and at every Door I found an Arm cover'd
with Gawſe, thruſt out thro a ſmall Loop-hole made on purpoſe.   At firſt
I fancied they were Arms of Wood or Braſs, to ſerve for Sconces, to
light up Candles in at Night; but it ſurpriz'd me when I was told, I muſt
cure the Perſons to whom thoſe Arms belong'd.

I T is a falſe Notion, that the Jewiſh Women can go into all the A-
partments of the Women of the Seraglio, to ſell their Jewels: they are
allow'd to come no farther than into a certain Hall, where they drive
their Trade, nor is the Door open'd to them, till the Eunuchs have ſearch'd
them heedfully; and a Man who ſhould be catch'd in a Woman's Habit,
would have his Throat cut in an inſtant, and a Chriſtian Woman would
be us'd very ſcurvily.   The Eunuchs alone paſs to and fro upon the Meſ-
ſages, and carry in the Jewels, and bring back the Money; and they
underſtand well enough how to pay themſelves for their pains.   After
all, what Uſe can theſe Eunuchs make of their Money, who have neither Relations nor Friends, and who can reap no other pleaſure from it
than to handle their Gold, and devour it with their Eyes? Their principal Aim, they ſay, is to ſecure their Lives at the Revolutions which
happen upon the Sultan's Death; but they are very ſeldom in danger,
who look to the Women.

T H E other Officers, who take care of the Seraglio, of whom I am to *The Surveyor*
ſpeak, are the Surveyor of the Baths; the Grand Falconer, whoſe Offi- *of the Baths,*
cers carry a Hawk upon their right Fiſt; the Grand Huntſman, who has &c.
under him above twelve hundred Dog-keepers; the Governour of the
Hounds and the Setting Dogs; the Governour of the Grey-hounds, the
Maſtiffs, and the Spaniels; the Grand Querry, who has two chief Quer-
ries under him, who command a great many Officers, and thoſe alſo an
infinite number of Grooms; for there is no Place, where Horſes are more
valued, than in *Turky*.   They feed them with a little Barley and minc'd
Straw, which they give them Evening and Morning in a ſmall quan-

tity, and the reſt of the Day they travel on briskly, and thereby become capable of holding out extraordinary Courſes: It is ſaid alſo, that the Horſes which come from *Arabia*, and from about *Babylon*, will travel thirty Leagues without reſting; they have admirable Legs, but no Hips nor Cheſt.

The Capigi's. I MUST not forget two other ſorts of Officers, who are of wonderful Uſe to the Grand Signior, as well within as without the Seraglio; and theſe are the *Capigi's* and the *Chiaus's*: The Body of the *Capigi's* or Porters conſiſts of about four hundred, commanded by four Captains of the Port, who are every one upon Guard in turn upon Council-Days. The Pay of the Porters is fifteen Aſpers a day: their Habit is like the Janizaries, but they have no Horns before their Bonnets. Fifty of theſe *Capigi's* are upon Duty every day at the Gate of the firſt Court of the Seraglio, and as many more at the Gate of the Court of the Divan. When the Grand Signior is diſſatisfied with the Conduct of a Viceroy or Governour, he ſends one of theſe *Capigi's* with an Order to demand his Head. The *Capigi* ſtrikes it off, after he has ſtrangled him; and ſeaſons it with Salt, to preſerve it, if the Road is very long, and carries it in a Sack to the Sultan: ſo that theſe *Capigi's* are perfectly Executioners.

The Chiaus's. THE *Chiaus's* are employ'd in more-honourable Commiſſions; they carry the Emperor's Orders over his whole Dominions, and are charg'd with the Letters he writes to Sovereign Princes: they are, as it were, Exempts of the Guard to the Grand Signior. Their Number is about ſix hundred Men, commanded by a Chief, who is call'd the *Chiaus-Bachi*. This Officer performs the part of Grand Maſter of the Ceremonies, and Introductor of Ambaſſadors. On the Days of the Divan, he places himſelf at the Door of the Grand Signior's Apartment, with the Captain of the Guard then in waiting. The Pay of the *Chiaus's* is from twelve Aſpers a day, to forty. They are at the Command of the Grand Viſier, the Viſiers, and the Beglerbegs, and even the Baſſa's: but the Rank of thoſe whom they ſerve, is diſtinguiſh'd by the Apple at the top of their Staff; which, for the principal Officers, is of Silver, but for others, of Wood. The greater part of the *Chiaus's* do the Duty of Serjeants, in citing Parties to appear before the Divan, and to meet and agree Matters among themſelves: they never lay down their Staff or their Bonnet:

net : the Bonnet is very large, and is like the Bonnet of Ceremony of the firſt Officers of the Empire.

I T is time, my Lord, that I ſhould inform you concerning the Officers who dwell out of the Palace, and who never come there, but when they are ſummon'd, or the Duty of their Place calls them. At the head of his Miniſters the Sultan places the Grand Viſier, who is, as it were, his Lieutenant-General ; with whom he divides, or rather to whom he leaves the Care of the whole Empire. The Grand Viſier is not only intruſted with the Finances, with foreign Affairs, and the Adminiſtration of Juſtice in Civil and Criminal Matters, but alſo with the Conduct of the War, and the Command of the Armies. A Man who is capable of ſuſtaining ſo great a Burden as he ought, is very uncommon : yet, there have been found Men, who have executed this Charge ſo skilfully, that they were the Wonder of their Age. The *Cuperli's*, Father and Son, were triumphant both in Peace and War, and, by a Policy almoſt unknown before, dy'd quietly in their Bed. *Cuperli*, their Relation, who was kill'd in the Battle of *Salankemen*, was alſo a great Man ; and, had he liv'd, would perhaps have protected the State from the Revolutions with which it is ſtill threatned. This Empire, which at this day ſeems to be declining, ſtands in need of ſuch Miniſters.

The Grand Viſier.

W H E N the Sultan names a Grand Viſier, he puts into his hands the Seal of the Empire, upon which is engraven his Name : this is the Badge of the firſt Miniſter ; he carries it always in his Boſom. He diſpatches all his Orders with this Seal, without conſulting or giving an account to any one. His Power is unlimited, unleſs with reſpect to the Troops, whom he cannot puniſh without the Concurrence of their Commanders. Excepting this, Affairs of all ſorts are brought before him, and are decided by his Judgment. He diſpoſes of all Honours, and all the Poſts of the Empire, except thoſe of Judicature. The Entry of his Palace is free to all the World, and he gives Audience even to the meaneſt of the Poor. Yet, if any one thinks he has great Injuſtice done him, he may make his way to the Grand-Signior himſelf, by putting Fire upon his Head; or elſe he fixes his Petition upon the end of a Reed, and ſo carries his Complaint to the Sultan.

THE

T H E Grand Vifier appears in his high Station with a world of Magnificence: he has above two thoufand Officers or Domefticks in his Palace, and never fhews himfelf in publick, but with a Turbant adorn'd with two Plumes of Feathers, charg'd with Diamonds and precious Stones: the Harnefs of his Horfe is fet with Rubies and Turquoifes, and his Houfing is embroider'd with Gold and Pearls. His Guard is compos'd of about four hundred *Bofnians* or *Albanians*, whofe Pay is from twelve to fifteen Afpers a day: fome of thefe attend him on foot, when he goes to the Divan; but when he marches into the Field, they are well mounted, and carry a Lance, a Sword, a Hatchet, and a pair of Piftols. They are call'd *Deli's*, that is, *Fools*, becaufe of their fantaftical Airs, and their Habit, which is ridiculous; for they have a kind of Seaman's Jacket.

T H E Grand Vifier is preceded by three Horfe-Tails, on the top of each of which is a gilded Apple: this is the Military Enfign of the *Ottomans*, which they call *Thou* or *Thouy*. For a certain General of this Nation, they fay, being at a plunge to rally his Troops, who had loft all their Standards, thought of this Device, to cut off a Horfe's Tail, and erect it on the point of a Lance: the Soldiers flock'd in to this new Enfign, and came off with Victory.

W H E N the Sultan honours the Grand Vifier with the Command of an Army, he takes out one of the Plumes of his own Turbant, at the head of the Troops, and delivers it to him to place in his own. And it is not till after this Mark of Diftinction, that the Soldiers acknowledge him for their General; and he has the Power to confer all vacant Pofts, even Viceroyfhips and Governments, upon the Officers who ferve under him. In a time of Peace, tho the Sultan difpofes of the chief Employments, yet the Grand Vifier continues to have a mighty Influence in procuring them to be difpos'd to whom he thinks fit; for he writes to the Grand Signior, and receives his Anfwer immediately: it is in this manner that he advances his own Creatures, or avenges himfelf upon his Enemies, whom he is able to get ftrangled, purely by the Reprefentation he makes to the Emperor about their ill Behaviour. He frequently vifits the Prifons by Night, and always takes an Executioner along with him, to put to death thofe he judges culpable.

T H E

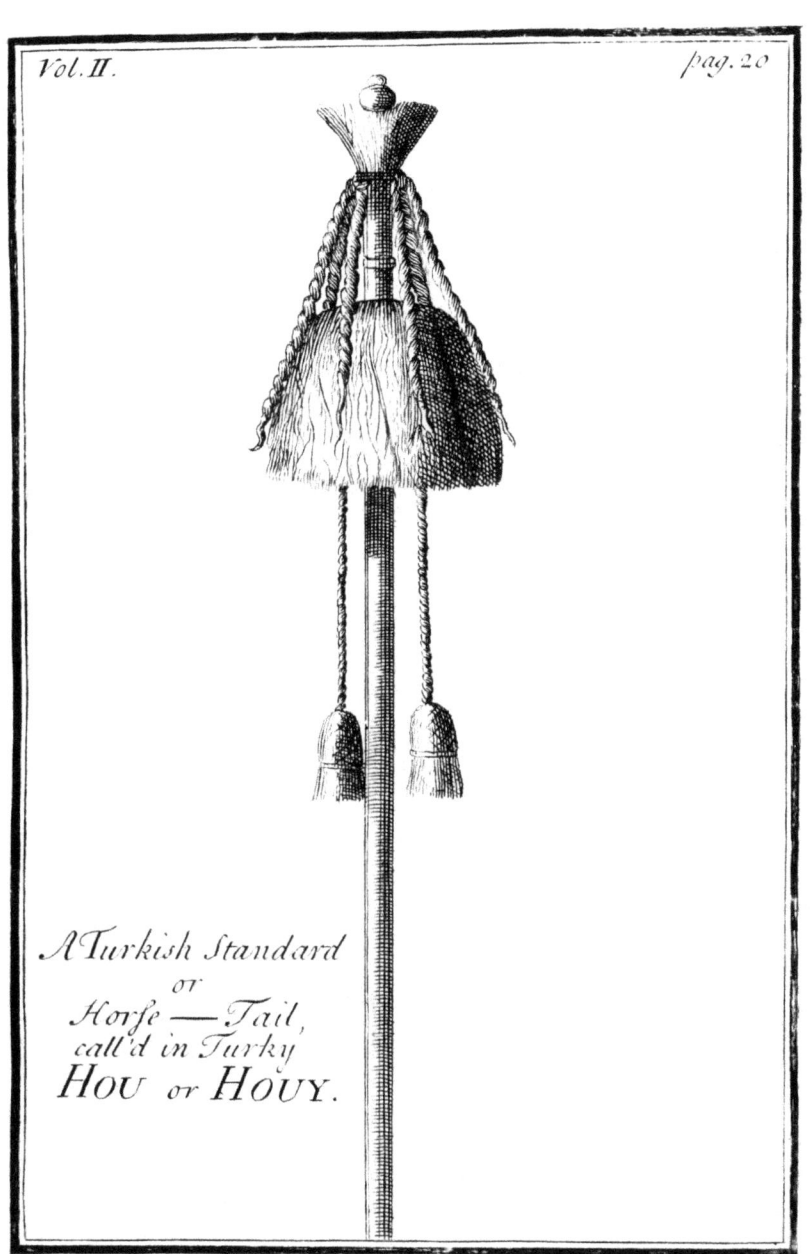

A Turkish Standard
or
Horſe — Tail,
call'd in Turky
HOU or HOUY.

THE ſtated Salary allow'd to the Grand Viſier is but twenty thou- ſand Crowns, yet he enjoys an immenſe Revenue. Not an Officer in this vaſt Empire, but makes him conſiderable Preſents, either to obtain or keep his Poſt: This is a ſort of indiſpenſable Tribute. The principal Enemies of the Grand Viſier are thoſe, who, next to the Sultan, command in the Palace, as the Sultana-Mother, the Chief of the Black Eunuchs, and the Favourite Sultaneſs: for theſe are ever contriving to ſell the great Offices; and that of Viſier being the higheſt, they watch him narrowly, even in his minuteſt Actions: and as much as he is truſted, he is encompaſs'd with Spies; and the Powers which are againſt him, ſometimes engage the Soldiery to mutiny, who under a pretence of ſome Grievances, demand this Miniſter's Head, or his Depoſal: upon which, the Sultan reſumes the Seal, and ſends it to him whom he chuſes to ſucceed.

THIS Prime Miniſter therefore is, in his turn, oblig'd to make rich Preſents, in order to preſerve himſelf in his Poſt. The Grand Signior is draining him perpetually, either by honouring him with Viſits, for which he pays very dear, or by asking of him from time to time conſiderable Sums. The Viſier alſo puts every thing to Sale, to furniſh himſelf for theſe Expences: his Palace is the Market where all Favours are ſold; but he uſes a world of Caution in managing the Traffick; for *Turky* is the only Place in the Earth, where Juſtice is often well obſerv'd in the midſt of the greateſt Injuſtice.

IF the Grand Viſier is of a Martial Genius, he finds his account better in War than in Peace. Tho his commanding the Army obliges him to be at a diſtance from the Court, he has his Penſionaries, who act for him in his Abſence; and a War with Strangers, provided it go not too far, is more favourable to him than a Peace, which may occaſion one at home. The Militia is then buſy'd in defending the Frontiers, and the War leaves them no time to think of an Inſurrection; for the moſt turbulent and ambitious Spirits, being eager to diſtinguiſh themſelves by extraordinary Actions, often fall in the Field: Beſides, this Miniſter has not a more proper way to win the Eſteem of the People, than by fighting againſt the *Unbelievers*.

AFTER

AFTER the firſt Viſier, there are ſix others, who are ſtil'd ſimply Viſiers, Viſiers of the Bench or of the Council, and Baſſa's of the three Horſe-Tails, becauſe three Horſe-Tails are carried before them when they march, whereas there is only one borne before the ordinary Baſſa's. Theſe Viſiers are Men of Wiſdom and clear Judgment, and knowing in the Law, and aſſiſt at the Divan; but they never deliver their Opinion upon the Affairs which are treated there, unleſs requir'd by the Grand Viſier, who often ſummons a ſecret Council of the Mufti alſo, and the Cadileſquers or Juſtices-General. The Stipend of theſe Viſiers is two thouſand Crowns *per annum*: The Grand Viſier commonly refers Matters of ſmall conſequence to them, as well as to the ordinary Judges; for as he is in a manner the Interpreter of the Law, in Points not regarding Religion, he generally follows only his own Opinion, either out of Vanity, or to ſhew the Credit he poſſeſſes.

THE Grand Viſier holds a * Divan in his own Houſe every day, except Friday, which is a Day of Reſt with the *Turks*. During the Remainder of the Week, he goes four times to the Divan of the Seraglio, *viz.* on Saturday, Sunday, Monday, and Tueſday: he is preceded by the Chiaus-Bachi, and ſome of the Chiaus's, and ſeveral Virgers, accompanied by the chief Lords of the Empire, and follow'd by his *Albanian* Guard, and above four hundred Perſons on horſeback, who march thro an infinite Croud of People, making a thouſand Acclamations for his Proſperity. On the Days of the Divan, an Hour before Sun-riſing, three Officers mounted, place themſelves before the Seraglio, to make certain Prayers there, while they wait for the Arrival of the Miniſters, whom they ſalute by Name, with a loud Voice, as they paſs along. At the ſight of the Palace, the Baſſa's forget their Gravity, and when they are thirty or forty Paces from the Gate, fall a galloping, and range themſelves on the right ſide of the firſt Court, to expect the Grand Viſier: the Janizaries and the Spahi's are planted in the ſecond Court under the Gallerys, the Spahi's on the left ſide, and the Janizaries on the right. The whole Train diſmounts in the firſt Court, and paſſes on to the ſecond; but the Gate of the Divan is not open'd till the Grand Viſier arrives, and after a Prieſt has made the Prayer for the Souls of the Emperors deceas'd, and for the Health of him who reigns.

THOSE

T H O S E who have Bufinefs at the Divan, enter the Hall in a throng; but the Vifiers and Juftices-General, out of Refpect, enter not, except with the Grand Vifier; and then the whole Affembly proftrate themfelves on the Ground. When this Minifter is feated, the two Juftices-General take their place on his left hand, which is the moft honourable with the *Turks* : he of *Europe* is the firft next to the Grand Vifier, and he of *Afia* the fecond: then the Treafurers-General of the Empire place themfelves, with whom is a Sub-Intendant, and two Artizans. The Vifiers are difpos'd on his right hand, according to their Rank, with the Guard of the Seals; and if there is any Beglerbeg or Viceroy newly return'd from his Government, the Grand Vifier does him the honour to feat him next to the Vifiers.

T H E Y begin with the Affairs of the Finances. The Chiaus-Bachi firft goes to the Door of the Treafury to take off the Seal, and brings it to the Grand Vifier, who examines whether it is whole and undefac'd. Then the Treafury is open'd, to put in or take out Money neceffary for paying the Troops, or to anfwer other Occafions; after which, the Grand Vifier delivers the Seal back, to be affix'd to the Door. From the Finances, they proceed to Matters of War, and confider the Demands and Anfwers of Embaffadors, and expedite the Orders of the Port, Patents, Grants, Paffports and Privileges. The Reis-Effendi, or Secretary of State, receives all the Difpatches from the Grand Vifier's hands, and fends them forward: If they are Orders of the Port, the Chancellor feals them; but for the Letters of the Signet, the Grand Vifier only fets the Emperor's Signet beneath, which he ftamps upon them, having firft dipt it in Ink. They go next upon Criminal Caufes; the Accufer appears with his Witneffes, and the Accufed is acquitted or condemn'd without delay. They conclude with what Civil Affairs are offer'd at the time.

I T is at this Tribunal, that the loweft Man in the Empire has the Confolation of having Reafon done him, even againft the greateft Lords of the Country: the Poor has the liberty of demanding Juftice; and Muffulmans, Chriftians, and Jews are equally heard. There is no brawling and fquabling, and one fees no Advocates or Proctors: the Clerks of the Secretarys of State read every one's Petition. If it is for a Debt, the Vifier fends a Chiaus to fetch the Debtor, and the Creditor produces his Evidence, and the Money is told out upon the fpot, or the Debtor is condemn'd to receive a certain Number of Blows with the Batoon. If it is a Queftion of Fact, two or three Witneffes decide it in an Hour; and let the Affair be of what nature it will, it never takes up above feven or eight Days. They have recourfe to the Alcoran, and the Vifier interprets the Law, if it be a Queftion of Right: In a Matter of Confcience, he confults the Mufti by a fhort Note, where he ftates the Cafe, without naming the Perfon. Concerning Affairs of the Empire, he fends an Abftract of the Petitions to the Grand Signior, and waits his Anfwer. The Secretary's Clerks write down all the Refolutions taken by the Grand Vifier: the Secretary is encompafs'd with Regifters, who draw up the Writing in as few Words as poffible, and he delivers out all the Decrees; and there being no Appeal, the Caufe is never reviv'd, either by annulling the Decree, or by a Writ of Review.

IT muft be allow'd on the other hand, that Law-Suits are much rarer in *Turky* than with us: for the Grand Signior's Subjects having only the Ufe of the Goods, which they hold merely by his Pleafure, leave very little ground of Contention when they die; whereas our Donations, Teftaments, and Marriage-Contracts, are Sources of infinite Difputes. An *Italian* told me one day at *Conftantinople*, that we fhould be very happy in *Europe*, if we could appeal from our Courts to the Divan: his Reflection made me fmile; for, added he, one might go to *Conftantinople*, and all over *Turky* too, if there were occafion, before one Suit would be finally decided in *Europe*. A *Turk* of *Africa* pleading before the Parliament of *Provence*, againft a Merchant of *Marfeilles*, who had led him a Dance for many Years from Court to Court, made a very merry Reply to one of his Friends, who defir'd to know the State of his Affairs: *Why, they are wonderfully alter'd,* fays the African; *when I firft arriv'd here, I had a Row of Piftoles as long as my Arm, and my Deed was compris'd in half a Sheet of Paper: but at prefent, I have a Writing above four times as long as my Arm, and my Train of Piftoles is but half an Inch.*

WITH all thefe Precautions, a great deal of Injuftice is done in *Turky*; for they admit the Evidence of all forts of Perfons: and People of the greateft Honefty are fometimes expos'd to lofe their Goods and

their

their Life, upon the bare Depofition of two or three falfe Witneffes. If Juftice is well executed in the Divan of *Conftantinople*, it is becaufe they confider the Sultan is always lift'ning at a Window juft above the Grand Vifier's Head, which is cover'd only with a Lattice and a piece of Crape : but do they not commit crying Injuftices in the Divans of other Towns, where the Cadi's fuffer themfelves to be corrupted by Money, and are only govern'd by their Paffions? One may appeal, it is true, from their Judgment to *Conftantinople*; but every Man is not in a condition to make the Journey. See here alfo another great Abufe.

T H E Religious among the *Turks*, by a particular Privilege, are exempted from common Juftice, infomuch that many who have enrich'd themfelves in the Adminiftration of Affairs, and apprehend they fhall be call'd to account, turn Dervifes or Santons. There is no Religious Order among Chriftians fo powerful as that would be, which fhould have a Liberty of receiving Perfons, who, after they had ruin'd a Province by their Extortions, fhould be permitted to imitate this *Turkifh* Practice, and affume the Habit.

T H E Soldiers have the Privilege of being judg'd only by their Commanders, or their Deputies. During the four Hours the Divan of *Conftantinople* is fitting, the Spahi's and the Janizaries attend in the fecond Court under the Galleries, where they keep a profound Silence, and every one holds in his Hand a Silver Staff gilded. The Colonel of Horfe, and he of the Foot, here difpenfe Juftice to their Soldiers; who, to prevent Diforder, are forbidden to ftir from their Place, without being call'd. If they have Petitions to prefent, they give them to two of their Comrades, who are appointed to go and come upon this Service. This Privilege encourages a world of Mifchiefs in the Provinces; for moft of the Rogues throw themfelves among the Janizaries to efcape the Punifhment of their Crimes.

I F O R G O T, my Lord, to tell you, that there is a Clofet on one fide of the Hall of the Divan, where feveral Officers wait during the Council; as the Keeper of the Rolls of the Grand Signior's Revenues, he who regifters every thing which enters the publick Treafury, or goes out, and he who is appointed to fee the Pieces weigh'd and prov'd.

Vol. II.　　　　　　　　　　E　　　　　　　　　　The

The Chiaus-Bachi, and the Capigi-Bachi paſs to and fro in the Court, to execute the Grand Viſier's Orders.

EMBASSADORS always have their Audience of the Grand Signior upon a Divan-day, and are introduc'd by the Captain of the Guard then on Duty. The Embaſſador is plac'd upon a Stool, over againſt the Grand Viſier, and diſcourſes of Buſineſs till Dinner is ſerv'd up: after which, the Preſents the Embaſſador is oblig'd to make, are brought into the Hall; and when the Grand Viſier and the Officers of the Divan have obſerv'd them, the Capigi's carry them along one by one, and expoſe them in the Court, that every one may judge of the Magnificence of the Prince who ſends them. During this, the Embaſſador has a Veſt preſented him, and ſome are diſtributed alſo to his Retinue. The Sultan ſhews himſelf in the Hall of Audience, which is near the Divan, ſitting upon his Throne: the Throne is erected upon Pillars, which ſupport a wooden Canopy, all cover'd with Plates of Gold, adorn'd with Numbers of Jewels, the Diamonds and Precious Stones of which are of wonderful Value. It is plac'd in a Corner of the Hall upon a Sofra rais'd a foot and a half high, and cover'd with a Carpet of the utmoſt Magnificence. The Sultan ſits with his Legs a-croſs, and is attended only by the Chief of the White Eunuchs, the Keeper of the Secret Treaſury, and ſome Mutes. It is impoſſible to have any other than a Side-view of his Face, becauſe the Door of the Hall does not anſwer directly to the Corner where the Throne is plac'd. Thoſe of the Embaſſador's Train, who were pre-ſented with Veſts, make their Salute to the Sultan firſt, and are con-ducted each of them by two Capigi's, who hold them under both Arms. The Embaſſador himſelf, who, according to the Cuſtom of the Coun-try, pays his Salute laſt, is led up in the ſame Poſture, by two Cap-tains of the Port; and it is order'd ſo, that in advancing and retiring, they never turn their Back to the Sultan. It was uſual once to kiſs his Hand; but it has been thought proper to lay aſide this Ceremony, ever ſince *Amurat* I. the Son of *Orcanes,* was ſtabb'd by a wretched Soldier, who deſign'd by it to revenge the Death of the Deſpot of *Ser-via,* his Maſter. For ſome time after this, they continu'd to kiſs a long Sleeve, which was faſten'd to the Emperor's Veſt on purpoſe.

*

Count

Count *Lefi* and *Marcheville*, Embaffadors of *France* had the honour to   Letter I.
do this: but this Practice is now abolifh'd; and at prefent Embaffadors
make a bare Salute, tho the Captains of the Guard endeavour as much
as they can to make them bow down, but without Succefs: for the
Embaffadors being appriz'd of what ought to be done, ftand firm, and
keep themfelves upright with all their Strength. When they have made
their Reverence, they are left alone in the Hall with the Secretary of
the Embaffy, and the Interpreter; to whom, after they have open'd them,
they deliver their Prince's Letters, and the Interpreter having explain'd
them, they withdraw. The Sultan falutes the Embaffador with a gentle
Inclination of his Head: he treats a Moment with the Vifiers concerning
the Subject of the Embaffy, and deliberates upon the Affairs in queftion, fuppofing they are of confequence. The Grand Vifier returns
from thence to the Divan, where he ftays ftill Noon, which is the Hour
when the Council breaks up; then he goes home, preceded by a Company of Janizaries, and another of Chiaus's on horfeback, by his Foot-
Guard, and follow'd with an infinite Croud, who form a very numerous Court.

UPON the Day of the Divan, the Emperor generally caufes the principal Officers to give him an account of all that pafs'd in the Affembly, and chiefly of the Duty of their Charge. They are call'd upon
for this fucceffively one after another. The Janizary-Aga, when he fees
the Capigi-Bachi and the Chiaus-Bachi coming to him, advances towards
them with four of his Captains, who accompany him as far as the
Prince's Apartments, at the Door of which he conjures them to pray God
to infpire the Sultan to forgive his Faults. He enters alone to undergo
the Examination; and if the Prince is fatisfy'd with his Conduct, he
returns in Peace: if the Sultan finds him to be guilty, he ftamps upon
the Ground with his Foot, at which Signal the Mutes enter, and ftrangle
the Aga without other Formality.

THE Spahi-Aga is alfo cited to the Grand Signior upon the fame
Occafion; but he commonly comes away with more Cheerfulnefs than
the reft; for what reafon, I can't tell. The other Great Men of the
Empire are afraid of falling under the Stroke, or, to fpeak more properly, under the String of the Mutes. The Juftices-General are the only

    Perfons

Perſons not ſubject to this melancholy Hazard, becauſe they belong to the Law. Sometimes the Sultan conſults the Mufti before he puts his Officers to death; and demands of him in *Writing* what Puniſhment a Slave would deſerve, who ſhould commit ſuch Faults. The Mufti, who knows well enough this is merely a Formality, and that the Honour would ſoon be diſpens'd with, if he did not give into his Maſter's O-pinion, ſeldom ſcruples to determine it is Death; and very often contrary to his better Sentiments.

THE Preſents the Grand Signior makes to the Prime Viſier are always ſuſpected; at leaſt he is oblig'd to make his Acknowledgment for them, by a Sum anſwerable to his Maſter's Grandeur. Sometimes, as a Mark of unuſual Diſtinction, this Prince in the Morning gives his firſt Miniſter a Veſt, which he had worn the Day before, and in the Afternoon he ſends for his Head, which is ſurrender'd with a perfect Reſignation; ſo true is it, that Nature in many caſes yields to Preju-dices. It is Prepoſſeſſion which makes Martyrs in all Religions except the Chriſtian, where Martyrdom is an Effect of Grace. If *Deſcartes* and *Gaſſendus* had ever gone to *Conſtantinople,* as they were once think-ing to have done, what a world of excellent Reflections would they have made upon the Morals and Politicks of the *Turks ?* The Great Ones of the Port die with Tranquillity a violent Death, and eſteem it a holy and glorious thing to die, if it is by the Sultan's Order; at leaſt, they act as if they thought ſo : it is their Policy alſo, to give them no time to conſider, by allowing them only to make one ſhort Prayer.

The Caimacan WHEN the Grand Viſier is not at *Conſtantinople,* the Caimacan ſupplies his room, and acts by his Direction. The Word *Caimacan* in *Turkiſh* ſignifies Lieutenant or Deputy. This Lieutenant holds a Divan, and gives Audience to Embaſſadors : but the happieſt Circumſtance be-longing to his Office, is, that he is not anſwerable for Events in Affairs of State; and if the Grand Signior finds fault with any thing, the Cai-macan excuſes himſelf by the Orders he receiv'd from the Prime Viſier. Beſides this, the Caimacan is Governour of *Conſtantinople,* where he exerciſes a ſurprizing Policy : If a Baker ſells Bread by falſe Weights, he is faſtned by the Ear for twenty four Hours to the Door of his Shop. They who ſell the firſt Fruits take Money firſt, but they don't ſell theſe

*

dearer

dearer than the next : for Novelty is not fo gainful in *Turky*, as it is Letter I.
in *France* ; and a Tradefman who fhould go to make a Profit of it, would
expofe himfelf to the Baftinado.   One may fend Children to Market
with Safety, if they do but know how to ask for what they want.
The Caimacan's Officers ftop the Children in the Streets, and examine
what they have got, and weigh it ; and if it is right, they let them pafs ;
but if they find there is a Cheat in the Weight, or the Meafure, or the
Price was too dear, they go back with them to the Man who fold it,
and he is condemn'd either to the Baftinado, or to a Fine.   It is the
Intereft of the Fruiterers that the Children be honeft, and able to govern
their Appetites ; fince if they fhould eat a Fig or a Cherry upon the way,
the poor Tradefmen would pay the Damage : For thirty Blows of the Batoon are generally given, if one Onion is found fhort, and twenty-five
for a Leek.   If any one is excus'd from the Baftinado, the common Punifhment for fhort Tale or Meafure, then they put about the Seller's
Neck two thick Planks bor'd hollow, and fill'd at each end with heavy
Stones.   In this Condition they lead the wretched Fruiterer all over the
Town ; and if he defires to reft himfelf in his Progrefs, he muft pay
down fuch a Number of Afpers.   Surgeons are alfo chaftis'd fometimes
after the fame manner ; but inftead of Stones they hang on, at the end
of the Planks, feveral fmall Bells, which make a lamentable Tinkling,
as they march along the Streets.   This fignifies they have fuffer'd fome
People to die thro their Neglect ; and the Defign of this Ceremony, fay
the Muffulmans, is only to warn Perfons not lightly to truft their Life
in the hands of fuch Murderers.

IF a dead Body is found in the Street, the next Neighbours are condemn'd to pay for the Blood, fuppofing the Author of the Murder is not
difcover'd.   The Terror that all are in of this Calamity, makes every
one ftrive as much as poffible to compofe Quarrels, and prevent any
Diforders in the Neighbourhood.   The Shops are fhut at Sun-fet, and
not open'd again till its Rifing.   Every one retires home in time, and
keeps good Hours : in a word, there is more noife made in one Day in
a Market of *Paris*, than there is in a whole Year in all *Conftantinople*.
The Grand Signior goes about fometimes difguis'd, with an Executioner, to fee what paffes in this great City.   *Mahomet* IV. who hated the

fmoking

fmoking of Tobacco violently, and was inform'd it often prov'd the Occafion of fetting Houfes on fire, was not content with publifhing fevere Orders againft this Cuftom, but frequently made the round, to catch fuch as fmok'd; and it is faid that he hang'd up all he found, having firft caus'd a Pipe to be thruft thro their Nofe, and a Roll of Tobacco to be ty'd about their Neck. The Watch all over *Turky* carries to Prifon thofe whom they find abroad in the Night, be they of what Nation or Religion they will: but they find very few; for the Dread of the Baftinado, or being amerc'd, keeps every one at home. It is a common Saying in *Turky, That in the Night the Streets are only for the Dogs :* and here indeed they are very full of thofe Animals; for every one throws them out Victuals, and it is very dangerous to walk on foot at fuch a time. Thefe Creatures, which are as fierce and ravenous as our Butchers Curs, make a terrible Bellowing, and howl lamentably at the leaft Noife they hear ; and fometimes the very Chiding of the Sea, fets them a yelling.

THE Soldiers there are very peaceable, excepting the Levanti's, who ferve on board the Gallies: but befide that they commit Diforders only in the Suburbs of *Conftantinople,* the Prejudice is inconfiderable, becaufe the Caimacan permits the Chriftians to defend themfelves ; which was granted them upon the Complaints Embaffadors were making every day, of the Infults the Subjects of their Nation receiv'd. *Janizaries.* As for the Janizaries, they live fairly enough in *Conftantinople* ; but they are very much fallen from the high Efteem the antient Janizaries were in, who contributed fo much to the Eftablifhment of this Empire. Whatever Precautions the Emperors have fometimes taken to preferve thefe Troops from degenerating, they are declin'd very much : and it feems likely, that in another Age, they will ftill be lefs regarded, for fear of their rendring themfelves too formidable.

THO the greater part of the *Turkifh* Infantry carries the Name of Janizaries, yet it is certain, in all this great Empire, there are not above five and twenty thoufand, who are true Janizaries, or Janizaries of the Port. This Soldiery was once compos'd only of Tributary Children, inftructed in the *Turkifh* Religion; but at prefent this is not obferv'd : and People

are

are not molested on this account, since the Officers take Money of the *Turks* themselves to be entred in this Body.

FORMERLY the Janizaries were not permitted to marry, the *Turks* being persuaded that the Cares of a Family render Soldiers less fit for the Exercise of Arms. Yet now-a-days they who will, marry with the consent of their Chiefs, who also at the same time give them a Sum of Money. The principal Reason which keeps the Janizaries from marrying, is, that Batchelors only arrive at Offices, the most desirable of which are to be Chiefs of their Chambers: for this Soldiery is all lodg'd in a large Quarter, divided into 162 Chambers. Every Chamber has a Chief, who commands in it; but out of it, he only performs the Office of Lieutenant of the Company, and receives Orders from the Captain.

EVERY Chamber has its own Ensign-Bearer, its Expenditor, its Cook, and its Water-Carrier. Above the Captains is only the Lieutenant-General of the Janizaries, who is subject to the Aga. Beside the common Pay, the Emperor gives the Janizaries every Year a compleat Suit of Cloth of *Salonica*, and every day allows them a quantity of Rice, Meat and Bread. The Chamber lodges them for one half *per Cent.* upon the Pay they receive in time of Peace, and seven *per Cent.* in time of War. This Pay is but from two to twelve Aspers a day, and is never rais'd but by little and little, in proportion to their Service; when they are disabled, they have an Allowance for Life. The Cap peculiar to the Janizaries is made like the Sleeve of a Coat; one end is put upon their Head, and the other hangs down upon their Shoulders: to this Cap before is fastned a sort of Spike half a foot long, of Silver gilded, and adorn'd with Bastard Stones. When the Janizaries march into the Field, the Sultan furnishes them with Horses to carry their Baggage, and Camels to carry their Tents; to wit, one Horse for ten Soldiers, and one Camel for twelve. At the Accession of every Sultan, their Pay is augmented one Asper a day.

THE Chambers inherit the Effects of those Members who die without Children; and the rest, tho they have Children, always leave their Chamber a Legacy. The *Solaes* and *Peyes* alone, among the Janizaries, are the Emperor's Guard; the others never go to the Seraglio, but to attend their Officers upon Divan-Days, and to prevent Disorders which might

happen

happen in the Court: they are generally plac'd Centinel at the Gates, and the Crofs-ways of the Town, to keep watch there. They are fear'd every where, and refpected, tho they carry only a Cane in their hand; for Arms are not deliver'd to them, but when they take the field. The greater part of the Janizaries do not want for Education, being taken from the Body of the Azamoglans; which, either thro Impatience or on fome other account, they frequently forfake. Thofe who are to be admitted among the Janizaries, pafs along in Review before the Officer, and every one takes hold on the bottom of his Companion's Veft. Their Names are entred in the Grand Signior's Regiftry, after which they all run up to the Mafter of their Chamber, who, to make them know they are under his Jurifdiction, gives them every one a Box on the Ear as they pafs by. At their Inrollment they take two Oaths, the firft is to ferve the Grand Signior faithfully ; the fecond, that they will follow the Will of their Comrades in Matters relating to the Body. There is no Set of Men in *Turky* fo united as that of the Janizaries : it is this ftrict Union which preferves their Authority, and gives them the Daring fometimes to depofe the Sultan. Tho there are but twelve or thirteen thoufand in *Conftantinople*, they are affur'd that their Brethren, what part foever of the Empire they are in, will not fail to approve their Conduct.

I F they think they have occafion to complain, their Difcontent begins to fhew itfelf in the Court of the Divan, at the time of the diftributing the Difhes of Rice to them, prepar'd in the Grand Signior's Kitchen: for they eat it quietly, if they are contented ; and on the contrary, they throw the Difhes on the ground, and turn them topfy-turvy, if they are out of humour at the Miniftry. There is no Infolence they fcruple to utter at fuch a time againft the principal Minifters, being well perfuaded they fhall obtain Satisfaction: For this reafon the moft favourable Opportunity is taken early to prevent their Rifing, efpecially the time when they give them feveral Days Pay together. The Mutinies of the Janizaries are much to be dreaded; for how often have they in an inftant chang'd the Face of the Empire? The fierceft Sultans, and the moft skilful Minifters have often found how dangerous it was to keep on foot, in time of Peace, a Militia who fo well under-

ftand

ftand their own Interefts.   They depos'd *Bajazet* II. in 1512. and promoted the Death of *Amurat* III. in 1595.   They threatned *Mahomet* III.
with Dethronement.   *Ofman* II. who had fworn to deftroy them, having
imprudently difclos'd his Defign, was difgracefully treated by them .
for they made him walk on foot to the Caftle of the Seven Towers,
where he was ftrangled in 1622.   *Muftapha* I. whom this impudent Soldiery put in *Ofman*'s room, was depos'd two months after by the fame
hands as advanc'd him.   They alfo put to death Sultan *Ibrahim* in 1649.
after they had dragg'd him ignominioufly to the Seven Towers.   His
Son *Mahomet* IV. was not fo unhappy indeed; but they depos'd him
after the laft Siege of *Vienna*, which mifcarried yet only by the Fault
of *Cara Muftapha* the Prime Vifier.   In this Sultan's ftead was prefer'd his
Brother *Solyman* III. a Prince of no merit, who was alfo depos'd in his
turn fome time after.

WITH refpect to the Sultanefs-Mother, the Vifiers, the Caimacan, the
firft Eunuchs of the Seraglio, the Grand Treafurer, and their Aga himfelf, the Janizaries value them not, and demand their Heads upon the
leaft Uneafinefs.   All the World knows, how they us'd, at the beginning of this Century, the Mufti *Fefullah-Effendi*, who had been Precep-
tor of Sultan *Muftapha*.   This Prince, who lov'd his Tutor blindly, was
not able to prevent his being drawn upon a Hurdle to *Adrianople*, and
thrown into the River.   The only Expedient which could ever be de-
vis'd to reprefs the Infolence of thefe Soldiers, was, to encourage the Spahi's
againft them, and thereby make them jealous one of another; but they
agree together too well upon certain Occafions.   It fignifies nothing to
change their Quarters; for as the abfent always ftand to what their Fellows have done, it is impoffible to avoid their Fury, when they have
once taken it in their head, that they have fuffer'd fome great Injuftice.   The Hiftory of the *Turks* can furnifh few Examples of their having been appeas'd without confiderable Largeffes, or without its cofting
the firft Officers of the Empire their Lives.

THEY have never dar'd to confifcate the Treafure of the Janizaries,
nor to fhare the Goods their Officers poffefs in property in feveral parts
of *Afia*, as at *Cataya*, at *Angora*, at *Caraiffar*, and in other Places.   When
the General dies, the Treafurer inherits his Goods: he is the only Offi-

cer whofe Effects are not feiz'd to the Emperor's Profit. This General has the Privilege of prefenting himfelf before the Sultan with his Arms at liberty, whereas the Prime Vifier, and the other Great Men of the Port, never appear in his Prefence, but with their Arms a-crofs their Breaft; which is rather a fervile, than a refpectful Pofture.

AFTER the Aga, the principal Officers of the Janizaries are, the Aga's Lieutenant, the Grand Provoft, the Captain of the Serjeants, who march by the Emperor's fide upon Days of Ceremony; the Captains of his Foot-Archers, and the Commander of his Pages on foot; thefe laft, as well as the Archers, march by the Grand Signior's Perfon when he walks thro the City. They are but threefcore, and wear Caps of beaten Gold, embellifh'd before with Milk-white Feathers. As for the Foot-Archers, or Archers of the Guard, they are in number three or four hundred; and in a day of Battel, they are about the Sultan, arm'd only with Bows and Arrows, that they may not frighten the Grand Signior's Horfe. Their Habit is a Coat of Cloth, tuck'd up at the Corners as high as their Waift, fo as to fhew their Shirts: their Cap is Cloth, and ends in a Point, and is adorn'd with Feathers in fafhion of a Plume. Thefe Archers fhoot with their left hand, as well as with the right, which they are taught, that fo they may never turn their back upon the Sultan: when he paffes the Rivers, they fwim by his Horfe, and found the Fordings with all the Diligence imaginable: as a Reward, the firft time the Sultan paffes a River, he caufes a Crown apiece to be given to every one who was up to the Knee in Water; and if they were as high as the Middle, they have two Crowns, and three if they were above the Waift.

OUT of the Body of the Janizaries are taken the Gunners, and thofe who take care of the Arms. The Gunners are about twelve hundred, and receive their Orders from the Grand Mafter of the Artillery: they live at *Topana* in Apartments divided into 52 Chambers; but it is very happy that they are as not dextrous as the Chriftians, in the cafting and managing Artillery. They who look to the Arms are fix hundred in number, divided into 60 Chambers: they lodge in Apartments near *Sancta Sophia*; they not only take care of the antient Arms which are in the Arfenal, but of thofe of the Janizaries and Spahi's, which they deliver out to them in good Order, when they are going into the Field.

BESIDE

BESIDE the Janizaries now mention'd, all the Provinces of this
vaſt Empire are fill'd at preſent with Foot Soldiers who bear the Name
of Janizaries: but theſe Janizaries of the ſecond Order are not inroll'd
in the Body of Janizaries of the Port, and have nothing of the antient
Diſcipline of the *Turks*.   All ill Perſons who would skreen themſelves
from the ordinary Courts of Juſtice, and honeſt Perſons alſo who are
willing to cover themſelves from the Inſults of the others ; they who
would eſcape the Taxes, and be excus'd from publick Offices, purchaſe
of the Colonels of the Janizaries, who are in the Towns of the Province, the Title of Janizaries.   They are ſo far from receiving Pay, that
they give ſeveral Aſpers a day to theſe Officers, to enjoy thoſe Privileges : ſometimes they paſs for Invalids, or Penſioners for Life, and live
quietly at home, without being oblig'd to go into the Army.   Is it
ſurprizing after this, that the *Turkiſh* Forces are ſo much diminiſh'd?
They never have had ſo many Soldiers, nor ſuch ſmall Armies : the
Officers who are oblig'd to take the field, paſs their own Domeſticks
for Soldiers, and put the Pay of thoſe who ought to bear Arms in the
Prince's Service, into their own Pockets.   The Corruption which is introduc'd into this great Empire, ſeems to threaten it with ſome ſtrange
Revolution.

NEITHER muſt we confound with the Janizaries, another ſort of
Infantry, call'd *Azapes* and *Arcangi's*.   The Azapes are the old Muſſul-
man-Bands, more antient than the Janizaries themſelves, but very much
deſpis'd.   They ſerve for Pioneers, and ſometimes are merely a Bridge
to the Horſe in marſhy Grounds, and ſo many Faſcines to fill up the
Ditches of a Place beſieg'd.   The Arcangi's have no more Pay than the
Azapes, but are appointed only to ravage the Frontiers of the Enemy.
Yet in full Peace (for the War is not eſteem'd to be declar'd, unleſs the
Artillery is drawn into the Field) they are perpetually making Incurſi-
ons, and pillaging their Neighbours.   If any one among theſe Troops
happens to become a good Soldier, after ſome vigorous Action, he is
entred in the Body of the Janizaries.

THIS, my Lord, is the State of the *Turkiſh* Infantry, nor is that
of their Cavalry at preſent one tittle better : It is compos'd of two ſorts,
known by the Name of *Spahi's*, but they muſt be carefully diſtinguiſh'd.

The

The one are upon the Emperor's Pay, and the others not. The Spahi's in Pay are divided into feveral Standards, the principal of which are the Yellow and the Red: thofe who have no Pay, are of two forts, the *Zaims* and the *Timariots*.

T H E Spahi's in Pay, are taken from among the Ichoglans and the Azamoglans, who have been bred up in the Grand Signior's Seraglio's. Their loweft Pay is twelve Afpers a day, and the higheft a hundred: Thofe who come from Ichoglans, generally begin with twenty or thirty, which are increas'd according to their Merit, or the Intereft of their Friends. In time of War, all the Spahi's in Pay, who bring in Heads of the Enemy, are advanc'd two Afpers a day. And they who firft acquaint the Sultan with the Death of any of their Comrades, are rais'd as much.

T H E Spahi's are pay'd in the Hall, and in the prefence of the Grand Vifier, or his Chiaia, in order to avoid all occafion of Complaint. Tho the Spahi's are born of unknown Parentage, they may yet be look'd on as the Nobility of the Country: their Education makes them more accomplifh'd than the other *Turks*; and in every place Good Manners ought to conftitute a real and true Nobility. Thofe of the Red Standard were heretofore only Servitors to the Yellow; but now they are all equal; and the Red have even overtopt their Mafters, under *Mahomet* III. who in a Battel, in which the Spahi's of the Yellow gave ground and fled, reftor'd the Fight by the Valour of the Red.

T H E Arms of both are a Lance and a Scymiter, and they make ufe of a Dart, which they manage with wonderful Dexterity: the Dart has a Steel Point at one end, and is about two foot and a half long. They alfo carry a Sword, but it is faftned to the Saddle, and hangs down upon the Horfe's Thigh, fo as not to hinder them in difcharging their Piftol and Carbine. Some likewife ufe Bows and Arrows, efpecially the Spahi's of *Anatolia*; for thofe of *Europe* or *Romelia* rather chufe the Arms in ufe with us. Thefe Troops however fight without Order, and in a Croud, inftead of throwing themfelves into Squadrons, and rallying regularly. *Mahomet Kuperli* the Grand Vifier, who was a great General, was fo far from bringing them to Difcipline, that he affected to humble them, and keep them ignorant, for fear of increafing their Infolence; fince which time, they have extremely loft their antient Reputation:

*

they

they baftinade them now on the Soles of the Feet, left if they fcourg'd them, they fhould be difabled from mounting their Horfe; and for a contrary Reafon, the Janizaries are fcourg'd, becaufe they are oblig'd to ufe their Feet in marching.

WHEN the Grand Signior goes to command his Army in Perfon, he caufes large Sums to be divided among the Spahi's. One Spahi and a Janizary are plac'd Centry at each Cord of his Tent, and the fame at the Chief Vifier's. The other Standards of the Spahi's are, the White, the White and Red, the White and Yellow, and the Green. The moft famous Spahi's are thofe call'd *Mutafaraca*, who receive forty Afpers a day. The Emperor is their Colonel; their Duty is to attend upon him : they are about five hundred.

AS to the other Cavalry, call'd *Zaims* and *Timariots*, they are Per- *Zaims and Timariots.* fons to whom the Grand Signior gives certain Commands, term'd *Timar*, for Life, on condition they maintain fuch a number of Horfe for his Service. The firft Sultans being Mafters of the Fiefs of the Empire, erected Baronies or Commands out of them, to reward any extraordinary Services, and principally for raifing and fubfifting a Body of Troops without iffuing Money. But it was *Solyman* II. who eftablifh'd the Order and Difcipline of thefe Baronies, and fettled by his Decrees the Number of Men each one fhould be oblig'd to find. This Body has been not only very powerful, but very celebrated alfo thro the whole Empire: But Avarice, the common Vice of the Eaft, has made them decline feveral Years ago. The Viceroys and Governours of Provinces prevail fo far by their Intrigues at Court, that even the Commands which lie out of their Government, are given to their Domefticks, or to them who offer the moft Money.

THE Zaims and the Timariots differ little more than in their Income. The Zaims have the moft confiderable Commands, and their Revenues make from 20000 to 818819 Afpers. If they produce even an Afper above this, it becomes the Property of fome Baffa. Alfo, when a Commander dies, his Command is divided, fuppofing the Income of it has been augmented under the deceas'd, as it commonly happens to be; for they are generally improv'd rather than leffen'd. The Zaims are oblig'd to maintain at leaft four Horfe, which is after the rate of one Man for five thoufand Afpers of Rent.   THERE-

THERE are two forts of Timariots, the one receive their Provifions from the Port, the other from the Viceroy of the Place; but their
Equipages are lefs than thofe of the Zaims, and their Tents are fmaller,
and proportion'd to their Revenue. They who receive their Patents
from the Court, have from 5 or 6000, to 19999 Afpers; if they fhould
receive one Afper more, they would pafs into the Rank of Zaims.
They who have their Patents from the Viceroys, have an Income from
three thoufand Afpers, to fix thoufand. Every Timariot is bound to
provide one Horfeman for every three thoufand Afpers his Income produces.

THE Zaims and the Timariots are oblig'd to march in Perfon to
the Army, at the firft Orders, and nothing can excufe them; the Indif-
pos'd are carried in Litters, and their Children in Baskets or Cradles.
The Timariots muft furnifh Baskets to their Troopers, to carry Earth,
for filling up Ditches and Trenches. Thefe are better difciplin'd than
thofe who are properly call'd the Spahi's, tho the Spahi's are more per-
fonable and lufty : and whereas the laft never engage but in a Croud, at
the head of the antient Cavalry; the Zaims and Timariots are divided
into Regiments, commanded by Colonels under the Baffa's. The Baffa
of *Aleppo* is Colonel-General of this Body of Horfe, when he is in the
Army, becaufe being Seraskier of the Army by his Place, it belongs to
him to command in Chief, during the Abfence of the Grand Vifier.

I SHOU'D now fpeak of the Militia of *Egypt* ; but as I have not
been there, I do not underftand it enough, my Lord, to offer you any
Account of it : I fhall therefore pafs to the Maritime Affairs, concerning which I have carefully inform'd my felf in *Conftantinople*, and the
Iflands of the *Archipelago*. It is not ftrange that the *Turks* are fo weak
at Sea, becaufe they want good Mariners, skilful Pilots, and experienc'd
Officers. The Pilots of the Grand Signior fcarcely know how to ufe
the Compafs; and thofe of the Saicks, which are their Merchant-Ships,
certainly underftand nothing of it. They fteer by their Knowledge of
the Coafts, which is very erroneous ; and they generally truft themfelves
in long Voyages, as to *Syria* and *Egypt*, to *Greeks* who have run the
Courfe with Chriftian Privateers, and have got the Track of the Countries of *Afia* and *Africa* by rote. However, if the *Turks* would apply

*

themfelves

themfelves to Navigation, they would eafily become Mafters of the *Me-* diterranean, and would chafe away the *Corfairs* who do fo much Mif- chief to their Traffick. Without reckoning the Supplies they might draw from *Greece*, the Ifles of the *Archipelago, Egypt*, and the Coaft of *Africk*, the *Black Sea* alone would furnifh them with more Wood and Rigging than are needful, even for a very formidable Navy. At prefent the Maritime Forces of this great Empire are reduc'd to twenty eight or thirty Men of War; and they arm out not above fifty Gallies. The *Turks* had much more powerful Fleets in the time of *Mahomet* II. of *Selimus*, and of *Soly-man* II. but they never made any great Expeditions. Since the War of *Candia*, they have mightily neglected the Sea, and perhaps would have done fo much more, if *Mizomorto*, the Captain-Baffa, had not in our days reftor'd and improv'd their Navy. The Advantage which arofe by the Sea to the Iflands of *Spalmadori* under the *Venetians*, made him fet a won-derful value upon the Ifland of *Scio*, and gave the Mahometans frefh Spirits. He was a Man of extraordinary Capacity for the Sea, and try'd all Methods to engage Chriftian Officers in the Grand Signior's Service. The Sultan may now have fix or feven Renegado Captains, who are well experienc'd ; but the Seamen know nothing of the Tackle, and the Gun-ners are miferable to the laft degree. The Succeffor of *Mizomorto* was but little efteem'd. *Adrama Baska*, who was nam'd for Admiral up-on the Death of the other, was able to have brought the Condition of the *Turkifh* Navy to Perfection, if fome who envy'd him, had not got him * ftrangled a little after his Promotion. He was known among the *Turks*, by the Name of the Baffa of *Rhodes*, and among the Chrifti-ans, by that of the Butcher's Son of *Marfeilles*. He was taken very young in a Ship belonging to that City, and was fo unhappy as to turn Mahometan. He had the Character among the *Turks* of a very upright Man, and very difinterefted. It is faid, that as he was going the Round one day, to execute Juftice at *Scio*, he ask'd to whom three or four She-Affes belong'd, who were loaded with weighty Stones, and were ty'd to the Door of a certain Houfe ; and underftanding their Mafters

---

* January 1706. *The Pretence was, That he had not been ready enough in extinguifhing a Fire which had hurt fome Houfes by the Arfenal.*

were

were hard by at Breakfaſt, he paſs'd on; but at his Return, being diſturb'd to find the poor Creatures were ſtill in the ſame condition, and that no care had been taken to feed them, he ſent for their Owners, and told them, it was but juſt that the Aſſes ſhould eat in their turn : the Peaſants readily aſſented; but were ſurpriz'd when he order'd each of them to bear one of the Stones upon his Back all the time the Aſſes were eating. The ſame Story is alſo told concerning Sultan *Morat*.

THE Poſt of Captain-Baſſa is one of the nobleſt in the Empire: He is great Admiral and General of the Gallies: his Power is abſolute, when he is out of the *Dardanelles*; ſo that he can ſtrangle the Viceroys and Governours who are on board, without waiting for the Sultan's Order. The Grand Viſier is the only Miniſter, who is above him: It is the ſecond Poſt in the Empire; and he is accountable to the Grand Signior alone. Not only the Sea-Officers, but all the Governours of the Maritime Provinces likewiſe, receive Orders from him. At *Conſtantinople* there are not above 28 or 30 Men of War.

THE Gallies are diſtinguiſh'd into two Claſſes; namely, thoſe of *Conſtantinople*, and thoſe of the *Archipelago*: thoſe of *Conſtantinople* are at Sea only in the Summer. At the cloſe of the Campaign they are diſarm'd, to be laid up in the Arſenal of *Caſſum Baſſa*: the greater part of the Beys or Captains are Renegades. Beſide the Body of the Gally, Artillery, and Bisket, the Emperor alſo allows them Pay, and the reſt of their Equipage, which conſiſts of 200 Oars, and a Boat to go on ſhore. If the Captains are rich enough to ſubſiſt their Slaves who row, they make a conſiderable Profit, for they are allow'd twelve thouſand Livres for Rowers, and make an Advantage alſo of the Journeys in which they employ their Slaves by Land, during the other part of the Year. When there are not Rowers enough, they preſs the Slaves of private Men at *Conſtantinople*: but very little Service is done by theſe poor Wretches, who have no Experience, and periſh moſt of them at Sea. You well know, my Lord, the Service of the Sea requires much more Practice than that of the Land. To reinforce the Soldiers of the Gallies, the *Turks* add ſome of the Janizaries.

THE Gallies of the *Archipelago* are oblig'd to be in a readineſs to put to Sea at all times. The Captains are paid by Aſſignments upon the Iſlands,

and

and are bound to find their own Slaves, and pay them; for the Grand Letter I.
Signior allows them only the Veſſel, Artillery, and Rigging.   They
avoid an Engagement all they can, in order to preſerve their Slaves;
and moſt of them have neither the number of Gallies they ought to
maintain, nor their Equipage compleat, becauſe the Captain-Baſſa, for a
Sum of Money (which the others know proper ways enough to hand
to him) often winks at it; conſequently, the Military Diſcipline is very
indifferently obſerv'd.

THE Beys of *Rhodes* and *Scio* ought to provide ſeven Gallies for
each Iſland: He of *Cyprus* ſix: Thoſe of *Mytelene, Negropont, Salonica,*
and *Caval,* one apiece: *Andros* and *Syra* only one; and *Naxos* and *Paros*
the like.   The Captain-Baſſa ſails round the *Archipelago* in the Summer,
to raiſe the Capitation-Tax, and learn the ſtate of Affairs which have
happen'd.   He commonly holds his Days of Audit in a Port of *Paros,*
call'd *Drio,* which is the Centre, as it were, of the *Archipelago.*   The
Officers of the Iſlands repair thither to make their Preſents to him, and
pay in the Sums at which each Iſland is tax'd.   Here alſo the Captain-
Baſſa finally judges all matters, as well civil as criminal.   I am,

My L o r d,

Your moſt Humble and

moſt Obedient Servant,

T O U R N E F O R T.

LETTER II.

To Monseigneur the Count de Pontchartrain, *Secretary of State,* &c.

My Lord,

Of the Religion, Manners, and Customs of the Turks.

I N my laſt, I had the honour to inform you concerning the Government and Polity of the *Turks*; and in this I ſhall ſpeak of their Religion, Manners, and Cuſtoms.

O F all falſe Religions, the Mahometan is the moſt dangerous, becauſe it not only ſtrongly flatters the Senſes, but in many Points alſo agrees with Chriſtianity. Mahometiſm is founded upon the Knowledge of the true God, the Creator of all things, upon the Love of one's Neighbour, the Purification of the Body, and a quiet peaceable Life. It abhors Idols, and the Worſhip of them is ſtrictly prohibited.

Birth of Mahomet.

M A H O M E T was born an Idolater among the *Arabs, An.* 570. He had naturally a Fund of good Senſe. God forbid, I ſhould deſire to make an Encomium on him here; but I know not how to avoid looking upon him as an extraordinary Genius, and admire how ſuch a Man could be able, without the Aſſiſtance of Grace, to recover himſelf from Idolatry. They ſay, *Sergius*, a Neſtorian Monk, who ran away from *Conſtantinople*, contributed to diſabuſe him from the Errors of Paganiſm; and *Mahomet* ſhook off ſo great a Prejudice, and open'd his Eyes to diſcern the Truth.

I T appears by the Alcoran, that theſe two Men have taken out of the Holy Scripture what they thought convenient to their Purpoſe; but as in their time there were far more Jews in *Arabia* than Chriſtians, they

follow'd

follow'd the New Testament less than the Old, that so they might engage the Jews in their Sect, without too far neglecting the Christians. If *Mahomet* had not had the Folly to affect to pass for the Messenger of God, his Religion had not differ'd from Socinianism; but he had a fancy to play an extraordinary part, in making People believe he had a Correspondence with the Superior Beings. As he had neither a Mission, nor a Gift of Miracles, he was oblig'd, in order to establish his System, to join Craft and Knavery to Reason. His Enthusiasms, whether they were dissembled, or really Fits of an Epilepsy, persuaded the Multitude, that he was infinitely above other Men, and inspir'd from Heaven. His Wife and his Friends boasted he was the Interpreter of the Lord, and was sent into the World on purpose to publish his Orders. The Pigeon which he had taught to flutter about his Head, contributed not a little to support the Mystery: this Bird pass'd for the Angel *Gabriel*, who came to whisper Messages in his Ear.

THAT he might not startle the Idolaters too much, he chose to appear neither a Jew nor a Christian; and to ingratiate himself with both the last, he adopted part of the Faith of each into his Doctrine. He taught there were three sorts of written Laws communicated to Men by the Lord, and in which they might be sav'd; because they were enjoin'd by all of them to believe in one only God, the Creator and Judge of all Men. The first Law, he said, was given to *Moses*; but as it was too burdensome, few Persons were able to fulfil it strictly. The second is that of Jesus Christ; which, tho it is full of Grace, is very difficult to be observ'd, by reason of its Opposition to corrupted Nature. On this account, continu'd he, the Lord who abounds in Mercy, has sent you, by my Ministry, a Law easy and proportion'd to your Weaknesses; that so by following this exactly, every one may be able to attain Happiness in this World, and in the next.

AS I do not understand the Genius of the *Arabian* Language, not its Delicacies, the Alcoran seems to me a Book very ill compos'd, which among some good things, contains a world of childish and frivolous Tales. Notwithstanding which, the Mahometan Religion, as to some Trifles, respecting the Care every one ought to take of his Body, seems very sensibly design'd. Perhaps to engage the Imagination of Idolaters,

which

which was accuftom'd to Figures of Wood and Stone, *Mahomet* thought it was neceffary to footh them with agreeable Images taken from the other World; and that, in order to come at them by Reafon, he muft enter into their Tafte, by promifing fenfual Pleafures after Death to People, who, in their Life-time, were acquainted with no others. This Book, fuch as it is, comprehends all the Laws Ecclefiaftical and Civil of the Mahometans, and teaches them whatever they ought to believe and practife. They never offer to open it, without having firft laid it upon their Head, which with them is the higheft Token of Veneration they can give; and their chief Employment is to read it, according to that Precept in it, *Apply your felves often to read the Book which is fent you, and pray inceffantly, for Prayer turns away Sin.* They are perfuaded that thofe who read it over fo many times, make fure of Paradife. In a word, they call it, *The Book*, by way of Excellence ; for *Alcoran* fignifies nothing but *The Scripture.*

IT is needlefs to relate here how this Book was compos'd at firft, and how it was reform'd after *Mahomet*'s Death : it is fufficient to remark that there are four Sects among the Mahometans. The moft Superftitious is that of the *Arabians*, who adhere to the Traditions of *Abubeker.* That of the *Perfians*, which was founded by *Hali*, is the moft refin'd ; but the *Turks* who follow that of *Omer*, treat them as Hereticks, and pronounce Anathema's againft them. The fimpleft of all is that of the *Tartars*, who follow *Odeman* or *Ofman*, the Chief Compiler of the Memoirs of *Mahomet.*

THE only Article of Faith the Mahometans have, is, that there is but One God, and that *Mahomet* is the Meffenger of God. As to the Commandments, the *Turks* reduce them to five. 1. To pray five times a day. 2. To faft in *Lent*. 3. To give Alms, and do Works of Charity. 4. To go in Pilgrimage to *Mecha*. 5. To fuffer no Filth upon their Body. There are four other Points added, but they are not abfolutely neceffary to Salvation. 1. To keep Friday a Sabbath. 2. To be circumcis'd. 3. To drink no Wine. 4. Not to eat Swine's Flefh, nor things ftrangled.

THE Mahometans regard Friday above the other Days of the Week, becaufe they believe it was upon a Friday that *Mahomet*, being perfecuted

ted

ted by the Idolaters, was forc'd to save himself by flying from *Mecha* to *Medina* in *Arabia*. It is from this Day the Mahometan Æra begins, which they call *Egire*; and this celebrated Friday fell upon *July* 22. in the Year 622, from the Death of Jesus Christ. They are oblig'd to go every Friday at Noon to the Mosque to Prayers; but the Women are excus'd, for fear they should occasion distraction to the Men. The Tradesmen keep their Shops shut this Day till Noon, and such as are pretty rich, do not open them till the Morrow.

CIRCUMCISION, and Abstinence from Swine's Flesh and things strangled, were perhaps inserted in their Law, merely in complaisance to the *Jews*, who were then as much courted by the Mahometans, as they have since been despis'd. The Publick Good led their Legislator to forbid the Use of Wine to his Disciples. *Abstain*, says he, *from Wine, and Games of Chance, and from Chess; these are the Invention of the Devil, to sow Hatred and Division among Men, to keep them from Prayer, and hinder their calling upon the Name of God.* Notwithstanding they confess Wine is an excellent thing, and that the Temptation of it is so inviting, that it makes the Sin very pardonable. They laugh at us who drink it with Water, and say, that since it is mix'd in drinking, one should satisfy one's Appetite, and not provoke it. With respect to Swine's Flesh, the *Turks* have it in abhorrence; but the *Persians* look upon Abstinence from it rather as a Counsel than a Command. They eat it, or forbear, as they also do by Wine, according to the Practice of the Sultan whose Taste is follow'd blindly by the whole Empire. It is a pleasure to Travellers, when they enter the King of *Persia*'s Territories, that they can then drink Wine, without making a Secret of it, and can see whole Herds of Swine in the Fields. The *Persians* who dwell upon the Borders, know the Christians so well, that they run out to them as fast as they can, with Bottles of Wine and Hams, when they spy a Caravan.

AS for Circumcision, the *Turks* esteem it rather as a Mark of Obedience to their Religion, than as an essential Law: there is nothing said of this Ceremony in the Alcoran, and it is rather a Tradition borrow'd from the *Jews*. The Mahometans are of opinion, that Children dying without Circumcision are nevertheless sav'd; and they break their little Finger, before they bury them, to denote they have not been circum-

cis'd,

cis'd. The moſt ſcrupulous (as there are ſome ſuch in all Religions) be-lieve the Circumciſion of their Father has an effect upon them: but thoſe who pretend to know the Fundamental Points of their Religion better, agree that Circumciſion had not been eſtabliſh'd, but to put the Mahometans in mind, thro the reſt of their Life, of what they promis'd to God by their Profeſſion of Faith, namely, that there is no God but God, and that *Mahomet* is the Meſſenger of God; and that for this rea-ſon, Children ought not to be circumcis'd till the Age of 12 or 14 Years, that ſo they may attend to what is done. Some of their Doctors believe Circumciſion was not taken from the *Jews*, but only for the better obſerving the Precept of Cleanneſs, by which they are forbid-den to let any Urine fall upon their Fleſh. And it is certain that ſome Drops are always apt to hang upon the *Præputium*, eſpecially among the *Arabians*, with whom that Skin is naturally much longer than in other Men. At preſent moſt Renegades are not circumcis'd; it is thought e-nough to make them lift up their Finger, and pronounce the Words which expreſs the Profeſſion of their Faith. Perhaps it is out of Con-tempt that they do not circumciſe them; for the *Turks* have a common Saying, that a bad Chriſtian will never make a good *Turk*.

The Ceremony of Circumci-ſion.

THE *Turkiſh* Girls are not touch'd by circumciſing; but in *Perſia* they cut off the *Nymphæ*. Upon the Day of the Circumciſion, in *Tur-key* a Feaſt is made for the Relations of the Child, who is to be circum-cis'd. He is dreſs'd as handſomely as may be, and is led upon a Horſe or a Camel, to the Sound of Inſtruments, thro the whole Town, if it is of a moderate Compaſs; or thro a quarter of it only, if it is very large. He holds an Arrow in his right Hand with the Point toward his Heart, to ſhew he would ſooner pierce that part, than renounce his Faith. His Comrades, his Friends, and Neighbours follow him on foot, ſinging his Praiſes with Tokens of Joy, to the Moſque, where the *Iman*, after a ſhort Exhortation, cauſes him to make a Profeſſion of Faith, and lift up his Finger: after which he orders the Surgeon appointed, to place him upon a Sopha, and perform the Operation. Two Servants hold a Napkin ſpread out before the Child; and the Surgeon having drawn the Foreskin as low as he can, without prejudice, he holds it with his Pincers, and cuts it with a Razor, and ſhewing it to the Aſſiſtants, cries

*

with

with a loud Voice, *God is great*. The Child roars out all the while, for the Pain is very acute: every one comes to congratulate him upon his being admitted into the Rank of *Muſſulmans*, that is, the Faithful.

IF the Relations are rich, they cauſe the Children of the Poor in their Neighbourhood to be circumcis'd at their own Charge. After the Ceremony, they retire in the ſame Order as they came, and march as in triumph to the Relations Houſe, who treat all who come for three Days. The Expence is only a large Kettle of Rice a day, ſome pieces of Beef and Mutton, and ſome Hens: nor is the Coſt much in Liquors; for the whole Company is ſatisfy'd with one great Jar of Water. The Rich entertain with Sherbet, Coffee and Tobacco, and the Relations make Preſents to the poor Boys who were circumcis'd with their Son ; they give Alms alſo to the Poor of their Pariſh. After they have well danc'd and ſung, the Gueſts, in their turn, make Preſents to the new Muſſulman. At the Houſes of Perſons of Diſtinction, they give Veſts, Arms, and Horſes. When one of the Grand Signior's Children is circumcis'd, there are publick Rejoicings, and all the Artillery of the Seraglio is diſcharg'd : Courſes are run in the *Atmeidan,* and other Places : Gambols are play'd in the Streets, and all the Diverſions of the Bairam renew'd.

IT is worthy remarking, that the Iman does not name the new-circumcis'd; but their Father gives them what Name he will, at the time when they are born. He holds the new-born Infant in his Arms, and lifting it up towards Heaven, to offer it up to God, he puts a Grain of Salt into its Mouth, and ſays, *God grant, my Son* Solyman, (for inſtance) *that his holy Name may always be as ſavoury to thee as this Salt, and that it may keep thee from taſting the things of the Earth.* Their Names are generally *Ibrahim* or *Abraham* ; *Solyman,* which ſignifies *Solomon* ; *Iſouph, Joſeph* ; *Iſmael,* hearing God ; *Mahomet,* Laudable ; *Mahmud,* Deſirable ; *Scander, Alexander* ; *Sophy,* Holy ; *Haly,* High ; *Selim,* Peaceable ; *Maſtapha,* Sanctify'd ; *Achmet,* Good ; *Amurat* or *Mourat,* Living ; *Seremeth,* Diligent.

FROM the Counſels, I paſs to the Commandments. The Muſſulmans are ſo convinc'd that their Prayers are the Keys of Paradiſe, and the Pillars of Religion, as they ſay, that they apply themſelves to them with a Care and Attention extremely edifying. Nothing can excuſe them from

praying;

praying; and it is enjoin'd them that when they are in the Army, they shall call up one another to pray, all the time their Comrades are under Arms. *Let them*, says the Alcoran, *who go to pray, not be drunk, but sober, and have their Mind free, that they may know what they ought to do, and what they ought to say.* It is said also in the same Book, that they who pray with a disorder'd Spirit, and without thinking what they are about, tho they seem to do a good Act, have nothing of the Love of God in them.

A S the *Turks* believe that what defiles the Body, is capable also of defiling the Soul; so they are persuaded, that what purifies the one, has a power in like manner to purify the other. Upon this Principle, which is directly contrary to that of many Christians, they prepare themselves for Prayer by Ablutions. *Good People*, says the Alcoran, *when ye would say your Prayers, ye must wash your Face, your Hands, your Arms, and your Feet.* In like manner, the married Persons who have lain together, must bathe. If the Sick and the Travellers can get no Water, let them rub their Face and Hands very clean with Powder; for God loves Cleanliness: He would have the Prayers we make to him perfect, that we should thank him for the Favours he bestows on us, and often call upon his holy Name.

The great Ablution of the Turks. T H E Mahometans have reduc'd the Duty of this Commandment to two Ablutions, the great and small. The first is of the whole Body, but this is enjoin'd only to married People, who have lain together; to those who have had any Pollution in their Sleep, or who have let some Urine drop upon their Flesh when they made Water. These are the three grand Defilements of the Mussulmans. That nothing may be cover'd from the Water which ought to purify their Body and their Soul, and that it may enter the better, they pare their Nails very carefully, and take off the Hair from all Parts, except the Chin. The great Ablution consists in plunging themselves three times under Water, let the Season be as severe as it will. I have seen *Turks* in the depth of Winter leave the Caravan to throw themselves stark naked into the Brooks which were on the side of the Road, without catching either the Cholick or the Pleurify: after which, they came and join'd the Company again with such an Air of Tranquillity, as is seen in the Face of Persons whose Conscience is at peace. When they find a warm Spring, they waſh themſelves in it with pleaſure. In moſt rich People's Houſes there are Tubs which are fill'd with Water every Morning, to make the Grand Ablution. In our Paſſage from *Scio* to *Conſtantinople*, there was an honeſt Muſſulman among us, who gave three pence a time to two Mariners, to take him down by the Ship's ſide, and plunge him thrice into the Sea, as cold as it was.

IN order to make the leſs Ablution, they turn their Face towards *Me-cha*, and waſh their Hands and their Arms as high as the Elbow, and rince their Mouth three times, and clean their Teeth with a Bruſh. After this, they are oblig'd to waſh the Noſe thrice, and ſquirt thro the Noſtrils ſome Water, which they drink up out of the hollow of the Hand: they alſo ſprinkle their Face three times; they are enjoin'd to rub themſelves from the Forehead down to the lower part of the Head with the right hand thrice; from whence they paſs to the Ears, which they muſt make very clean within and without; and the Ceremony concludes with the Feet.

MAHOMET might ſay, if he pleas'd, that his Law was eaſy to be practis'd; but, for my part, I thought it troubleſome enough, and make no queſtion but moſt of the Renegadoes break thro theſe Trifles. When they make water, they ſquat down like Women, for fear ſome Drops of Urine ſhould fall into their Breeches. To prevent this Evil, they ſqueeze the part very carefully, and rub the Head of it againſt the Wall; and one may ſee the Stones worn in ſeveral Places by this Cuſtom. To make themſelves ſport, the Chriſtians ſmeer the Stones ſometimes with *Indian* Pepper, and the Root call'd *Calfs-foot*, or ſome other hot Plants, which frequently cauſes an Inflammation in ſuch as happen to uſe the Stone. As the Pain is very ſmart, the poor *Turks* commonly run for a Cure to thoſe very Chriſtian Surgeons, who were the Authors of all the Miſchief: they never fail to tell them it is a very dangerous Caſe, and that they ſhould be oblig'd perhaps to make an Amputation: the *Turks*, on the contrary, proteſt and ſwear they have had no Communication with any ſort of Woman that could be ſuſpected. In ſhort, they wrap up the ſuffering Part in a Linen dipp'd in Oxicrat, tinctur'd with a little Bole-Armenic; and this they ſell them as a great Specifick for this kind of Miſchief.

WHEN they go to ſtool either at home or in the field, they furniſh themſelves with two large Cloths, which they carry at their Girdle, or acroſs their Shoulders juſt as a Butler carries a Napkin ; they alſo take a Pot of Water in their hand, which ſerves to make the *Taharat,* that is, to waſh themſelves below with their Finger. The Grand Signior himſelf cannot diſpenſe with this Cuſtom ; it is the firſt Leſſon his Governour teaches him : we may preſume, that after this Operation the *Turks* muſt waſh and ſcour the tops of their Fingers frequently. Nor is this the only Inconvenience ; for there are a great many things which annul this Ablution, and oblige them to begin it anew : as for inſtance, if they happen to break wind ; but it is an inſufferable Misfortune if a Man has a Looſeneſs, for in that caſe this Ablution, which muſt be perpetually repeated, becomes an exceſſive Burden. I have heard the *Turks* ſay, that one of the principal reaſons which hinders them from travelling into Chriſtian Countries, is becauſe they cannot have Conveniences to perform theſe Duties.

AS to a particular Ablution, that muſt be done for the leaſt Fault ; as, for having blown their Noſe with the right hand ; for having waſh'd the Parts of the Body more than three times ; for having us'd on this occaſion Water warm'd in the Sun. It is the ſame alſo, if they happen to throw the Water upon their Face with too much Violence ; if Blood or any Ordure falls upon their Body, if they vomit, if they fall into a Swoon, if they drink Wine, or ſleep at Prayers : in a word, if they touch a Dog, or any other unclean Animal. All theſe reaſons cauſe them to build Reſervatories and Fountains, and Turn-Cocks about their Moſques, or in their Houſes. Upon want of Water, they are permitted to make uſe of Sand, Powder, *B. 1. c. 13.* or ſome Plants proper to cleanſe themſelves with. *Rablais*'s Chapter, which carries a pleaſant Title, would be a wonderful Relief to them, if it was tranſlated into their Language.

AFTER they have purify'd themſelves, the *Turks* fix their Eyes on the Ground, and retire ſeriouſly inward, in order to diſpoſe themſelves for their Prayer, which they make five times a day. 1. In the Morning, between the Break of Day and Sun-riſing. 2. At Noon. 3. Between Noon and Sun-ſet. 4. At Sun-ſet. 5. About an hour and a half after the Sun is down. All theſe Prayers are accompanied with many Bowings,

* and

and some Prostrations. They may make their Prayers either at home or in the Mosques; and they have notice given them of the Hours appoint-ed for this Exercise, by Men hir'd on purpose, who guide themselves by the Course of the Sun, or by an Hour-glass. These Fellows are a sort of speaking Clocks, for at set Hours they go up to the Galleries of the Pinacles, and stopping their Ears with their Fingers, bawl out as loudly as they are able, the following Words; *God is great, there is no other God but God; come to the Prayer, I summon you with a clear Voice.* They re-peat these Words four times, turning themselves first to the South, then to the North; after that to the East, and lastly to the West.

AT this Signal every one makes his Purification, and then goes to the Mosque, at the Door of which they put off their Shoes, unless they chuse to take them with them in their hand, for fear they shou'd be mix'd with those of others who come there. All this is done with a profound Silence. They salute with a deep Reverence the Nich where the Alcoran is plac'd; and this Place is directed toward *Mecha.* After this, every one lifts up his Eyes, and puts his Thumbs into his Ears before he sits down: the very manner of sitting down is also the most humble among them as can be, for they sit upon the Calf of the Leg; they continue thus for some time, and cast down their Eyes, and kiss the Earth thrice: after this they take their Seats, and wait for the Priest to begin, whom they follow, and make the same Inclinations as he does. It is at this time, that their Decency is most admirable: they salute nobody, nor dare to hold discourse, nor take notice of any one whoever it is, nor mind what pas-ses. The whole Assembly is unmov'd, no one either spits or coughs: in fine, they give no token of Life, but by some profound Sighs, which are rather the Aspirations of the Soul towards God, than mechanical Mo-tions of the Body. Amidst these Sighs the Priest stands up, and spreads his Hands upon his Head, stops his Ears with his Thumbs, and lifting his Eyes towards Heaven, sings with a loud and distinct Voice, *God is great*; *Glory to thee, O Lord: May thy Name be blessed and praised: may thy Greatness be acknowledged; for there is no other God beside thee.*

THIS is the Prayer which they commonly repeat with their Eyes turn'd down, and their Hands across their Stomach. They also use the following Prayer, which is the same to them, as the Lord's Prayer is to us.

H 2　　　　　　　　　　　　　　　　　*IN*

IN the Name of God, full of Goodnefs and Mercy! Praifed be God, the Lord of the World, who is one God, full of Goodnefs aud Mercy. Lord, who fhalt judge all Men; we worfhip thee, we place our whole Truft in thee. Preferve us, who call upon thee, in the right way, which thou haft chofen, and doeft favour with thy Acceptance. It is not the way of the Infidels, nor of thofe againft whom thou art juftly incens'd. So be it.

AFTER this, they make the Inclinations, and reft their Hands upon their Knees, which are half bent, and make this Prayer, *God is great : Glory to thee, O Lord,* &c. or elfe they fay three times, *Let the Name of the Lord be glorified.* Then they proftrate themfelves again, kiffing the Ground twice, and crying out as often, *O great God, may thy Name be glorify'd.* They alfo recite that Prayer, *In the name of God, full of Goodnefs and Mercy,* &c. To which they add the following Article out of the Alcoran, *I acknowledge that God is God, that God is eternal, that he neither begot, nor is begotten, and has none who is like him or equal to him.* After having made the Inclinations which the Hour of Prayer requires, they raife themfelves half up, refting ftill upon their Feet; and cafting their Eyes upon their Hands, fpread open like a Book, they pronounce the following Words.

ADORATION and Prayers are due only to God. Salvation and Peace be to thee, O Prophet. The Mercy, the Bleffings, and the Peace of the Lord be upon us and upon the Servants of God. I declare there is but one God, that he has no Companion, and that Mahomet *is the Meffenger of God.*

THEY clofe their Prayers with the Salutation of the two Angels, who, they believe, are at their fide. In performing this Duty, they take hold on their Beard, and turn to the right hand and to the left. One of thefe Angels, they imagine, is white, and the other black: The white, as they believe, excites them to do Good, and keeps a Regifter of their good Actions; and the black rules over their evil Actions, to accufe them for them after their Death. In faluting each Angel, they fay, *The Salvation and the Mercy of God be upon thee.* They believe alfo that their Prayers will not be heard, unlefs they firft refolve firmly to forgive their Enemies. It is for this reafon, that they never let a Friday pafs without making a hearty Reconciliation; and hence it is that we never hear of any Detraction or Injury among the *Turks.*

THE

THE Friday-Prayers are defign'd for invoking the Grace of God upon all Muffulmans. On Saturday they pray for the Converfion of the *Jews*; and on Sunday for that of the Chriftians; on Monday for the Prophets; on Tuefday for the Priefts, and for them who honour the Saints in this World; on Wednefday for the Dead, and for the Muffulmans who are in Slavery among the Infidels; on Thurfday for the whole World, of whatever Nation, and of whatever Religion. The Mofques are moft frequented upon Friday, and are better illuminated, and the Prayers are made with the greateft Solemnity.

WE never faw them at Prayer in the Mofques, becaufe the Chriftians are not fuffer'd to enter while any Muffulman is there; but we have feen them at Prayer in the Caravans. The Chief of the Caravans, knowing what Hour it is by the Elevation of the Sun, ftops them, and calls them to Prayers, exactly like the ordinary Chanters: the Chriftians and the Jews wait by on horfeback, if they pleafe, or elfe ride out during the time. Every Muffulman fpreads his Carpet on the Ground, and makes the Inclinations, and fays over the Prayers. Very often the Chief of the Caravan fupplies the Place of the Prieft; but if they light upon a Dervife, as they commonly do in the Caravans of *Afia*, he exercifes the Function. All this is done in the middle of the Field, with the fame Attention and Decency, as if they were in a Mofque. When there are but two or three *Turks* in a Caravan, one fhall fee them ftep afide out of the Road to pray, and then put on full fpeed to get up to the Company. Nothing can be more exemplary than thefe Exercifes; and it has rais'd the utmoft Indignation in me againft the *Greeks*, who commonly live like fo many Brutes.

BESIDE the daily Prayers I have mention'd, the *Turks* refort to the Mofques at Midnight in *Lent*, to make the following Prayer.

LORD God, who paffeft by our Faults; thou who alone oughteft to be lov'd and honour'd; who art great and victorious; who ordereft the Night and the Day; who pardoneft our Offences, and cleanfeft our Hearts; who fheweft Mercy, and difpenfeft thy Benefits to thy Servants: Adorable Lord, we have not honour'd thee as thou oughteft to be honour'd. Great God, who deferveft that we fhould fpeak of nothing but thee; we have not fpoken of thee fo worthily as we ought. Great God, whom we ought to thank con-

tinually,

tinually, we have not given thee sufficient Thanks. Merciful God, all Wisdom, all Goodness, all Virtue come from thee: it is of thee we must seek Forgiveness and Mercy. There is no God but God. He is one only. He has no Companion: Mahomet *is the Messenger of God. My God, let thy Blessing be upon* Mahomet, *and upon the Race of* Mussulmans.

The Lent of
the Turks.
THE *Turkish Lent* takes its Name from the Month in which it falls, which is the Moon of *Ramazan* or *Ramadan,* for they always reckon by Moons. Their Year consists of 354 Days, divided into twelve Moons or Months, which begin upon the new Moon: these Months contain alternately 30 Days and 31. The first of them, which has 30 Days, is call'd *Muharrem*; the second *Sefer,* and contains but 29 Days; the third *Rebiul-euvel*; the fourth *Rebiul-ahhir*; the fifth *Giamazil-euvil*; the sixth *Giamazil-ahhir*; the seventh *Regeb*; the eighth *Chaban*; the ninth *Ramazan* or *Ramadan*; the tenth *Chuval*; the eleventh *Zouleudé*; the twelfth *Zoulhigé.* These Months do not follow the Seasons, because they do not agree to the course of the Sun; and their Years have twelve Days fewer than ours: the *Ramazan* falls higher every Year the same number of Days: from whence it comes that in some Years it runs thro all the Seasons.

THE *Lent* was appointed in the Month of *Ramazan,* because *Mahomet* declar'd the Alcoran was sent to him from Heaven at that time. The Fast which it ordains, is different from ours, in that it is absolutely prohibited, during the whole Course of that Moon, to eat or drink, or take any thing into their mouth, or even to smoke, from Sun-rising till its setting. To make amends, while the Night continues, they are allow'd to eat and drink without distinction of Meats or Drinks, excepting only Wine; for it would be a high Crime to taste this, and formerly the Crime could be expiated only by pouring melted Lead down the Offender's throat: at present they are not so severe, tho they still punish it corporally. In the Night also they never spare *Aquavitæ,* during this time of Penitence; and much less the Sherbet and Coffee: and there are some, who under a pretence of Penitence indulge themselves more deliciously than all the rest of the Year. Self-Love, which is always ingenious, prompts them at this time to enjoy Good-Cheer, in a Season appointed for Mortification: the Devotees comfort their Stomachs with Sweetmeats, tho they are

made

made ordinarily of Honey and Rofin. The Rich obferve *Lent* as ftrictly as the Poor, and the Soldiers as the Religious, and the Sultan himfelf as the meaneft private Man. In the day-time they take their Repofe, and mind nothing but to fleep, or at leaft to fhun the Exercifes which occafion Drought; for it is an intolerable Punifhment not to be able to drink Water amidft fuch exceffive Heats. Labourers, and Travellers, and Country-People, fuffer very much; it is true, they are excus'd in breaking *Lent*, provided they keep an account of the Days, and faft the fame number afterwards, when their Affairs permit. Upon the whole matter, *Lent* with the Mahometans is only living differently from their ufual manner. When the Moon of *Chaban*, which immediately precedes that of *Ramazan*, is pafs'd, they watch very carefully for the New Moon. An infinite Croud of People of all Conditions poft themfelves upon the high places, and run away to give notice of its Appearance; fome do it out of Devotion, and others to obtain a Reward. The very moment they are affur'd of the Fact, they publifh it through the whole Town, and begin to faft. In places where there are any Cannons, they fire one Round at Sun-fet. They light up fuch a prodigious number of Lamps in the Mofques, that they look like fo many Chappels on fire: they take care alfo to make great Illuminations upon the Pinacles in the night.

THE *Muezins* at the Return of the Moon, that is, at the Clofe of the firft Day of the Faft, proclaim with a loud Voice, it is then time to pray and eat. The poor Mahometans, who are choak'd with Thirft, begin then to fwill off huge Draughts of Water, and fall greedily to their Plates of Rice. Every one refrefhes himfelf with the beft Provifion; and as if they apprehended they fhould die with hunger, they go out to eat abroad after they have ftuff'd themfelves at home: fome run to Coffee, others to Sherbet; and the more Charitable give Victuals to fuch as come. One may hear the Poor cry in the ftreets, *I pray God fill their Purfe, who give me fomething to fill my Belly.* They who think to improve their Pleafures, fatigue themfelves in the Night as much as they can, that they may reft the better in the Day, and pafs the time of the Faft without trouble. They fmoke then during the Darknefs, after they have eat fufficiently, and play upon Inftruments, and have Puppet-fhews by Lamplight. All thefe Diverfions continue till the Morning is clear enough for

them

them to diftinguifh, as they fay, a white Thred from a black : then they repofe themfelves, and the Name of a Faft is given to undifturb'd Slumber, which continues till Night.   None but fuch as are forc'd by Neceffity, go about their ordinary Work.   Where is then, according to them, the Spirit of Mortification, which ought to purify the Souls of Muffulmans!   Thofe who love a diforderly Life, wifh this Penitential Seafon were to laft half the Year ; and the more, becaufe it is follow'd by the Grand *Bairam*, in which, by an agreeable Alternative, they fleep all night, and rejoice all the Day.

The Bairam.   ABOUT the end of the Moon *Ramazan*, they look out heedfully for that of *Chuval*, and proclaim the *Bairam* as foon as they perceive it. One hears then nothing but the Sound of Drums and Trumpets in the Palaces and Publick Places.   If the cloudy Weather hinders their difcerning the New Moon, they keep back the Feftival one day ; but if the Clouds continue, they fuppofe there ought to be a New Moon, and kindle Bonfires in the ftreets.   The Women who are fhut up all the year, have the liberty of going abroad the three days this Feaft continues ; and every where are feen Muficians, Flying-Chairs, and Wheels of Fortune.   In thefe Chairs they are carry'd aloft in the Air, by means of Cords which the Men pull with more or lefs Violence, as the Perfon chufes. The Wheels of Fortune are like thofe of a Water-mill, and are turn'd round, without thofe who are feated in them fo much as touching one another, tho every one finds himfelf in his turn at the top and at bottom of the Wheel.

THE firft Day of *Bairam* the Muffulmans make a general Reconciliation with one another, and join Hands mutually in the Streets ; and having kifs'd thofe of their Enemies, they lay them upon their Head. They wifh one another a thoufand Profperities, and fend Prefents as we do at the beginning of the Year.   The Preachers explain in the Mofques fome Points of the Alcoran ; and after the Sermon, is fung the following Prayer : *Salvation and Bleffing upon thee,* Mahomet, *Friend of God. Salvation and Bleffing upon thee,* Jefus Chrift, *the Breath of God.   Salvation and Bleffing upon thee,* Mofes, *the Familiar of God.   Salvation and Bleffing upon thee,* David, *the Monarch eftablifh'd by God.   Salvation and Bleffing upon thee,* Solomon, *the Faithful of the Lord.   Salvation and Blef-*
fing

sing upon thee, Noah, *who wert saved by the Favour of God.   Salvation and Blessing upon thee,* Adam, *the Purity of God.*

THE Grand Signior appears more magnificent upon this Day than ordinary; and receives the Compliments of the Great Ones of the Port, and gives them a sumptuous Repast in the Hall of the Divan. At his Return from *Sancta Sophia,* they say, he mounts his Throne, having the Chief of the White Eunuchs at his left side. If the Sons of the Cham of *Tartary* are at Court, they come first of all to prostrate themselves before him, and withdraw not till they have kiss'd his Hands, and wish'd him a happy Festival. Then the Grand Visier presents himself at the head of the Viceroys and Bassa's, who are in Town; and having made his Compliment to the Sultan kneeling, he kisses his Hand, and takes the Place of the Chief of the white Eunuchs. The Mufti, accompanied by the Chief-Justices, the Head-Cadi's, and the most celebrated Preachers, and, in a word, by all those who are call'd the principal Officers of the Faith, and by him who terms himself the Chief of the Race of *Mahomet:* the Mufti, I say, bowing his Head to the Ground, and with his Hands in his Girdle, goes to kiss the Sultan's Shoulder; and they say the Prince advances one Step to receive him. The Janizary-Aga makes his Compliment last of all, after the Officers who attended the Mufti have made their Reverence. At the Repast, the Grand Signior distributes Vests of Sable to the prime Officers of the Port. All this is transacted at the Entrance of the Seraglio. In the inner Rooms of the Palace the Sultan receives the Compliments of the Chief Eunuchs and Waiters. The Sultanesses also come out of their Apartments, and are carried abroad with the Grand Signior in Coaches; but they fasten the Coaches up as carefully, as if they were carrying out so many Prisoners. I am inform'd, that during the three Days the Women are permitted to come to the Sultan, he is serv'd only by black Eunuchs; the Pages and white Eunuchs, and in short, all whose Complexion is not black, being remov'd for all that time. The Women also visit one another, after they have paid their Homage to the Emperor.

THE Mahometans likewise observe other Festivals during the rest of the Year. I have given your Lordship an Account of the smaller *Bairam* in my third Letter: this is solemniz'd the 70th Day after the other, *viz.* upon the 10th of the Moon af *Zoulhigè*; and the Pilgrims who go to *Mecha*, order their Journey so skilfully, as to arrive there the Evening before. The *Turks* celebrate with Joy also the Night of *Mahomet*'s Birth, which is from the 11th to the 12th of the third Month. They make the usual Illuminations in the Mosques and Pinacles of *Conftantinople*. The Emperor goes to the new Mosque, where he gives a Collation after Prayer, and orders Sweet-meats and Drinks to be distributed. *Mahomet*, as the Mussulmans believe, was carried to Heaven upon *Alborac*, the Night from the 26th to the 27th of the 4th Month, which is a Day of a high Festival with them. Two Months before the *Ramazan*, they celebrate the Night from the 4th to the 5th of the 7th Month, to put them in mind that *Lent* is at hand. They never fast on account of these Feasts; but on the contrary, after having pray'd by night in the Mosques, they go in the day-time to make merry at home, or with their Friends.

THE *Turks* do not wait for Festival Days to do Works of Charity; for as Alms-giving is an indispensable Commandment with them, they esteem it the most certain Means to increase their Store, and draw down the Blessing of Heaven upon their Estates. *They who read the Alcoran*, says Mahomet, *who pray, and who give of the Goods which God has given them, either in publick or in private, may rest assur'd they shall lose nothing thereby. They shall be amply re-imburs'd for all they have given. God, whom we ought always to glorify, pardons the Sins of them who do Charity, and pays with Interest whatever is given in his Name.* They are enjoin'd to give Alms only in view of pleasing God, and not from a Principle of Vanity. *Lose not the Gain, ye Rich, of your Alms, in seeking to have them seen: for he who bestows them in order to have them seen, and not with an Intention of rendring himself accepted of the Lord in the Day of Judgment, is, with respect to the things of Heaven, as a Field full of Stones cover'd with a little shallow Dust, which the least Rain washes away, so that nothing remains but the Stones.*

THE Mahometan Casuists are not agreed by what Rules every one ought to proportion his Alms. Some think it is sufficient to give One in the Hundred of all one's Goods; others pretend, they ought to lay by a fourth Part for the Poor: but the most Severe oblige them to give a tenth. Beside private Alms, there is no Nation which expends more

upon

upon publick Foundations than the *Turks*. Even they who have but a Letter II. moderate Fortune, leave something after their Death, to maintain a Man to give Water in the Summer-Heats to drink to Paſſengers, as they go along by the Place where they are bury'd : Nor do I queſtion but they would have alſo order'd Veſſels of Wine, if *Mahomet* had not forbidden the uſe of it. The manner of giving Alms is very well explain'd in the following Precept: *Help your Father and Mother, your next Relations, the Orphans, your Neighbours, them who travel with you, the Pilgrims, and thoſe who are under your power; but do nothing out of Vanity, for that is abhorr'd of God. I will ſeverely puniſh* (ſays the Lord) *and will cover with Confuſion the Covetous, who not content to impart nothing to others out of the Goods which I entruſted them with only as Stewards, on the contrary affirm, they ought not to give. Let them who have Faith give Alms, and pray before the Day of Judgment comes; for there will be no time for obtaining Paradiſe after that terrible Day.*

T H E R E are no Beggars to be ſeen in *Turkey*, becauſe they take care to prevent the Unfortunate from falling into ſuch Neceſſities. The Sick viſit the Priſons, to diſcharge thoſe who are arreſted for Debt: they are very careful to relieve Perſons who are baſhfully aſham'd of their Poverty. How many Families may one find, who have been ruin'd by Fires, and are reſtor'd by Charities? They need only preſent themſelves at the door of the Moſques. They alſo go to their Houſes to comfort the Afflicted. The Diſeas'd, and they who have the Peſtilence, are ſuc-cour'd by their Neighbours Purſe, and the Pariſh-Funds; for the *Turks*, as *Leunclavius* obſerves, ſet no bounds to their Charities. They lay out Mony for repairing the Highways, and making Fountains for the benefit of Paſſengers; and build Hoſpitals, Inns, Baths, Bridges, and Moſques.

T H O the fineſt Moſques are at *Conſtantinople*, at *Adrianople*, *Burſa* or *Pruſa*; yet there are ſome Conveniences provided in thoſe of the princi-pal Towns, and Receptacles of Water for making the Ablutions. The Body of the Moſque is generally a very handſome Dome, the Inſide is very plain, and upon the Walls is written the Name of God in *Arabick*. The Nich where the Alcoran lies, is always fronting the ſide towards *Mecha* ; and the Dedication of the moſt celebrated Moſques, is made by fixing there a piece of ſtuff which had ſerv'd to ſupport the Moſque at *Mecha*.

The fmalleft Mofque has its Pinacle ; and thofe which make any tolerable figure, have two : if there is none, the Muezim places himfelf at the Door, and putting his Thumbs into his Ears, turns himfelf to the four Quarters of the World, and proclaims the Hours of Prayer. This Chanter ferves inftead of a Clock, a Quadrant, and a Dial ; for there is no fuch thing as a Watch in all *Turkey.* Their Service is uniform in all the Churches. All the Officers are under the Curate, who, as the chief Minifter, preaches and fays the Prayers. As good as the Pavement of the Church is, it is always cover'd with a Carpet or a Mat. As for the Revenues of the Mofques, it is certain none of them are poor ; the greater part are very rich : and they fay the Church poffeffes a third part of the Lands of the Empire. *Orchan,* the fecond *Ottoman* Emperor, chang'd the *Greek* Churches into Mofques ; his Succeffors did the fame, and augmented their Revenues, fo far were they from leffening them. This Emperor alfo was the firft who caus'd Hofpitals to be built for the Poor, and the Pilgrims ; he founded Colleges, and endow'd them, for the Education of Youth. There are few confiderable Mofques, but have their Hofpitals and Colleges. The Poor, of whatever Religion they are, are reliev'd in thefe Hofpitals ; but they admit none into the Colleges except *Mahometans,* who are taught to read and write, and interpret the Alcoran. Several apply themfelves there to Arithmetick, Aftrology, and Poefy ; tho the Colleges are principally appointed to breed them to the Law.

THE Inns upon the publick Foundation are large Buildings, long or fquare, and in appearance like a Barn. On the Infide there is only a Bench fix'd to the Wall, about three foot high, and fix broad : the reft of the Place is for lodging the Mules, Horfes, and Camels. The Bench ferves the Men for a Bed, a Table, and a Kitchin. They have fmall Chimneys there, feven or eight foot wide, one with another, where they hang on the Pot. When the Broth is ready, they fpread a Napkin, and fit round with their Legs acrofs, like Taylors. The Bed is foon made after Supper, for they only fpread their Carpet, and lay on their Baggage and Clothes ; a Saddle fupplies the place of a Pillow, and their Clothes ferve them for a Coverlid. The greateft Convenience is, that in the Morning they mount their Horfe without getting down from the Bench they flept on, for it is even with the Stirrups. The Carriers hold the

Stirrup

Stirrup while the Paffengers mount ; thefe Fellows never fleep, but fpend Letter II
the moft part of the Night in feeding the Horfes, and making them ready.

AT the Door of thefe Inns are Bread, Eggs, Hens, Fruits, and fometimes Wine; and if any thing is wanted, they procure it at the neighbouring Town.   If there are any Chriftians there, then one may get fome
Wine ; if not, one muft go on without it.   They pay nothing for Lodging: Thefe publick Hoftries in fome degree keep up the Hofpitality fo
praife-worthy in the Antients.

THE private Inns in the Towns are more convenient and better built;
they are very like Monafteries, for a great many have a little Mofque
belonging to them.   The Fountain is commonly in the middle of the
Court, and the neceffary Houfes are round about ; the Chambers are
rang'd along a large Gallery, or in very lightfome Dormitories.   In the
publick Inns the whole Entertainment cofts a fmall Piece of Mony to the
Keeper, and Provifions are cheap enough in the others : the beft way to
make one eafy there, is to get a Room where a Man may drefs his own
Diet ; the Market is juft at hand, for you have Meat, Fifh, Bread, Fruits,
Oil, Butter, Pipes, Tobacco, Coffee, Candles, and Wood, at the door.
One muft apply to the Jews or the Chriftians for Wine, and for a fmall
matter they will bring you fome privately ; the Jews have the beft, and
the worft is the *Greeks* : we had generally very good, becaufe our People,
who had an Intereft there, took care to give out through the Quarters
that we were Phyficians.   They came about us to ask for Medicines, or
to beg us to look upon their Sick, and the Fee was commonly fome Bottles of excellent Wine.   There are feveral of thefe Inns, where Straw,
Barley, Bread, and Rice are provided at the Founder's charge.   Thofe of
Europe are better built, and better endow'd and accommodated, than
thofe of *Afia*; for in the great Towns they are cover'd with Lead, and
embellifh'd with feveral Domes : but it raining very feldom in *Afia*, it is
more pleafant to walk in the Fields, in fine Weather, along the Brooks
fide, where one may catch admirable Trouts ; there are Partridges alfo
almoft in every corner.

AS Charity and Love of one's Neighbour are the moft effential Points
of the *Mahometan* Religion, the Highways are generally kept mighty
well; and there are Springs of Water common enough, becaufe they are

wanted :

wanted for making the Ablutions. The Poor look after the Conduit-Pipes, and those who have a tolerable Fortune repair the Causeys. The Neighbourhood joins together to build Bridges over the deep Routs, and contribute to the Benefit of the Publick, according to their power. The Workmen take no Hire, but find Masons and Labourers *gratis* for the several sorts of Work. You may see Pitchers of Water standing at the doors of the Houses in the Towns for the use of Passengers; and some honest Mussulmans lodge themselves under a sort of Sheds, which they erect in the Road, and do nothing else during the great Heats, but get those who are weary to come in and rest themselves, and take a Refresh-ment. The Spirit of Charity is so extensive among the *Turks,* that the Beggars themselves, tho there are very few to be seen, think they are oblig'd to give their Superfluities to other poor Folks; and carry their Charity, or rather Vanity, to such an Extreme, that they give their Leavings even to sufficient Persons, who make no scruple to receive their Bread and to eat it, to shew how highly they esteem their Virtue.

THE Charity of the *Mahometans* is extended also to Animals, and Plants, and to the Dead. They believe it is pleasing to God, since Men who will use their Reason, want for nothing; whereas the Animals, not having Reason, their Instinct often exposes them to seek their Food with the loss of their Lives. In considerable Towns, they sell Victuals at the Corners of the Streets, to give to the Dogs; and some *Turks* out of cha-rity have them cured of Wounds, and especially of the Mange, with which these Creatures are miserably afflicted toward the end of their Life: and one may see Persons of good Sense, out of mere Devotion, carry Straw to lay under the Bitches which are going to whelp; and they build them small Huts, to shelter them and their Puppies. One would hardly believe there are Endowments settled in Form by Will, for main-taining a certain number of Dogs and Cats, so many Days in the Week; yet this is commonly done: and there are People paid at *Constantinople,* to see the Donor's Intention executed, in feeding them in the streets. The Butchers and Bakers often set aside a small Portion to bestow upon these Animals. Yet with all their Charity the *Turks* hate Dogs, and ne-ver suffer them in their Houses; and in a time of Pestilence they kill as many as they find, thinking these unclean Creatures infect the Air.

ON

ON the contrary they love Cats very well ; whether it be for their natural Cleanliness, or because they sympathize with themselves in Gravity, whereas the Dogs are wanton, sporting, and noisy. Besides, the *Turks* believe, from I know not what Tradition, that *Mahomet* had such a love for his Cat, that being consulted one day about a Point of Religion, he chose rather to cut off the Skirt of his Garment upon which the Cat lay asleep, than to wake her in getting up, to go and speak with the Person who was waiting. The *Levant* Cats however are not more beautiful than ours, and the fine Cats of a Tabby-grey Colour, are very scarce there : they bring them from the Island of *Malta*, where the Breed is common enough. Among the Birds, the *Turks* look upon Turtle-Doves and Storks as sacred, and it is not lawful to kill them ; on the contrary, the *Greeks* of the *Archipelago* are great Eaters of the Turtle-Doves, and count them a delicious Dish : they are in short the best Wild-Fowl of the *Levant*, and yield to a Francolin only in bulk ; but they must be eaten roasted, for those which are salted in Barrels, like Anchovies, lose all their Taste. The *Turks* think they do a Work of Charity in buying Birds in a Cage, in order to set them at liberty ; tho at the same time they make no scruple to keep up their Women in a Prison, and our Slaves at the Chain. Those who catch Birds by Bird-lime, or any other way, believe they do no harm, because their Intention is to furnish them to those who are able to redeem them, in order to release them, and thereby have an occasion to do good Works ; so that every one hopes to find his account in it before God : so true is it, that the Direction of the Intention is natural to all these Men.

AS to Plants, the most Devout among the *Turks* water them out of charity, and cultivate the Earth where they grow, that they may thrive the better. Sultan *Osman*, they say, seeing a Tree at a distance, which had the Figure of a Dervise, settled a Salary of an Asper a day for a Man to take care of it. Tho it was Simplicity, not to say Folly, to follow this Emperor's Example, yet the good Mussulmans believe they do in it a thing agreeable to God, who is the Creator and Preserver of all things. They are also weak enough to imagine they do a Pleasure to the Dead, in pouring Water upon their Tombs ; for this, say they, may be a Refreshment to them : and there are several Women, who go to eat

and

and drink in the Cemeteries upon a Friday, believing that by this they appeaſe the Hunger and Thirſt of their deceaſed Husbands.

BEFORE I entertain you, my Lord, with an Account of all the Practices of the *Turks* with reſpect to the Dead, it will be proper to explain the two Commands which are remaining; namely, that concerning the Journey to *Mecha*, and that concerning Cleanlineſs. The Pilgrimage to *Mecha* is not only difficult becauſe of the Length of the Way, but on account of the Dangers alſo in *Barbary*, where Robberies are frequent, Water ſcarce, and the Heats exceſſive. It is true, the *Mahometans* may have a Diſpenſation, and ſubſtitute a Man to run theſe hazards in their ſtead. They look upon the Temple of *Haram*, which is that of *Mecha*, as the Work of *Abraham*. *Cauſe all the World to know*, ſays the Alcoran, *that God has commanded them to follow the Religion of* Abraham, *who was neither Idolatrous nor Unbelieving: That it is* Abraham *who built the Temple at* Mecha, *which is the firſt that was built for praying to the Lord. The Honour which is paid to this, is well-pleaſing to God; who wills that all who are able to go thither, ſhould go.* The Muſſulmans never trouble themſelves about the falſe Chronology, and would condemn any one to the flames, who ſhould dare to deny there was ſuch a Town as *Mecha* in *Abraham's* time.

THE four Places of Rendevouz for the Pilgrims, are *Damaſcus*, *Cairo*, *Babylon*, and *Zebir*. They prepare themſelves for this toilſome Journey, by a Faſt which ſucceeds that of *Ramazan*, and aſſemble in Troops at the places appointed. The Subjects of the Grand Signior, who are in *Europe*, reſort generally to *Alexandria* on board Ships of *Provence*, the Maſters whereof are obliged to give the Pilgrims paſſage *gratis*. At the approach of the ſmalleſt Veſſel, theſe good Muſſulmans, who think of nothing but falling into the hands of the *Malteſe*, run to kiſs the Banner of *France*, wrap themſelves up in it, and regard it as their Aſylum. From *Alexandria* they paſs to *Cairo*, to join the Caravan of *Africans*. The *Turks* of *Aſia* aſſemble at *Damaſcus*; the *Perſians* and *Indians* at *Babylon*; the *Arabians*, and thoſe of the adjacent Iſlands, at *Zebir*. The Baſſa's who go, embark at *Suez*, a Port of the *Red Sea*, three days Journey and a half from *Cairo*. All theſe Caravans take their meaſures ſo well, that they arrive the Eve of the leſs *Bairam* at the Hill *Arafagd*, which is

one

one day's March from *Mecha*. It is upon this celebrated Hill, they believe, that the Angel appear'd the first time to *Mahomet*, and here is one of their principal Sanctuaries. After having kill'd some Sheep to give to the Poor, they go to make their Prayers at *Mecha*, and from thence to *Medina*, where is the Tomb of the Prophet, upon which they spread every year a very rich and magnificent Pall, which the Grand Signior sends thither as a Present of Devotion : the antient Pall is worn away by pieces, for the Pilgrims tear off a piece of it, be it ever so small, and keep it as an invaluable Relique.

THE Grand Signior also sends, by the Super-Intendant of the Caravans, five hundred Sequins, an Alcoran cover'd with Gold, several rich Carpets, and a great many Pieces of black Cloth for the Hangings of the Mosques of *Mecha*. The noblest Camel in the Country is chosen to carry the Alcoran ; at his Return this Camel is hung with Garlands of Flowers, and cover'd with Benedictions, is richly fed, and excus'd from Labour all the rest of his days. They kill him with Solemnity when he is very old, and eat his Flesh as holy Flesh ; for if he should die of Age or Sickness, his Flesh would be lost, and be subject to Putrefaction. The Pilgrims who have made the Journey to *Mecha*, are held in great Veneration the remainder of their Life ; and being absolv'd of all sorts of Crimes, they commit them anew with Impunity, since, according to the Law, they are not to be put to death : they are reputed incorruptible, irreproachable, and sanctify'd from this World. Some *Indians*, they say, are foolish enough to put out their Eyes after they have seen what they call the Holy Places of *Mecha*, pretending that their Eyes ought not after that to be prophan'd by the sight of worldly things.

THE Children who are conceiv'd in this Pilgrimage, are esteem'd as so many little Saints, whether the Pilgrims beget them upon their lawful Wives, or upon strange Women ; for there are such waiting upon the Road, who offer themselves very humbly for so pious a Work. These Children are kept cleaner than others, tho it be very difficult to add any thing to the Neatness with which Children are generally kept over all the *Levant*.

MAHOMET would have deserv'd to be commended, if he had advis'd Cleanliness, as comely and useful to the Health ; but it was ridiculous in him to make it a point of Religion. Yet the Muſſulmans are ſo fond of it, that they ſpend a great part of their Life in waſhing. There is not a Village among them, which has not a publick Bath. Thoſe in the Towns are the chief Ornament of the Place, and are allotted for all ſorts of People, of whatever Quality and Religion they are. But the Men never bathe with the Women; and there is ſo much Modeſty obſerv'd, that any one would be reprov'd who ſhould ſee any thing thro Inadvertency; and if he did it by deſign, he would be baſtinado'd. There are ſome Baths which are for the Uſe of the Men in the Morning, and for the Women in the Afternoon; and others are frequented one day in the Week by one Sex, and the next by the other. One is ſerv'd very well in theſe Baths for three or four Aſpers; the Strangers commonly pay handſomer, and every one is welcome there from Four in the Morning to Eight in the Evening.

THE firſt Entrance is into a fine Hall, in the middle of which is the principal Fountain, the Baſin of which ſerves for waſhing the Linen of the Houſe: All round the Hall is a ſmall Bench about three foot high, cover'd with Mat; they ſit down upon this to ſmoke, and pull off their Clothes, which are folded up in a Towel. The Air of this firſt Hall is ſo temperate, that one can bear to have nothing upon one's Body but an Apron about the Waiſt, to cover one before and behind. In this Condition a Man paſſes into a ſmall Hall, which is a little warmer, and from thence into a larger, where the Heat is more ſenſible. All theſe Halls are generally cloſ'd above with ſmall Domes, which let in light at the top thro a round Glaſs, like thoſe our Gardiners put over their Melons. In the laſt Hall there are Marble Baſins with two Cocks, one of hot Water, and the other of cold, which every one mixes to his own Fancy, and laves upon his Body with little Buckets of Braſs belonging to the Place. The Pavement of this Chamber is heated by Furnaces beneath, and every one walks there as long as he thinks proper.

WHEN a Man deſires to be ſcour'd, a Servant of the Bath cauſes you at once to lie along upon your Back, and ſetting his Knees then upon your Belly, without further Ceremony preſſes and ſqueezes you violently, and makes every Bone crack. The firſt time I fell into one of theſe Fellow's hands, I thought he had put out all my Limbs: they handle after

*

the

the fame manner, the Joints of the Back and the Shoulder-blades. In Letter II.
brief, if you would be fhav'd, he fhaves you, or gives you a Razor to
fhave yourfelf, if you chufe it; but for this, you muft withdraw into a
Clofet, at the door of which you hang up a Towel as a Signal for no
body to enter; and when you come out, you take it away again, and
go into the great Hall, where another Servant preffes your Flefh all over
with his Hands fo dextroufly, that having kneaded it, as I may fay, without doing you any harm, he forces out a furprizing Quantity of Sweat.
The little Camelot-Bags they make ufe of here, are inftead of the Stri-
gils of the Antients, and are much more convenient. To clean the
Skin the better, they pour a world of hot Water upon the Body; and
if you have a mind to it, they ufe a piece of perfum'd Soap: in a word,
they wipe you with Linen very clean, dry, and warm; and the Ceremony concludes with the Feet, which the fame Man wafhes very carefully,
when you are come back into the great Hall, where you left your Clothes;
it is there that you are accommodated with a fmall Mirror, and pay your
Money, after you are drefs'd, and have reftor'd the Linen you had for
your Ufe. In this Hall they fmoke, drink Coffee, and have Collations;
for after this Exercife a Man finds himfelf very hungry. By difcharging the
Glands, the Bath certainly facilitates Perfpiration, and by confequence
the Circulation of Juices which fupply the Body. A Man perceives
himfelf very light when he has been well purify'd; but he muft be accuftom'd to the Bath from his Youth, for otherwife the Breaft is very
much affected by thefe warm Rooms.

THE Women are very happy when they are permitted to go to the
publick Baths; but moft of them, efpecially fuch whofe Husbands are
rich enough to build them Baths at home, have not this liberty. In the
publick Baths, they entertain one another without any Conftraint, and
pafs their time more agreeably than in their own Apartments. The
Men who have any Complaifance for their Wives, do not refufe them thefe
innocent Diverfions. Too much Conftraint makes them fometimes feek
Reafons for a Divorce.

MARRIAGE among the *Turks* is only a Civil Contract, which
the Parties have in their power to break; and nothing feems more convenient: yet as they are frequently weary of Marriage here, as well

as

as elsewhere, they have wisely provided that frequent Separations shall be chargeable to the Family. A Woman may demand to be separated from her Husband if he is impotent, or given to unnatural Pleasures, or if he does not pay his Tribute upon Thursday and Friday Night, which are the times consecrated to the conjugal Duties. If the Man acquits himself well, and supplies her with Bread, Butter, Rice, Wood, Coffee, Cotton, and Silk to spin her Garments, she cannot be parted from him. A Husband who denies his Wife Money to go to the Bath twice a Week, is subject to a Separation ; for if the Woman turns her Slipper upside down in presence of the Judge, it is a Sign her Husband would force her to consent to things forbidden. Then the Judge sends to look for the Husband, and bastinades him, and dissolves the Marriage, unless he brings some very good Reasons in his Defence.

A HUSBAND who would be parted from his Wife, wants Pretences as little in his turn, tho the thing is not so easy among the *Turks* as People imagine. The Husband is not only oblig'd to settle a Dowry upon his Wife for the rest of her Days ; but supposing that in a return of Tenderness towards her, he should desire to take her again, he is condemn'd to let her lie for twenty-four Hours with some other Man, whom he shall think fit. He generally chuses one of his Friends, whom he knows to be most discreet : sometimes also he takes the first Comer ; and it often happens, they say, that some Women who are pleas'd with their Change, refuse to return to their first Husbands again. This is practis'd only toward such Wives as are espous'd. The *Turks* are permitted to keep two other sorts, namely, such as they have in pay, and their Slaves. They espouse the first, the second they hire, and the last they purchase.

WHEN a Man would marry a Woman in form, he makes his Address to the Relations, and signs the Articles, after they are all met in the Presence of the Cadi, and before two Witnesses. It is not the Father and Mother, but the Husband, who endows the Woman : when the Dowry is fix'd, the Cadi delivers to the Parties the Copy of the Marriage-Contract ; the Woman, on her part, brings only her Partition of Goods. Against the Nuptial Day, the Bridegroom has his Marriage bless'd by the Curate ; and to draw upon himself the Favour of Heaven, he distributes Alms, and sets some Slaves at liberty. Upon the Wedding-day, the Bride mounts on

horse-

horseback, cover'd with a large Veil, and rides thro the Streets under a Letter II.
Canopy, accompany'd by several Women and some Slaves, according
to the Quality of her Husband. The Men and Women, who play on Instruments, assist in the Ceremony: After this, are carried along the Goods,
which make not the least Ornament of the Procession. As this is all the
Profit which accrues to the Husband, they affect to place upon Horses
and Camels a great many Coffers, which make a fine Appearance, but
are commonly empty, or have nothing in them but the Habits and Jewels. The Bride is also led home in triumph by the farthest way to her
Husband's, who receives her at his Door. Then these two Persons who
have never seen one another, nor chang'd a word but by the Interposition
of some Friends, join Hands, and make the tenderest Protestations that
a sincere Passion can inspire. They forget not also to make a Speech,
which is eloquent at least, for it is impossible the Heart should have much
share in it.

T H E Ceremony being perform'd in presence of the Relations and
Friends, they spend the Day in Feasting and Dances, and seeing Puppet-
Shows. The Men make merry in one Company, and the Women in
another, till at last, Night comes on, and Silence succeeds to this tumul-
tuous Joy. Among the Rich, the Bride is conducted into the Chamber
by an Eunuch; but if there is no Eunuch, some Woman-Relation takes
her by the Hand, and delivers her into her Husband's Arms. In some
Towns of *Turky*, there are Women whose Profession it is to instruct the
Bride what she ought to do when she approaches her Spouse, who is oblig'd to undress her piece by piece, and to put her to bed. During this
time, they say, she repeats a long Prayer and takes care to tye her Girdle in several Knots, so that the poor Bridegroom exercises himself for
whole Hours, before he can finish the disrobing. It is only by the Report of another, that a Man understands whether the Woman he espouses
be handsome or ugly. There are a great many Towns, where the next
Day after the Wedding the Relations and Friends go to the House of
the new-married Couple, to take a bloody Cloth, and shew it in the
Streets as they ride along, with Instruments playing before them. The
Mother or the Relations forget not to prepare such a Cloth, both for
that end, and to shew; in case of need, that the Parties were satisfied one

with

with another. If the Women live prudently, the Alcoran requires them to be treated well, and condemns the Husbands who use them otherwise, to make amends for their Offence by Alms, or by other Works of Piety, which they are oblig'd to do before they lie with them.

IF the Husband dies first, the Woman takes her Dowry, and nothing more; and the Children, when the Mother dies, can oblige the Father to give the Dowry to them. In case of a Divorce, the Dowry is lost, if the Husband's Reasons are sufficient; if not, he must continue it, and maintain the Children.

THIS is the Condition of lawful Wives. As for them who are hir'd, there is not so much Formality about them. After the Father and Mother's Consent, who are willing to deliver their Daughter to such a Man, they repair to the Judge, who draws a Writing, that such a Man is willing to take such a Woman to serve for a Wife, that he undertakes to maintain her and the Children they shall have together, upon condition he shall be able to dismiss her when he thinks fit, paying her a certain Sum, in proportion to the Number of Years they shall live together. To colour over this evil Practice, the *Turks* throw the Scandal of it upon the Christian Merchants, who having left their Wives behind in their own Country, hire others in the *Levant.* As for Slaves, the *Mahometans,* according to the Law, may use them as they please; they give them their liberty when they will, or hold them in Servitude for their whole Life. What is commendable in this Libertine Way of Living, is, that the Children which the *Turks* have by all their Wives, equally inherit their Fathers Goods; with this difference only, that the Children of the Slaves must be declar'd free by Testament. If their Father does not do them this favour, they follow the condition of the Mother, and are at the discretion of the Eldest of the Family.

THO the Women in *Turky* do not shew themselves in publick, they are yet very magnificent in their Habits; they wear Breeches like Men, which reach as low as the Heel in manner of a Pantaloon, at the end of which is a very neat Sock of *Spanish* Leather. These Breeches are of Cloth, Velvet, Sattin, Brocade, Fustian, or fine Linen, according to the Season, and the Quality of the Wearer. There are Women at *Constantinople* debauch'd and profligate to such a degree, that under a shew of adjusting

justing their Clothes, they discover in the open Street all that which
Modesty enjoins them to conceal, and get their Living by this detesta-
ble Trade.   The *Turkish* Women wear upon their Shift a Waist-coat, and
upon that a kind of Cassock of very rich Stuff; this Cassock is button'd
down below the Breast, and girt about with a Girdle of Silk or Leather,
with some Plates of Silver enrich'd with Jewels.   The Vest they wear
upon the Cassock, is of a Stuff which is more or less thick, according to
the Season; and the Fur of it is more or less costly, according to the
Person's Condition.   They often fold one part of the Vest over the
other, and the Sleeves reach to the Fingers-Ends ; and they commonly
carry their Hands thrust in at the Slits in the side of the Vest.   Their
Shoes are exactly like the Mens, that is, embellish'd with a Border of Iron
about the Heel.   To give their Stature the best Advantage, instead of a
Turbant, they wear a Bonnet of Pasteboard, cover'd with Cloth of
Gold, or some handsome Stuff.   This Bonnet, which is very high, resem-
bles, in some manner, a certain sort of inverted Basket, which is seen
in the antient Medals upon the Heads of *Diana*, *Juno*, and *Iris*.   This
Fashion is observ'd in the *Levant*; but as the Women among the *Turks*
are oblig'd to cover themselves all over, they have a Veil upon the Bonnet, which hangs down to the Eye-Brows; the rest of the Face is cover'd with a fine Handkerchief, ty'd so strait behind, that the Women look just as if they were bridled.   Their Hair hangs in Tresses
upon their Back, and is a wonderful Grace to them ; and those who have
not good Hair of their own, wear artificial.

T H E *Turkish* Women, according to the Report of our Countrymen
at *Constantinople* and *Smyrna*, who see them at the Bath with liberty
enough, are generally handsome and well-made.   They have a delicate
Skin, regular Features, an admirable Chest, and above all, black Eyes;
and several of them are compleat Beauties.   Their Habit indeed is no
Advantage to their Shape; but among the *Turks*, the thickest Women
pass for the best made, and slender Shapes are not esteem'd.   Their Breasts
are at full liberty under their Vest, without any restraint of Stays or Bo-
dice; in a word, they are just as Nature has made them, whereas with
us, by endeavouring by Machines of Iron and Whalebone to correct Nature, who sometimes at a certain Age discovers Faults in the Back-Bone

and

and the Shoulders, the fine Women are frequently mere Counterfeits Befides, their Diet is fweeter and more fimple than that of our Women, who eat Ragous, and drink Wine and ftrong Liquors, and fpend a great part of the Night at Play : Is it furprizing then that they have Children crooked, or with falfe Shapes? The Blood of the *Levant*-Women is alfo much purer ; their Cleanlinefs is extraordinary ; for they bathe twice a Week, and fuffer not the fmalleft Hair or the leaft Soil to be upon their Body : all which conduces extremely to make them healthy. But they might fpare the Care they take of their Nails and their Eye-brows; for they colour their Nails of a dark red, with a Powder which comes out of *Egypt*, and ufe another Drug for their Brows to make them black.

A S to the Qualities of the Mind, the *Turkish* Women want neither Wit, Vivacity nor Tendernefs; and it is owing to the Men of this Country, that they are not capable of more beautiful Paffions : but the extreme Conftraint with which they are guarded, makes them go a great way in a little time. The more brisk among them fometimes caufe their Slaves to ftop a comely Man, as he paffes along the Street. They commonly faften upon Chriftians, and we may eafily believe they do not chufe thofe who feem the leaft vigorous. We were told at *Conftantinople*, that a handfome *Greek*, as he was returning from an Adventure of Gallantry, unhappily fell into a Trap-door, by the fault of the Slave who conducted him : the Trap-door was at the end of a Spout, which difcharg'd itfelf into the Town-ditch. One may imagine how heartily the poor *Greek* curs'd the Adventure, and how fpeedily he ran to the Bath to wafh himfelf clean. The Slaves of the *Jews*, who are the *Turkish* Womens Confidents, enter their Apartments at all Hours, under a pretence of carrying them Jewels, and often take with them fome jolly young Fellows difguis'd in Womens Apparel; they fpread them out with a Fardingale, to make them look bulky. The Hour of Morning and Evening Prayer is the common time for intriguing in *Turky*, as well as in many parts of *Spain*; but this can be practis'd only in great Towns, where the diforderly Women, and fuch whofe Husbands are conveniently goodnatur'd, are very ftrict at their Devotions, while their Husbands are in the Mofque. The Meeting is made in the Houfes of the *Jeweffes*, where the *Turkish* Women love a good Company; and there Strangers have all

*

the

the Liberty with them that can be. Love is ingenious in every Coun-try; but whatever Precautions are taken to conceal the Game, they are often surpriz'd in those Places where they thought themselves secure. Adultery is rigorously punish'd in *Turky*; and in that case the Husbands are Masters of the Life of their Wives; for if they are revengeful, the wretched Women who are caught in this flagrant Offence, or convicted in Form, are put into a Sack fill'd with Stones, and drown'd: but most of them know how to manage their Intrigues so well, that they seldom die this death. When their Husbands give them their Life, they are more happy sometimes than they were before; for then they oblige them to marry their Gallant, who is condemn'd to die or turn *Turk*, supposing he is a Christian. The Gallant is often condemn'd also to ride thro the Street upon an Ass, with his Head towards the Tail, which they make him hold in his Hand like a Bridle, with a Crown of Garbage, and a Cravat of the same Stuff. After this Triumph they entertain him with a certain Number of Blows of the Battoon upon the Reins and upon the Soles of the Feet; and for the last Punishment, he pays down a Fine proportionable to his Estate. The Savages of *Canada* are not so rigorous; for tho they condemn the Adulteress, yet they agree that the Frailty being so natural to the two Sexes, they should mutually forgive one another, if the Faith is broken which is plighted in so delicate a Matter.

THE Alcoran detests Adultery, and ordains, that he who shall accuse his Wife, without being able to prove it, shall be condemn'd to fourscore Strokes of the Battoon. As the thing is difficult to be prov'd in *Turky*, where there must be Witnesses, the Husband is oblig'd to swear four times before the Judge, that he speaks the Truth, and protests five times, that he desires to be accurs'd of God and Men if he lyes. The Woman laughs in her heart, for she is believ'd upon her Oath, provided she prays to God five times that she may perish, if what her Husband says is true. Does it not seem that every Woman in such a condition ought to be dispens'd with from speaking the Truth?

JEALOUSY excepted, the *Turks* are a well-natur'd People, and take all possible measures to avoid the occasions of it; for they never suffer their Wives Faces to be seen by the dearest Friend they have in the whole World. They are alfo well made, and of a manly Stature: the Blood changes lefs with them than with us, perhaps becaufe they are more fober, and their Nourifhment is more wholefome and light; and there are fewer crooked People, or lame, or Dwarfs. It is true, their Habit hides many Defects, which ours difcovers. The firft part of the Habit is a pair of Breeches in manner of Pantaloons or Drawers, which reach to the Heels, and end with a yellow *Spanifh* Leather Sock, which goes into Slippers of the fame Leather. Inftead of a Heel, the Slippers are adorn'd with a fmall Iron, only one Finger and a half broad, and four high, bent like a Horfe-fhoe, which keeps them from wearing out in that part: the Tip is curv'd in a Bow, and they are fow'd more neatly than our Shoes. Tho they have only a fingle Sole, they laft a long time; efpecially thofe of *Conftantinople*, where they ufe the beft and lighteft Leather of the *Levant*. The Sultan is no better fhod than others. The Chriftians who are Strangers, are not fuffer'd to wear yellow Slippers; for the Subjects of the Grand Signior, Chriftians or Jews, have them either red, violet, or black. This Order is fo well eftablifh'd, and obferv'd with fuch Exactnefs, that one may know what Religion any one is of by the Feet and the Head. The great Convenience of thefe Slippers is, that one puts them on and off without Trouble; but I loft mine feveral times in the middle of the Street, when I firft wore them, and never mifs'd them till the aking of my Feet gave me notice.

OUR Shoes are of a much better Fafhion, tho the *Turks* think them heavy and clumfey. Their Slippers are good only in fine Weather, for the leaft Drop of Water foils them: they are by no means fit for Perfons who love to go a fimpling. There is no walking in the Fields in thefe Slippers without being hurt by the fmalleft Pebble: it is true, they fometimes put on *Spanifh* Leather Buskins as light as Cloth, and border'd at the Heel with Iron, like the Slippers: the Muffulmans alone, and privileg'd Chriftians, wear them of yellow.

THE *Turkifh* Breeches are faften'd together at top by a Band three or four Inches wide, which goes into a linen Loop few'd on to the Cloth. They are not made to open more before than behind, becaufe the *Mahometans* do not urine after that manner. Their Shirts are made of fine foft Callicoe, and the Sleeves are wider than thofe of a Woman's Shift:

they

they turn up the Sleeves in their Ablutions as high as the Elbow, and very eafily, becaufe they have no Wrift-bands. Upon the Shirt they wear a fort of Caffock of Fuftian, or Satin, or Stuff of Gold, and reaching to the Heels. In the Winter it is lin'd with Cotton, and fome *Turks* have it of the fineft *Englifh* Cloth : it is juft fit acrofs the Breaft, and is button'd with Buttons of Silver gilt, or of Silk, as big as a Pepper-corn. The Sleeves are alfo made very fit, and are faftned with Buttons of the fame fize, which go into a Loop of Silk, inftead of Button-holes ; and the Caffock is the fame. For Quicknefs in dreffing, they button only two or three Buttons here and there; fometimes the Sleeves have at the end a fmall Band which covers the upper part of the Hand. They wear a Girdle upon the Caffock ten or twelve feet long, and one foot and a quarter wide : the beft Girdles are made at *Scio* : they go twice or thrice round the Waift, fo that the two ends, which are handfomely toffel'd, hang down before.

THEY wear a Dagger, and fometimes two in this Girdle ; thefe are merely Cafe-Knives, and the Handle is adorn'd with Gold or Silver, and precious Stones. As they have no Pockets, they alfo carry their Handkerchiefs under the fame Girdle, and their Tobacco-Box, Letter-Cafe, *&c.* they thruft into their Bofom, which makes them look very big. The great Veft comes over the Caffock, and during the Heats they wear it like a loofe Coat, without putting their Arms into the Sleeves; but it would be the higheft Indecency to prefent themfelves in this Pofture before Perfons of Diftinction. The Sleeves of thefe Vefts are ftrait enough, and not lin'd with Furs, becaufe they would then be of an ungraceful Bignefs, and would hinder them from ufing their Arms freely : Thefe Sleeves come down to the Wrift, and are turn'd up with a broad Facing of the fame Fur as the Veft is lin'd with. The ordinary Furs are the Fox-skin, the Martin, and the fmall Badger ; and the better are the Sable-Tail very dark, or the Breaft of the *Mufcovian* Fox bleach'd very bright : thefe laft are very dear, becaufe a great many Martins Tails or Foxes Breafts go to line one Veft ; they coft from five hundred Crowns to a thoufand, and the deareft rife to four or five thoufand Livres. The Vefts are of Cloth of *England, France,* or *Holland,* of

a Scara Scarlet, Musk, or Coffee Colour, or Olive-Green ; and they reach to the Heels like the Garments of their Antients.

THE Turbant or *Saric* is compos'd of two pieces, namely, a Bonnet, and the Linen which is wrapp'd about it. The *Turks* call the Linen *Tulbend*, from whence comes our *Turbant*. The Bonnet is a kind of Cap red or green, without Brims, pretty flat, tho somewhat rising at the top, quilted, as I may say, with Cotton, but it does not cover the Ears : about this Cap they roll several Folds of Callicoe. It is a particular Art to know how to give a Turbant a good Air; and it is a Trade in *Turky*, as selling Hats is with us. The *Emirs*, who boast of their being descended from the Race of *Mahomet*, wear a Turbant all green; but that of other *Turks* is red, with a white Border. It must be chang'd often, to keep it clean. Upon the whole matter, this Habit is convenient enough, and I found it better than my own.

THE *Turks* take a world of care of handsome Beards, and value them highly. One of the greatest Marks of Friendship with them, is, to kiss one's self, holding one's Beard ; as it is a flagrant Injury to pull any one by the Beard, or cut it off. When they swear, it is by their Beard; and a Lawyer who had no Beard, would be despis'd. Those who follow Arms, are content with wearing one noble Mustachio, and are very proud of fine Whiskers. The manner of Saluting among the *Turks*, is, to make a light Inclination of the Head, and at the same time lay their Hand upon their Heart, wishing a thousand Benedictions, and calling those whom they salute Brethren. When it is a Person of Distinction, they advance toward him without bowing; and when they are come up within reach, they stoop down, and taking up a corner of his Vest before, lift it about a foot and a half high ; they kiss it with respect, or else let it fall, according to the Quality of the Person : when they have made their Compliment, or spoke of their Business, they withdraw, after having observ'd the same Ceremony.

IN ordinary Visits, they only lay their Hand upon their Heart, and sit cross-legg'd upon a Sofa, which is a low-rais'd Bench : they commonly bring in Pipes of Tobacco ready lighted; the Pipes are very clean and neat, and two or three feet long, and consequently the Smoke comes very mild into the Mouth, and has none of that stinking Oil which burns the

Tongue,

Tongue, and inflames the Throat when one uses short Pipes. The To- bacco also which is smok'd in the *Levant*, is the best in the World; it is commonly the Tobacco of *Salonica*, but that of *Asia* is better, and especially that of *Syria*, which they call Tobacco of *Ataxi* or *Ataquie*, because they plant it about the antient Town of *Laodicea*. The *Turks* mix Wood of Aloes, or other Perfumes, among the Tobacco; but this spoils it. The Bowls of their Pipes are bigger and more convenient than ours. The Pipes of *Negropont* and *Thebes* are made of a natural Clay, which they cut with a Knife as it rises out of the Quarry, and which grows hard afterward of itself. After Tobacco, Coffee and Sherbet are brought in; the Coffee is excellent, but they never put Sugar in it, whether it be out of Avarice, or because they think it better without Mixture. Beside Tobacco, People of Quality treat also with Perfume: One Slave burns Drugs under your Nose, while others hold a Cloth over your Head, to hinder the Fumes from being dissipated too soon: a Man must have been us'd to these Scents, otherwise they are noisome.

MOST Visits are perform'd with these Ceremonies. There is no need of much Wit to transact Business well; for a good Mien and Gravity are instead of Merit in the *East*, and much Gaiety would spoil all: not that the *Turks* are not Men of Wit, but they speak little, and pride themselves in Sincerity and Modesty more than Eloquence. It is not thus with the *Greeks*, who are unmerciful Talkers. Tho these two Nations are born under one Climate, their Tempers are more different than if they liv'd very remote from each other; which can be imputed only to their different Education. The *Turks* use no unnecessary Words, and the *Greeks* on the contrary talk incessantly. In Winter they spend whole days in the *Tendours*; and there it is they have their great Chats, and the Neighbour is never spar'd. These *Tendours* are Tables boarded round the sides, and in which they shut themselves up Waist-high, Men and Women, Maids and Batchelors, after they have set a small Stove there to keep them warm. Our Missionaries may declaim against these *Tendours* as much as they please, the Custom is too convenient to be suppress'd. The *Turks* practise what their Religion enjoins, but the *Greeks* do not; and their Misery causes them to play a thousand Fooleries, authoriz'd by bad Example, and perpetuated from Father to

*   Son.

Son. In fhort, the *Turks* make profeffion of Candour and good Faith, whereas how long the Faith of the *Greeks* has been fufpected, one may eafily fee by their own Hiftorians.

AN Uniformity runs thro all the Actions of the *Turks*; and they never change their manner of Life. There is no fuch thing as making great Feafts with them; they are fatisfy'd with a little, and you never hear of a *Turk*'s being undone by feeding too high. Rice is the ftanding Difh in their Kitchens; and they drefs it three feveral ways: That which they call *Pilau* is dry Rice, fat, and which melts in the Mouth, and is more agreeable than the Hens and Rumps of Mutton they boil with it: they boil it over a fmall Fire, with a little Liquor, and never ftir it, nor uncover it; for by expofing it to the Air, it would turn to a thick Milk. The fecond way of dreffing it they call *Lappa*, it is boil'd up to the fame Confiftence as with us, and may be eat with a Spoon; but the *Turks* ufe their Fingers, and the Hollow of their Hand ferves them for a Trencher. The third way is *Tchorba*; this is a fort of Rice-cream, which ferves them for a Broth. * This feems to be that Preparation of Rice which the Antients gave to fick Perfons.

* Sume hoc
Ptifarearium
Oryzæ. *Hor.*

THE *Levant* Hens are very good, but the Butcher's Meat is not extraordinary in a great many Places. They fell there Buffaloe's Flefh for Beef, which is violently tough. The Mutton is very fat, and taftes of the Suet, efpecially the Rump, which is perfectly a Roll of Fat of a prodigious Thicknefs: the *Turks* never kill it till juft as they hang the Pot upon the Fire. As they value only the Broth, they cut the Flefh out in Morfels before they put it into the Kettle, and then boil it with all forts of Game. When they roaft, they chop it ftill fmaller, and draw all the pieces upon a very long Spit, putting a piece of Meat and then an Onion alternately. There is good Beef at *Conftantinople*, and excellent Hares; and upon the Coafts of *Afia* the Heat-cocks are admirable, and fo are the Partridges. The beft Fifh in the World is taken in the *Levant*. Befide the forts we know, the Black Sea furnifhes a number of others which are unknown to us. Sometimes the *Turks* have a Ragou of Meat hafh'd with a little Fat, and ftrew'd over with curdled Rice; they make up Rice alfo in Rolls, which they wrap over with Vine-leaves or Cabbage, according to the Seafon, after having boil'd it

in

in an Earthen Pan cover'd clofe.　All thro the *Levant* they make abominable Bread with very good Wheat, for their Dough is neither kneaded nor le-
ven'd ; yet, for all this there is fometimes good Paftry enough, and made
with very fine Puff-Pafte.　Their Difhes are of Porcelain, fine Earth, or
Pewter.　The moft common are Copper tinn'd ; for *Afia the Lefs* abounds
with Copper-Mines.　They tin it very neatly, and very quick ; for
they make the Difh red-hot, and ftrew Sal Armoniac upon it, and then
rub the Tin over it, and polifh it with a Burnifher.　This Tin adheres
to the Copper fo well, that their Veffels do not lofe it fo eafily as ours.

WHEN the Hour of Eating is come, they fpread a piece of black,
Spanifh Leather upon the Ground or the Sofa, according to the number,
who are to eat.　They who love Neatnefs lay it on a Table of Wood,
half a foot high, upon which they fet a great wooden Bowl with Plates
of Rice and Meat,　The Mafter of the Houfe fays the ordinary Prayer,
In the Name of God Almighty and Merciful, &c.　One Napkin of blue
Linen is handed round the Table, and ferves all the Guefts ; and they
have one wooden Ladle among them with a long Handle, which helps
to fharpen their Appetite to the Rice.　Meat and Fruits are alfo produc'd, and cold Water is never omitted at the Clofe of the Treat.　We
have rofe from Table fometimes with our Belly perfectly frozen ; but
to make us amends, they gave us Coffee boiling hot : and we fmok'd
like the reft of the Company, but it was more out of Complaifance than
Pleafure.　Tobacco in Smoke, taken medicinally, is good for an Afthma,
for Pains in the Teeth, and for feveral Maladies occafion'd by Serofi-
ties, to which fome are very fubject : In this fenfe, Tobacco is proper
enough for the *Turks* ; for their Habit of the Turbant expofes them to
Defluxions, becaufe its Thicknefs hinders Perfpiration, and it does not
cover the Ears.　Tobacco alfo humours their Lazinefs ; they fwallow
their Spittle out of Cuftom and out of Cleanlinefs, and without any Prejudice.　When I went to bridle my felf before Perfons of Fafhion, and for-
bore to fpit, it made me heart-fick : Decency however requires one to fpit
into a Handkerchief, in order to fave the Carpet upon the Floor, or elfe one
muft fit at one Corner, and take up the Carpet, and fpit upon the Boards.

THE firft time we were oblig'd to lodge among the *Turks*, we were
puzzled fufficiently to know where we fhould lie.　Our Hoft had only
one

one Hall where we eat, one fmall Kitchen juft befide it, and another Chamber which belong'd to his Wife; this was evidently not intended for us: and befides, there was neither Bed, Couch, Bench or Chair to be feen; for the *Turks* of all People in the World encumber a Room the leaft with Moveables: when at once a Slave drew out of a Cup-board in the Wall all the Materials for making our Beds. To make three Beds, he fpread three Quilts, very fcanty and very hard, upon the Board we had eat upon, and upon thefe he laid three Cloths, and then a fecond Cloth upon every one; but, according to the Fafhion of the Country, the laft Cloth was few'd to the Counter-pane, left it fhould flip off in the Night. Every Bed had alfo its Pillow; and when we rofe, the fame Slave folded up the Baggage in a moment, and put it into the Cup-board; and all this was done as fwiftly as one can fhift the Decoration of an Opera.

THE Idlenefs in which moft part of the *Turks* live, obliges them to feek out for Amufements, which is the propereft Term on this occafion; when they play together, it is only to pafs the time, as they fay, and not to win Money. *Mahomet*, who had nothing in view but the Peace of Families and the publick Tranquillity, has given them good Principles about this Subject. *Abftain*, fays he, *from playing at Games of Hazard and at Chefs; thefe are the Inventions of the Devil to caufe Divifion among Men, to divert them from their Prayers, and hinder their calling upon the Name of God.* As to Chefs, they do not obey his Injunction, but they underftand neither Cards nor Dice: they play fometimes at Drafts. The *Mancala* is their Favourite-Game; it is a Table with two Leaves like a Draft-board, and fix Spots on a fide; they play two at a time, and each has 36 Men, which he ranges on the Spots on his fide.

THE moft ingenious Muffulmans employ themfelves in reading the Alcoran, and the Commentators upon it. Others take to Poetry, in which they are faid to do very well; nor am I furpriz'd at it, for the Blood of the fineft Genius's *Afia* and *Greece* has formerly produc'd runs in their Veins, or at leaft they are under the Influences of the fame Heaven. Some *Turks* delight in Mufick, and fpend the whole day in playing upon an Inftrument without being tir'd, tho they only repeat the fame Tune. The *Dervifes* are great Muficians and great Dancers; but I muft firft mention the Lawyers, before I fpeak of the Religious.

THE

T H E Mufti, who is at the head of the Lawyers, is chief of their Religion, and the Interpreter of the Alcoran. He is nam'd by the Sultan, and feldom depos'd: the Sultan chufes a Man of Probity, learn'd in the Knowledge of the Law, and whofe Reputation is eftablifh'd. By this Choice he becomes the moft refpected Officer of the Empire: he is the Oracle of the Country, and they ftand to all Decifions, which he makes only by *Yes* or *No*, which he writes under the Queftion propos'd. For this he has three Officers; one who ftates the Queftion well, after having difentangled it from the Difficulties which might obfcure it; the other copies it out, and the third applies his Mafter's Seal to it, when he has given his Anfwer. This Anfwer removes all Difficulties; there is no Appeal, and the Matter is ended for ever. When it is about Peace or War, the Death of great Officers, or fome Affairs relating to the Good of the Empire, the Sultan propofes the Point to him in Writing in form of a Doubt, and without naming the Perfon; as thus, *What ought to be done in fuch a Cafe?* It concerns the Mufti to be circumfpect, for many times he is confulted only out of Formality, and is depos'd if he does not anfwer according to the Prince's Pleafure. Sultan *Morat* having to do with a ftubborn Mufti, demanded of him fiercely, *Who was it made thee Mufti? Your Highnefs,* he reply'd. *Very well,* faid the Sultan; *fince I was able to clothe thee with that Dignity, am I not able to ftrip thee of it?* It is not faid what the Mufti return'd; but he was degraded. There have been feveral Mufti's who have fign'd the Depofition and the Death of the Emperors, who put them into their Places.

T H O they perfuade People that the Alcoran is a perfect Book, they do not forbear to give different Interpretations to the Law, according to the time and the occafion. The Grand Signior prefents the new Mufti with a Veft of great Price furr'd with Sable, and with his own Hand puts into his Bofom a Handkerchief full of Sequins. This Prefent and the Veft are valued at two thoufand Crowns. He alfo affigns him a Fund of about twenty five Crowns a day, which is generally rais'd upon fome Mofque. The Baffas who are at the Court, and the Ambaffadors and Refidents alfo, make him a confiderable Prefent, when they go to wifh him Joy of his Promotion. In a word, the Mufti is the only Officer whom the Grand Signior falutes with Refpect. He never refufes him Audience, and advances several Steps to receive him. The Grand Visier rises up to none, nor goes to any Person beside the Mufti. The Visier takes the left Hand of him, which is the Sword-side, and the most honourable Place among the Professors of Arms, because, they say, those who are on their right hand are under their Sword; but the Mufti and the Cadilesquers are very well content to take the right Hand, which is he Place of Honour among the Men of the Law: there is also never any Dispute between them. See here, how the Fancies of People are satisfy'd. If the Mufti is depos'd by the Intrigue of his Enemies, in order to place one of their own Faction in so advantageous a Post; he has the Disposal given him of some Employments of Judicature, which bring him in a very noble Revenue. But if the Mufti was guilty of High Treason, or any enormous Crime, it would be in vain for him to say the Law forbids him to be put to death; for he would be degraded, and sent to the *Seven Towers*, and there be pounded alive in a Mortar.

A F T E R the Mufti, the Cadilesquers are the Officers of Justice the most honour'd in the Empire. Next are the *Moula* or *Moula-Cadi's*, call'd Grand Cadi's, and the Cadi's or ordinary Judges. Among the Cadilesquers, or chief Judges, he of *Europe*, or *Romania*, is the First; he of *Asia* or *Anatolia* the Second, and he of *Egypt* the Third. The Cadilesquers do the Business of the Cadi in his absence: they very often come to be Mufti's, and apply themselves strictly to the Study of the Alcoran, which is their Civil and Religious Code: They are also stil'd Judges of the Army, because the Soldiers are judg'd only by them. Their Place at the Divan is at the Grand Visier's side, and they appeal sometimes to them from the Sentence of the Secular Cadi; in short, their Employment obliges them to have an eye upon all the Officers of Justice in the Empire. They give out the Cadi's Commissions, and those of the Moula-Cadi's; but for the last, they must have the Grand Seignior's Consent. Upon considerable Complaints well grounded, they depose the Cadi's, and condemn them to a Fine, after they have suffer'd the Bastinado.

T H E Judges of the great Towns are call'd *Moula* or *Moula-Cadi's*, those of small Towns, and of Burroughs and Villages, *Cadi's*. The administration of Justice lies wholly in the hands of this sort of Men in *Turkey*; and as all is corrupted at present, the Mufti is Pensioner to the

Cadilefquers, the Cadilefquers to the Moula, the Moula to the Cadi's, and the Cadi's to the People. Every Cadi has his Serjeants before him, to summon with a loud Voice those who are accus'd : If he who is summon'd, fails at the Hour appointed him, they grant to the other Party all he desires. It is commonly to no purpose to appeal from the Sentence of the Cadi; for a Process is never form'd over again : the Sentence would also be perpetually confirm'd, because the Cadi form'd it, as he understood it; and it is by this he commits horrible Abuses. However, the Cadi's are sometimes cashier'd; and if the Injustices they have acted are flagrant, they are punish'd, but the Law forbids the putting them to death. These Officers have been known at *Constantinople* ever since about 1390. for *Bajazet* I. oblig'd *John Paleologus* the *Greek* Emperor, to admit them into that City, to judge the Affairs which happen'd between the *Greeks* and the *Turks* who were settled there.

T H E Priests and the Religious among the *Turks* have the good luck to die in their Beds, as well as the Cadi's. The Priests commonly begin with proclaiming the Hours of Prayer in the Galleries of the Pinacles. If they carry themselves well, and have a fair Reputation, the People of the Parish present them to the Grand Visier, upon the Vacancy of the Cure; who dispatches their Presentment, after having made them read some Passages of the Alcoran, or after having laid this Book upon their Head. The Employment of the Priests is to say Prayers, to read in the Mosques, to bless Marriages, to assist the Dying, and accompany the Dead. To comfort the Dying who have Debts which they are unable to pay, the Curate calls the Creditors together, and exhorts them to forgive them to the dying Person, or to declare before Witnesses that they will never demand any thing of him. The Creditors who are hard-hearted enough to refuse this Favour, are reputed very ill Men.

T H E Y wash the Dead with a great deal of Care in *Turky*; they shave them all over, and burn Incense about them, to drive away the evil Spirits : they bury them in a Cloth, open at top and bottom; for they imagine that when the dead Person is laid in the Ground, two Angels come and make him get upon his Knees, to give an account of his Actions; for which reason most of the *Turks* leave a Lock of Hair upon their Head, for the Angel who makes them thus change their Posture, to

take

take hold on. That the Dead may be the more at eafe, they make a kind of Arch in the Grave, of light Planks, upon which they lay them all along. If the Dead liv'd a good Life, two Angels, white as Snow, fucceed to thofe who came to examine him, and entertain him with nothing but reprefenting the Pleafures he fhall tafte in the other World ; but if he was a great Sinner, two other Angels, black as Jet, torment him horribly : one, they, fay, ftrikes him into the Earth with a Club, and the other pulls him up again with an Iron Hook ; and they divert themfelves with this cruel Exercife even to the Day of Judgment, without difcontinuing it one moment.

MAHOMET, who had it upon his hands to manage the *Arabs,* has treated them according to their Tafte. As their Soil is an arid dry Defert, to comfort them, he has provided them a Paradife full of Fountains, and Gardens, and Groves impenetrable by the Sun, Parterres abounding with Flowers, and Orchards loaded with all forts of admirable Fruits. In this charming Place flows Milk, Honey, and Wine, but it is a Wine which never touches the Head, nor difturbs the Reafon. The moft accomplifh'd Beauties are up and down in the Walks, and are neither too eafy nor too cruel. A Man fhall efpoufe what Women he pleafes, for there all forts are to be found : their Eyes, which are as large as an Egg, fhall be always faftned upon their Husbands, who love them to Dotage. Their Daughters, according to this Prophet, are all pure and unfpotted; and the Maladies peculiar to the Sex are never heard of there ; nor are Savine, Mercury, *&c.* known among them. The beft thing that *Mahomet* has faid concerning the other World, is, that they muft not be reckon'd in the number of the Dead, who die in the ways of God, becaufe they live in God, and enjoy his Bleffings and his Love. The Damned, on the contrary, are precipitated into a devouring Fire, in the midft of which their Flefh is continually renew'd, in order to augment their Punifhment; they fhall fuffer an incredible Thirft, without being able to cool themfelves with one drop of Water ; and if by chance any thing is given them to drink, it will be a poifon'd Liquor which will fuffocate them without killing them : and to compleat their Miferies, they fhall have there no Women.

I FOR-

I FORGOT to mention, that before they bury their Dead, they expofe them in the Houfe upon a Bier, under a Pall of different Colours, according to the Quality of the Perfons; this Pall is red for Men of the Army, black for Citizens, and red for an Emir or a Cherif; the Turbants which are laid upon the Bier, are of the fame Colour with the Pall. The Priefts go before the Train, and pray for the deceas'd; the Poor follow with the Slaves and Horfes. There are alfo Mourners, as well as in the Interments of the *Greeks*: thefe make a mad fort of Mufick along the Street while the Body is burying, and after it is bury'd; they cover the Grave with certain Planks, upon which they throw on what Materials they find thereabout. After this, the Men retire, and the Women ftay there fome time; then the Priefts advance to the Grave to liften, in order to inform the Relations if the deceas'd makes a good Defence when the Angels queftion him; they take care enough not to fay he was confounded, becaufe they are well paid when they tell good News. The Women often go to pray upon their Husbands Graves, but it is always in open Day, and never by Night, for fear fome Adven-ture fhould befal them, like that of the *Ephefian* Matron. They fome-times carry Victuals to eat in the Cemeteries, efpecially on a Friday: fome believe this eafes the Dead; but the more reafonable fay it is done to draw the Paffengers thither, to pray to God for the Deceas'd.

ONE of the principal Reafons which caufes the *Turks* to bury the Dead in the Highways, is, to excite Paffengers to wifh them well; and the Wifh is generally, *That God would deliver them from the Torments which the black Angels make them fuffer.* They fet up two great Stones at each end of the Grave for Perfons of Diftinction: that at the head fhews the Difference of Sex, by a Turbant or by a Bonnet; and it is in this fort of Work that the Carvers of *Conftantinople* and the chief Towns of the Empire are employ'd: the Epitaph is engrav'd upon the other Stone. The Mafter-piece of the chief Artifts is to make a Tomb for the Grand Signiors; in which notwithftanding, they fucceed very ill, for they beftow Pains and Labour without any Skill or Tafte. They com-monly dig among the Ruins of the antient Towns to fearch for pieces of Pillars, or fome old Marbles, to make Grave-ftones of. They who take pleafure in Infcriptions, fhould not neglect to vifit the Cemeteries, be-caufe the *Turks*, the *Greeks*, and the *Armenians*, carry the fineft Marbles thither;

thither: the Cemeteries are of a prodigious Extent, for they never bury two Perfons in the fame Grave; and the Ground they take up about *Conftantinople*, if it were till'd, would bear Corn enough to feed that great City for half the Year; and there is Stone enough in them to build a fecond Wall round it.

I AM not acquainted well enough with the *Turkifh* Religious, to make a particular Defcription of the different Orders among them, for I have feen none but thofe they call *Dervifes*. Thefe are the chief Monks, who live in a Body in Monafteries under a Superior, who applies himfelf principally to Preaching: they make a Vow of Poverty, Chaftity and Obedience; but they eafily give themfelves a Difpenfation from the two former, and quit their Order alfo without Scandal, to marry when the Humour takes them. It is a Maxim with the *Turks*, that a Man's Head is too light and giddy to continue long in the fame Difpofition. The General of the Order of the Dervifes refides at *Cogna*, which was the antient *Iconium*, the Capital of *Lycaonia* in the fmaller *Afia*. *Ottoman*, the firft Emperor of the *Turks*, erected the Superior of the Convent of this City into Chief of the Order, and granted great Privileges to this Houfe. They fay it holds above five hundred Religious, and that their Founder was a Sultan of the fame Town, call'd *Melelava*, from whence they came to be call'd *Melelevi's*: they have this Sultan's Tomb in their Convent.

THE Dervifes who wear Shirts, have them, by way of Penitence, of the coarfeft Cloth they can get; and thofe who wear none, have a woollen Veft next their Skin of a brown Colour, made at *Cogna*, and which reaches a little below the Calf of the Leg: they button it when they have a mind, but moft part of the Year they go open to their Skin as low as their Girdle, which is generally of black Leather. The Sleeves of this Veft are as large as thofe of our Women's Shifts in *France*; and upon this they wear a fort of Caffock or Cloke, the Arms of which come no lower than the Elbow. Thefe Monks go bare-legg'd, and fometimes they ufe the common Slipper: upon their Head they have a Bonnet of Camel's Hair of darkifh white, without any Brims, and made in the Form of a Sugar-Loaf, but rounded at top like a Dome: fome roll a piece of Linen about it, to make a Turbant of it.

IN

IN the prefence of their Superiors and Strangers, thefe Religious ob-ferve an affected Modefty, turning down their Eyes, and keeping a profound Silence: but in other Points they are faid not to be fo modeft, for they are great Drinkers of *Aqua Vitæ*. The Ufe of Opium is more familiar to thefe than to other *Turks*. This Drug, which is Poifon to them who are not accuftom'd to it, and a fmall Dofe of which would kill other People, throws the Dervifes, who take it by Ounces at a time, into a Gayety equal to what Men have who drink a plentiful Quantity of Wine. A pleafing Fury, which one may ftile Enthufiafm, fucceeds this Gayety, and makes them pafs for extraordinary Perfons, if one is ignorant of the Caufe: but as their Blood is too much attenuated by this Drug, it occafions a confiderable Difcharge of Serofity in the Brain, and fo cafts them into a Slumber, and they lie a whole day without ftirring a Hand or Foot. This kind of Lethargy feizes them every Thurfday, which is their Day of Fafting; during which they dare not eat, according to their Rules, tho it be after Sun-fet.

THE Dervifes value themfelves much upon their Politenefs; their Beards are very clean and well comb'd; and their Verfes never turn upon the Women, unlefs it is upon thofe whom they hope to fee one day in Paradife. They are no longer fuch Fools, as to cut and flafh their Bodies, as they did formerly; for now they fcarcely raze the Skin: however, they burn themfelves fometimes on the fide of their Heart with fmall Wax-Candles, as a Mark of their Tendernefs to the Object of their Love. They draw the Admiration of the People by handling Fire without being burnt; they will hold it alfo in their Mouth a good while, like our Mountebanks. They perform a thoufand Feats of Activity, and play with the Jugler's Box furprizingly. They pretend to charm Vipers by a Specifick Virtue adhering to their Clothes, and are the only *Turks* who travel into the *Eaftern* Countries: They go into the Mogul's Dominions, and thereabouts, picking up Alms in abundance, and always take care to make their Meals at their Religious Houfes which lie in their way. Mufick is one part of their Study: their finging feem'd to us to be fad, and yet harmonious; and tho it is forbidden by the Alcoran, to praife God with Inftruments, yet they have fet it on foot in fpite of the Edicts of the Sultan, and the Perfecution of the Bigots.

THE.

THE principal Exercifes of the Dervifes, are to dance upon Tuefdays and Fridays; and this Comedy is preceded by a Preachment by the Superior of the Convent, or his Sub-delegate. Their Morals, they fay, are good, and may be of excellent Ufe to Perfons of any Religion. The Women, who are banifh'd from all publick Places where the Men refort, are permitted to attend thefe Preachments, and never fail to be prefent. During the time, thefe Religious fit within a Balluftrade, upon their Legs, with their Arms acrofs, and their Hands turn'd down: After the Sermon, the Singers, who are plac'd in a Gallery, which ferves for an Orcheftre, ftrike up with their Voices to the Fifes and Tabors, and fing a very long Hymn. At the fecond Stanza, the Superior, in a Stole and a Veft with hanging Sleeves, claps his Hands; at which Signal the Monks get up, and having faluted him with a profound Reverence, begin to turn round one after another, and whirl about fo fwiftly, that the Doublet they have upon their Veft, flies out, and fpreads juft like a Tent, in a furprizing manner. All thofe Dancers form a great Circle as merry as can be; but at the firft Stroke or Signal of the Superior, they give over, and return to their firft Pofture, as calmly as if they had never mov'd. They repeat this Dance at the fame Signal three or four times, the laft of which is much the longeft, becaufe the Monks are then well in Breath; and by a long Habitude, they finifh this Exercife without being giddy. As much Veneration as the *Turks* have for thefe Religious, they don't fuffer them to have many Convents, becaufe they never efteem fuch Perfons as do not beget Children. Sultan *Morat* defign'd to extirpate the Dervifes, as a fort of Men ufelefs to the Republick, and for whom the People had too much Confideration; but he contented himfelf with confining them to their Convent of *Cogna.* They have alfo a Houfe at *Pera,* and another upon the *Thracian Bofphorus.* We heard their Preachment in their Convent at *Prufa* in *Bythinia,* and faw them dance with a great deal of Pleafure thro the Rails of the Mofque.

THE *Armenian* Merchants in our Caravan, who fpoke *Italian,* explain'd to us part of the Sermon. The principal Subject was upon Jefus Chrift: The Preacher declaim'd againft the *Jews,* but coolly, for they are never in a Tranfport; and found fault with the Chriftians extremely, for believing the *Jews* had put to death fo great a Prophet, affuring us on the contrary,

A Dance of Dervices

trary, that he afcended into Heaven, and that the *Jews* crucify'd another Letter II.
Perfon in his ftead.

I KNOW not how to conclude more nobly, than by obferving the Efteem the *Turks* have for Jefus Chrift; fo far is it from being true, that they vomit out Blafphemies againft him, as fome Travellers have told us. If the *Turks* have the misfortune not to believe the Divinity of Jefus Chrift, they reverence him at leaft as a great Friend of God, and efpecially as a great Interceffor before the Lord. They confefs he was fent from God, ·to deliver a Law full of Grace; and if they treat us as Infidels, it is not becaufe we believe in Jefus Chrift, but for not believing that *Mahomet* came after him, to publifh another Law, lefs oppofite to corrupted Nature.

I am, My LORD, &c.

LETTER III.

To Monseigneur the Count de Pontchartrain, *Secretary of State, &c.*

MY LORD,

A Description of the Canal of the Black Sea.

BEFORE I engage in the Description of the *Black Sea*, I beg you to allow me the Honour to give you an account of what we observ'd as to the Canal whereby it discharges itself into the Sea of *Marmora*, which makes part of the *White Sea*, according to the Language of the *Turks*.

Βόσπορος Θράκιος. Po-lyb. & Strab. Βόσπορος της Χαλκηδονίης. Herod. lib. 4. ¹ *On the word* Χαλκηδών.

THE Canal of the *Black Sea*, or the *Bosphorus* of *Thrace*, begins properly at the Point of the Seraglio of *Constantinople*, and ends towards the Column of *Pompey*. *Herodotus, Polybius, Strabo,* and *Menippus* quoted by ¹ *Stephanus Byzantinus*, make it 120 Stadia in length, which come to fifteen Miles: but they place the Beginning of that Canal between *Byzantium* and *Chalcedon*, and the End at the Temple of *Jupiter*, where the new Castle of *Asia* stands at present. Tho this Difference be arbitrary, yet after Inspection of the Places, every body would, I believe, agree in my Measures. This Canal is very far from being in a right Line; its Entrance, which on the side of the *Black Sea* has the Form of a Tunnel, looks to the North-East, and is to be taken from the Column of *Pompey*, whence we reckon about three Miles to the new Castles. That

² Αργυρόνιον Ακρα. ³ Jupiter Urius, Ούρεος.

of *Asia* is built upon a Cape ² where the Temple of ³ *Jupiter the Distributer of good Winds*, is thought formerly to have been; upon which account that Place is still call'd *Joro*, by corruption, from *Jeron*, which

signifies

fignifies a *Temple*. The Caftle of *Europe* is on an oppofite Cape, near Lett. III. which ftood in times paft the Temple of *Serapis*, mention'd by *Polybius*. From thefe Caftles the Canal forms a great Elbow, in which are the Gulphs of *Saraia* and *Tharabia*; and from this Elbow it runs South-Eaft towards the Seraglio call'd *Sultan Solyman Kiofc*, five Miles diftant from the Caftles. After this, by another Elbow fhap'd like a *Zig-zag*, the fame Canal crimps by little and little to the South, till it comes to the Point of the Seraglio, where in my Opinion it ends. From this laft Elbow, to the Old Caftles, is reckon'd two Miles and a half; and thence to the Seraglio, or Point of *Byzantium*, fix. Thus, according to this Computation, the whole Canal is fixteen Miles and a half long, which is not very different from the Account of the Antients, who gain'd on the fide of *Chalcedon*, where they plac'd the Beginning of the Canal, what they loft between the Temples of *Jupiter* and *Serapis*, and the Column of *Pompey*.

THE Breadth of the Canal at the new Caftles, where thofe Temples ftood, is a Mile; and a Mile and a half, or two Miles, in fome other parts. The narroweft part of all is at the Old Caftles, whereof that of *Europe* is upon the Rifing, on which the Antients, as *Polybius* informs us, had built a Temple to *Mercury*; for which reafon it was nam'd the *Hermean Cape*. This Cape lay half way in the Canal, according to the Antients, who, as we have already faid, terminated it on one fide between *Chalcedon* and *Byzantium*, and on the other, at the Temple of *Jupiter*. This part is not more than 800 Paces broad, and the Canal is very near as narrow a little lower at *Courichifme*, a Village built at the foot of the Cape, which the Antients call'd *Efties*, whence it widens to the Seraglio for the length of a Mile, or a Mile and half. Thus the Waters of the *Black Sea* enter with fufficient Swiftnefs into the Canal of the new Caftles, and have free room to extend themfelves in the Gulphs of *Saraia* and *Tharabia*. From thence, without running at all fafter, they wind toward the Kiofc of Sultan *Solyman*, where they are forc'd to turn towards the South, without any vifible Augmentation of their Motion, except between the Old Caftle, where the Channel grows ftraiter.

IN this part (as *Polybius* remarks) befides that the narrowing of the Canal encreafes the Swiftnefs of the Water, it is reflected obliquely from

the

the Cape of *Mercury*, on which is the Old Castle of *Europe*, against the Cape of *Candil-bachesi* in *Asia*, and returns towards *Europe* about *Courichisme* at Cape *Esties*, whence it flows thro by the Point of the Seraglio. This is what *Polybius* observ'd in his time, that is, in the time of *Scipio* and *Loelius*, with whom he was intimately acquainted. For my part, I own I could not observe this Indentedness of Motion of this side the Castles, tho I pass'd the Canal four or five times ; but it is certain that upon a North Wind the Rapidity is so great between the two Castles, that no-Vessel can stop itself, nor get back again, without a Wind contrary to the Current : yet the Swiftness of the Waters diminishes so sensibly, that you may go down and up, without any Difficulty, when the Winds are not violent.

INDEPENDENTLY of the Winds, there are some very particular Currents in the Canal of the *Black Sea*; the most apparent is that which runs all along it, from the opening of the *Black Sea*, to the Sea of *Marmara*, which is the *Propontis* of the Antients. Before this Current enters the Canal, it beats in part against the Point of the Seraglio, as *Polybius, Xiphilinus*, and after them M. *Gilles* have observ'd; for one part of these Waters (tho the least considerable) flows into the Port of *Constantinople*, or the antient *Byzantium*, and following the Western Windings, runs into the Nook which goes by the Name of the *Fresh Waters* : nay, *Polybius* and *Xiphilinus* had a notion that these Waters reflected, form'd that celebrated Port, which the Antients admir'd by the name of the *Golden Horn*, upon account of the Riches it brought to that powerful City. That Portion therefore of the Canal which goes into the Port of *Constantinople*, makes a Current that follows the Turn of the Walls of the City : all the rest discharges itself into the Sea of *Marmara*, between the Seraglio and *Chalcedon*.

MONSIEUR *le Comte Marsilly* hath observ'd, that the two little Rivers of the *Fresh Waters* form'd a Current in the Port of *Constantinople*, from the North-West to the East, which, as it were, sweeping the Coasts of *Galata* and *Topana*, proceeds along those of *Fondoxli*, quite to *Arnautcui*, going up the Canal on the side of the Castles, in a course opposite to the great Current. When we know this, we shall not be surpriz'd that some Boats go up under favour of this little Current, while others

*

go down by keeping in the great one. It is likely, the Stream that goes
out of the Port glancing fide-ways againft the great Current, flides towards the North; whereas if it run againft it in any other line than fide-
ways, it would bear it along with it, or beat it back. M. *le Comte
Marfilly* has alfo obferv'd, that there is a little Current in the corner of the
Coaft of *Scutari*; fo that the Waters of the great Current that ftrike
againft Cape *Scutari*, are reflected back towards the North. According
to the Obfervations of that Learned Man, the Waters of the great Current being arrived at Cape *Modabouron*, afcend again along the Coaft of
Chalcedon towards Cape *Scutari*, and make another fort of Current.

THIS Diverfity of Currents has nothing in it very extraordinary.
It is eafy to conceive, that a Cape which juts out too far muft ftrike
back the Waters that run againft it in a certain line; but it is hard to
account for another hidden Current, which we fhall henceforth call *the
Under Current*, becaufe it is obfervable only in the great Canal beneath
the great Current, which we may call *the Upper Current*, which flows
quite from the Caftles to the Sea of *Marmara*. We are therefore to take
notice, that the Waters which poffefs the Surface of this Canal to a certain depth, run from the Caftles to the Seraglio. This is inconteftable;
but it is alfo certain, that beneath thefe Waters there is one part of the
Water of the fame Canal, which moves in a contrary Direction; that is
to fay, goes back up towards the Caftles.

PROCOPIUS of *Cefarea*, who lived in the fixth Century, informs
us, that the Fifhermen took notice that their Nets, inftead of finking
perpendicularly to the bottom of the Canal, were dragg'd from the
North towards the South, when they came to a certain Depth; while
the other part of the fame Nets, which defcended beyond that Depth to
the bottom of the Canal, were bent a contrary way. There is alfo
great likelihood that this Obfervation is ftill more antient, for the *Bofphorus* has in all times been very famous for fifhing. This Canal is call'd
Fifhy, in the Infcription which *Mandrocles* caus'd to be fet under the Picture wherein he had reprefented the Bridge over which *Darius* march'd
with his Army, when he went to fight the *Scythians*. *Procopius* tells us,
that according to the Remarks of the Fifhermen, the two oppofite Currents, one upper and the other under, are very perceptible in that part of
the

the *Bofphorus*, which is call'd the *Abyfs*. Perhaps thereabouts may be a deep Gulph form'd by a Rock, in fhape hollow like the Bowl of a Spoon, the hollow part looking towards the Caftles: for according to this Suppofition, the Waters that are to the bottom of the Canal fhocking violently againft this Rock, muft by fuch Reflection take a Determination contrary to what they had before; that is to fay, they muft run back towards the Caftles, and confequently flow in a line oppofite to that of the upper Current. The fhort abode we made at *Conftantinople*, would not allow us to examine into this Wonder. M. *Gilles* fpeaks of it as of a very extraordinary thing, and M. *le Comte Marfilly* obferv'd it with great attention; and indeed I think nothing can be more worthy of Obfervation. That skilful Philofopher would not venture to give his Opinion as to the Explication of fo fingular an Effect; and I propofe mine, only to fpur on the Learned to fearch into the true Caufe of this Phenomenon.

NEITHER is it eafy to give a reafon why when the *Bofphorus* difcharges fo little Water, the *Black Sea*, which receives fo prodigious a quantity, fhould not become larger. That Sea, whofe Extent is fo confiderable, befides the *Palus Meotis*, another Sea well worth notice, receives more Rivers than the *Mediterranean*. Every body knows that the greateft Collections of Water in *Europe* fall into the *Black Sea* by means of the *Danube*, into which run the Rivers of *Suabia*, *Franconia*, *Bavaria*, *Auftria*, *Hungary*, *Moravia*, *Carinthia*, *Croatia*, *Bofnia*, *Servia*, *Tranfylvania*, *Wallachia*. Thofe of *Little Ruffia* and *Podolia* run into the fame Sea by means of the *Niefter*. Thofe of the Southern and Eaftern Parts of *Poland*, of *North Mufcovy*, and of the Country of the *Coffacks*, come into it by the *Nieper* or *Borifthenes*. Do not the *Tanais* and *Copa* pafs into the *Black Sea* by the *Cimmerian Bofphorus*? The Rivers of *Mengrelia*, whereof the *Phafis* is the chief, empty themfelves alfo into the *Black Sea*, as do likewife the *Cafalmac*, the *Sangaris*, and the other Rivers of *Afia Minor*, whofe Courfe is to the North. And yet the *Bofphorus* of *Thrace* is not comparably equal to any one of the great Rivers we have here named. It is alfo certain, that the *Black Sea* does not increafe; tho according to the Rules of Phyficks, a Refervoir fhould grow fuller when its Difcharge is not anfwerable to the quantity of Water it receives. The

Black

Black Sea muſt therefore empty it ſelf as well by ſubterranean Canals, Lett. III. which perhaps may run through *Aſia* and *Europe*, as by the continual Expence of its Waters, whick ſoak into the ground, and flow far away from the Coaſts.   This kind of Tranſpiration is like that of the Body of Animals, which according to *Sanctorius*'s Computation is much more conſiderable than any made by the moſt ſenſible Evacuations.

SUPPOSING the *Black Sea* to have been a mere Lake without any Diſcharge, form'd by the Concourſe of ſo many Rivers; it could not poſſibly empty it ſelf, according to the Conformation of the Place, any otherwiſe than by the *Thracian Boſphorus*: the Mountains that are between the *Black Sea* and the *Caſpian*, oppos'd its Paſſage to the Eaſt.   The Waters of the *Palus Meotis* fall into the *Black Sea* on the ſide of the North, inſtead of allowing thoſe of the *Black Sea* to fall in upon them. The Rivers of *Aſia* repel the *Black Sea* from the South to the North.   The *Danube* drives it from its Mouths on the Weſt.   There was therefore no place but this Corner, which is to the North-Eaſt above *Conſtantinople*, where it could work away the Earth without oppoſition, between the Light-houſe of *Europe* and that of *Aſia*.   Neither could it diſcharge it ſelf on the ſide of either of thoſe Light-houſes, the Coaſts there being dreadfully ſteep: ſo that the Waters of the *Black Sea* were forced through a place, which conſiſted of nothing but Soil ; and through this Soil it was that they began to dig themſelves a Canal, by pouring upon it in front with a Column that ſoak'd through the Earth, and carry'd it away at ſeveral ſhakes.   According to this Hypotheſis, the Waters firſt made themſelves a paſſage in a ſtrait line between the two Rocks where the new Caſtles now ſtand, and ſoften'd the ground of the firſt Elbow, where now we ſee the Gulphs of *Saraia* and *Tharabia*, and were then compell'd to remain ſome time in a Baſon edg'd with very high Rocks ; but their natural Diſpoſition afterwards made them deſcend to the Kioſc of *Solyman* II. and from thence their Determination being alter'd by the Interruption of new Rocks, they form'd the ſecond Elbow of the Canal, the Earth whereof gave way to the South.

THIS Route was certainly traced out by the Author of Nature; for according to the Laws of Motion by him eſtabliſh'd, the Waters always throw themſelves that way where they find leaſt oppoſition.   Thoſe of

the

the *Black Sea* continu'd then to wash away the Earth that lay between the two Rocks where the old Castles are, and by this means carry'd their Canal quite to the Point of the Seraglio, the bottom of which is a living Rock, not by any means to be shaken. This large Heap of Waters did probably throw down at once the Dike of Earth that remain'd between *Constantinople* and Cape *Scutari*, and so discharg'd it self into the Sea of *Marmara*.

AT this time, if we may judge by appearances, happen'd the great Inundation spoken of by *Diodorus Siculus*, one of the most faithful Historians of Antiquity. That Author informs us, that the People of *Samothracia*, a considerable Island situated to the left of the Entrance of the *Dardanelles*, perceiv'd the Irruption that the *Pontus Euxinus* made in the *Propontis* by the Aperture of the *Cyanean* Islands; for the *Pontus Euxinus*, which was then look'd upon to be a great Lake, was so swell'd by the discharge of the Rivers which run into it, that it overflow'd into the *Propontis*, and drown'd part of the Cities on the Coast of *Asia*, which undoubtedly was lower than that of *Europe*. But notwithstanding this Situation, the Waters mounted to the very tops of the highest Mountains of *Samothracia*, and chang'd the Face of the whole Country. The Islanders had still the Tradition of it among them in the time of our Historian, who thereby has preserv'd us one of the finest Observations in all Antiquity; for it is certain this Alteration happen'd long before the Voyage of the *Argonauts*, and those Heroes undertook that Voyage but 1263 Years before Christ. This being so, what we just now propos'd, as a Philosophical Conjecture, becomes an Historical Truth, and must convince us that the great Passage of the *Propontis* into the *Mediterranean*, was made long before by the same Mechanism.

IT is very probable, that the Waters of the *Propontis*, which antiently might be nothing but a Lake form'd by the *Granicus* and *Rhyndacus*, finding it more easy to work themselves a Canal by the *Dardanelles*, than any other way spread themselves into the *Mediterranean*, and wash'd away the Flesh of the Rocks (if we may be allow'd such an Expression) by melting the Earth from them. The Islands of the *Propontis* are no more than the Remains of the Rocks which the Waters could not dissolve; as also were those which made so much noise antiently by the name of

the

Biblioth. Hist. lib.5. p.322.

Sanmandraki.

the *Cyanean* Iflands of *Europe* and *Afia* at the Mouth of the *Black Sea.* Lett. III.
The Iflands feem to be fo many Nails drove into the Globe of the Earth,
and of which the Mountains are as it were the Heads.

BUT what Changes did not the Iflands of the *Egean Sea* under- ' Archipelago.
go, by the overflowing of the *Euxine,* and more efpecially thofe which
lie, as it were, in a right line? fince that of *Samothrace,* adjacent to the
Canal, was fo overwhelm'd with its Inundation, that the Inhabitants were
at their wits end. The Fifhermen, when the Waters were abated, would Diod. Sic. Bib-
frequently draw out with their Nets Chapiters of Pillars, and other Limbs liotn. ibid.
of Architecture. Confidering what violent work the Waters made in the
Sea of *Greece,* can it be thought ftrange in the Hiftorians and Poets of
old to give out that feveral Iflands of the *Archipelago* funk to the bottom,
and new ones fprung out of them? Peradventure the famous *Delos* appear'd then for the firft time, and the People of the neighbouring Iflands
gave it that Name, which fignifies *Manifeft.* And yet moft of the antient Authors are look'd upon as fo many Dotards, and Tellers of old
Wives Fables. How many Colonies muft needs have been fettled after
fuch a Devaftation? and how do we know whether the Works of thofe
who gave an account of thefe Revolutions are extant, as well as thofe
of *Diodorus?* Thofe Paffages in *Pliny* which feem to us to be moft incredible, are perhaps the beft Pieces of many Authors who wrote of thefe
Matters, and whofe other Writings are loft.

I ASK your Lordfhip's Pardon, if I dwell a little longer on the Subject of Philofophy: The Example of a learned Minifter, to whom the
World is beholden on many accounts, has put me out of my way; not
that I mean to follow him in every thing; for as great an Admiral as he
was, and as much us'd to the Sea, I can't help thinking he took the Formation thereof in a Senfe diametrically oppofite to what is confonant to
Nature. He was of opinion, that the Ocean, by its Impetuofity having
difmember'd the Mountain of *Calpe* from the Lands of *Africa,* pour'd it
felf into that vaft Space now the *Mediterranean*: that this Sea afterwards
penetrating northerly, produc'd the *Propontis* or Sea of *Marmara,* the *Black
Sea,* and the *Meotick* Lakes. But independent of *Diodorus*'s Obfervation,
if we confider the Formation of things *gradatim,* is it not more reafonable to look upon the *Meotick* Lakes, the *Black Sea,* the *Propontis,* and

the *Meditteranean*, as so many huge Lakes of Water form'd by multitudes of Rivers disburdening themselves into them, than to fancy them the Expansions of the Ocean? What could become of the Waters which were gathering day and night in the same Basins? Doubtless they form'd Lakes of a prodigious Extent, which at length would have cover'd all the adjoining Lands, had they not broke down their Dykes in the manner before-mention'd.

'T I S there for certain that the Waters of the North do fall into the Mediterranean thro the *Bosphorus Cimmerius*, the *Bosphorus Thracius*, and the Canal of the *Dardanelles*, which, according to the Idea of the Antients, is another sort of *Bosphorus*; that is to say, an Arm of the Sea narrow enough for an Ox to swim over. The *Mediterranean* discharges itself into the Ocean at the Straits of *Gibraltar*, where by good Fortune it was easier for the Water to scoop itself a Canal, than to overspread the Lands of *Africa*. The All-wise God had left this Opening between Mount *Atlas* and that of *Calpe*; the Plug, as one may say, only wanted to be pull'd out. Perhaps the terrible Irruption which was then made into the Ocean, either sunk or carried away that famous Isle of *Atlantis*, which *Plato* describes beyond the Coast of *Spain*, and [2] *Diodorus Siculus* beyond that of *Africa*. The *Canary* Islands, the *Azores*, and *America*, may be (for ought we know) in the same Predicament: where then is the Wonder they should be peopled by the Descendants of *Adam* and *Noah*, or that their Inhabitants should use the same Weapons as the antient *Asiaticks* and *Europeans*, namely, Bow and Arrow?

P L I N Y had therefore better stick to the Opinion of some Authors who were not unknown to him, and who, as he himself confesses, brought into the Ocean the Waters from the North to the South. How shall we judge of the course of a stagnant Water, the *Saone* for instance, or *la Marche*, but by their Currency under the Arches of their respective Bridges? Now, in the *Bosphorus*'s before-mentioned, this Currency is apparent. There is but one Circumstance which can favour *Pliny*'s Opinion, and that is, the Saltness of the Water in all these Seas: it is impossible to account how these large Lakes we are speaking of, and which are form'd by nothing but the Accession of fresh-water Rivers, should be endu'd with this brackish Quality. But besides the Ocean's communicating

nicating

In Tim. tom. pag. 24. Edit. Hehric. Steph.
[2] *Bibliot. Hist. lib. 5.*

nicating with the *Mediterranean*, it is certain, that the Water of the *Black* Lett. III.
Sea is far lefs briny than that of our Seas; befides, all round the *Black Sea*,
the Land is full of foffile Salt, which is continually melting into it: this
Salt, mix'd with a certain Portion of Sulphur accruing from the Oil of
the Fifhes, which are there conftantly putrefying, heightens this degree
of Saltnefs, and imparts that tang of Bitternefs fo fenfible in Sea-watei.
The *Cafpian* Sea, for the fame reafon, is as falt as other Seas, tho it looks
to be only a Pool, which receives nothing but frefh Water continually
running into it.

BEFORE we return to the Canal of the *Black Sea*, it will not be a-
mifs to take notice, that *Polybius*'s Prophecy is not fulfill'd. He, good
Man, fancy'd that the *Euxine Sea* would one day become a Morafs, and
that very fuddenly too; becaufe, faid he, the Mud and Sludge which is
carried thither by the Rivers, muft form a Bar capable of choking up the
Mouth of it, as happen'd to the *Danube* in his time. 'Tis well for the
Turks, who enjoy great Advantages from their Trade to the *Black Sea*,
that the *Bofphorus* is ftill open, and perhaps wider than formerly it was.
Come what will come, there's no need to fear any fuch thing: a Bar never comes but at the Mouth of fuch Rivers whofe Waters are beaten
back to Land by the Surges of the Sea, and by the Tides. There's nothing in this Canal to give the Waters of the *Black Sea* a retrogade
Motion: on the contrary, 'tis an evacuating Paffage, thro which the
Water glides of itfelf; and being ever and anon pinch'd, as it were, and
contracted by the Defilees of the Land, acquires a Velocity, and fweeps
away whatever may oppofe its Progrefs. As for the Tides, *Strabo* has
obferv'd there was none at all in the *Bofphorus*; and Count *Marfilly* takes
notice that they were not perceptible. As rapid as this *Bofphorus* is, it
is fometimes frozen over. *Zonaras* writes that in the Reign of *Conftantine Copronymus*, there happen'd fo fevere a Winter, that People walk'd
upon the Ice from *Conftantinople* to *Scutari*; nay, that it bore Carts too.
In 401, the *Black Sea* itfelf was frozen for twenty days; and when the
Weather broke, fuch Mountains of Ice pafs'd by *Conftantinople*, as frighted
the Inhabitants.

IN the Summer-time both Sides of the *Bofphorus* afford a delicate Profpect. The Villages and Pleafure-Houfes difpers'd among the Forefts

O 2   make

make a very delightful Landskip, diversify'd with little Hills cover'd over with Coppices. The Letter I wrote containing an account of *Con-stantinople*, concludes with a Description of the Pavilion call'd *Fanari-Ki-osc*. I am now going to give a Description of the *Asiatick* Coast, from the Canal of the *Black Sea*, up as far as the Light-house beyond its Mouth: after which, I shall pass over to the Light-house and *Pompey*'s Pillar, on the side of *Europe*; and so coasting along the said Canal, return again to *Constantinople*.

I COULD no where have met with better Guides upon this Canal, than *Dionysius Byzantinus* a *Greek* Author, and another that was a *French-man*. The Description which the former has given of the *Thracian Bos-phorus*, is exact to a nicety. An Edition of it from the Manuscripts in the *Vatican*, and the King's Library, has been promis'd us by *Holstenius* and M. *du Cange*; but they have not had leisure to be as good as their words. M. *Gilles*, my other Guide, and a *Frenchman*, has with wonder-ful Accuracy confirm'd upon the spot the Description made by *Dionysi-us*, not forgetting the Name of the smallest Rock. I hope your Lord-ship will approve of the Plan of the *Bosphorus* I send you; it is drawn according to the Rules, the Distances well mark'd, and no considerable Fault, that I know of, in the Position of Towns. I thought it necessa-ry to add to the old *Greek* Names, those given them by the *Turks*, in order to illustrate the Observations made by *Dionysius* and *Gilles*. The first is thought to have liv'd about the time of *Domitian:* the other was of the Diocess of *Alby*, and dy'd at *Rome* in 1555. after he had tra-vell'd into *Asia* and *Africa* by Order of *Francis* I. to make Collections of Manuscripts and antique Monuments.

TO begin a Description of the Canal of the *Black Sea*, we must re-sume that of *Constantinople*, which concludes at *Fanari-Kiosc* built on the Cape of *Chalcedon*. To the East of this Cape is one of the Ports which the Antients call'd *Eutrope*, where the Children of the Emperor *Maurice* were put to death by order of *Phocas*, who dethron'd him in the beginning of the 7th Century. The Emperor's Widow and her three Daughters had their Heads struck off five Years after. It looks as if this Port was pre-ordain'd for the Butchery of this unhappy Family.

*

The

The Emperor *Juftinian* caus'd it to be repair'd in a manner becoming his Lett. III. Greatnefs of Soul.

Paffing the Port of *Eutrope*, you double the ¹ Cape of *Modabouron*, which terminates the Peninfula, on whofe Ifthmus the famous City of *Chalcedon* ftood. I am prone to believe that this Cape went heretofore by the name of *Herea*; for *Stephanus Byzantinus* places it over againft that Town, and quotes fome Verfes of *Demofthenes* of *Bithynia*, who affigns it the fame Situation. The Coaft of *Calamoti* extends beyond the Cape, and is fo call'd from a Church of St. *John Chryfoftom* built in a Morafs full of Rufhes. The other Port of *Chalcedon* is on the fame Coaft on the bending part of the Ifthmus facing the Weft, and confequently the City of *Conftantinople*. The Emperor *Juftinian* had expended immenfe Sums in forming Jettees, to hinder the entring of more than one Ship at a time: of thefe Works there's nothing now left but the Foundations. This fhews how injudicious they were, who made choice of this place for the building of *Chalcedon*, fince they were forc'd to make two artificial Ports, whereas the Port of *Byzantium* is by Nature the fineft Port in the World. This ill Choice occafion'd the Oracle of *Apollo*, and *Megabizes* General of *Darius*'s Troops, to call the Founders blind Buzzards: *Pliny* too gives it the Appellative of the Blindmens City.

CONSTANTINE the Great, had it not been for an aftonifhing Prodigy, had committed the like Overfight, if we may credit *Cedrenus*. The *Perfians* having deftroy'd *Chalcedon*, and that Emperor having order'd it to be rebuilt, as they were going to work upon't, feveral Eagles came, and with their Talons took away the Stones from the Workmen, and carried them to *Byzantium*. This Miracle being feveral times repeated, the whole Court was alarm'd; *Euphratas*, one of the Emperor's Chief Minifters, affur'd him it was the Will of God he fhould build a Church at *Byzantium*, in honour of the Virgin. *Chalcedon* feems to have been built on purpofe to embellifh *Byzantium*; for when the Emperor *Valens* had caus'd the Walls of *Chalcedon* to be level'd with the Ground, to punifh the Inhabitants for fiding with *Procopius*, he order'd the Materials to be fent to *Conftantinople*, to be us'd in that beautiful Aqueduct call'd the *Valentinian Aqueduct*. 'Tis afferted by *Ammianus Marcellinus*, that the Burghers of *Chalcedon*, among other Affronts which they pretended to

¹ *Port of* Irene.
² *Port of* Chalcedon *or* Calamoti.

⁰ Κάλαμος, *Rufh.*

Sabaia, *Beer.* TO put upon *Valens,* call'd him, while he befieg'd their City, *Beer-bibber.*
Solyman II. made ufe of nothing but the Ruins of *Chalcedon* to repair
the *Valentinian* Aqueduct, and to build *la Solymania.* The fettling of Pofts
feems to have been more antient than is generally believ'd : *Procopius*
fpeaks this of it with relation to *Chalcedon.* The Emperors, fays he, fettled Pofts, with intent to gain timely Information of whatever paft in
the Empire. There were no fewer than five Pofts a day, and fometimes
eight, with forty Horfes to each Poft, and as many Poftilions and Grooms
as were neceffary. *Juftinian* abolifh'd thefe Eftablifhments in many Places, efpecially thofe between *Chalcedon* and *Diacibiza,* which is the antient
Town of *Lybiffa,* fam'd for *Hannibal's* being bury'd there, and fituated
in the Gulf of *Nicomedia.* The fame Author, the more to expofe *Jufti-*
nian, advances, that he fet up an Afs-poft in divers Parts of the *Le-*
vant.

Chalcedon at this time is a poor beggarly Place, confifting of between
feven and eight hundred Houfes : it goes by the name of *Cadiaci,* or the
Judges Town ; but the *Greeks* continue to call it by the old Name. Here
a General Council was held, *anno* 451. in St. *Euphemia's* Church, where
the Fathers condemn'd *Eutyches,* who deny'd there were two Natures in
Jefus Chrift. There's no likelihood that that Church is what the *Greeks*
now make ufe for their Parochial Church, fince we are told by *Eva-*
grius that it was in the Suburbs ; and M. *de Nointel* Embaffador of *France*
to the *Porte,* avers, that the Remains of St. *Euphemia's* Church were a
Mile from the Town, where he met with an Infcription mentioning the
faid Council. The Coaft of *Chalcedon* abounds with Fifh : *Strabo* and
Pliny muft have been impos'd upon by thofe who made them believe that
the *Pelamides* or young Tunnies turn'd afide, and fheer'd off towards *By-*
zantium for fear of the white Rocks conceal'd under Water. On the
contrary, the Tunny-fifh of *Chalcedon* were fo much in vogue with the
Antients, that *Varro,* cited by *Aulus Gellius,* ranks them among the Delicacies of the Table ; and at this day you fee nothing but Tunny-fifh
Nets round the Town.

FROM *Chalcedon,* you go to Cape *Scutari,* call'd antiently the *Ox,*
or the *Ox Paffage :* from whence 'tis plain, that Place muft be confider-
ed as the beginning of the *Bofphorus. Polybius* fpeaking of the Rout from

Chalcedon

Chalcedon to *Byzantium*, obferves very juftly, that there's no croffing the Sea directly, becaufe of the ftrong Current between thefe two Cities. So again, when the fame Author defcribes the Current of the *Bofphorus*, he fays it comes from Cape *des Efties*, where *Courouchifme* now ftands, and fo proceeds to a Place call'd an *Ox* or *Cow*; for the Poets likewife gave out that *Io*, *Jupiter*'s Miftrefs, pafs'd over that Strait in fhape of a Cow. The Fleet of *Philip* of *Macedon*, who was befieging *Byzantium*, was beaten by the *Athenian* General *Chares*, near this Cape.

T H A T General's Wife *Damalis* was bury'd there: fhe dy'd during the Siege; and the *Byzantins*, in acknowledgment of the Services done 'em by her Husband, erected likewife an Altar in honour of her, and her Statue ftanding on a Pillar. The Place ftill retains the Name of *Damalis*, which fignifies a Cow. *Codinus*, the Reporter of this Story, took it out of *Dionyfius Byzantinus*, who has an old Infcription mentioning the Fact. The Seraglio of *Scutari* now takes up the fame fpot, call'd the *Cow-Cape*: I think it was *Solyman* II. built it. The Fountain of *Hermagora*, fpoken of by the fame Author, muft be within its Compafs.

C A R E muft be taken not to confound this Cape with the Beef-Market Place of *Conftantinople*, often call'd by Hiftorians fimply *the Ox*, and which was in the eleventh Precinct of the City. This Market-place took its Name from a brazen Stove fhap'd like an Ox, according to *Zonaras*, and brought from among the Ruins of *Troy*. In this Place it was that *Phocas*, by order of *Heraclius*, was burnt, after being beheaded, and depriv'd of thofe Parts which had been inftrumental in deflowering the Ladies of the firft Quality in *Conftantinople*. *Zonaras* likewife takes notice, that at the time of the Grand Revolution, when the *Comnenii* affum'd the Throne, and fhut up *Nicephorus Botaniates* in a Cloifter, their Party, who fpar'd not even the moft Sacred Things, carried on their Diforders as far as the Place call'd *the Ox*; which Place, by the way, has been the Theatre of many illuftrious Martyrdoms. *Codinus* tells us, that *Julian* the Apoftate caus'd feveral Chriftians to be burnt in the faid Stove or Furnace, the top whereof was form'd like a Bull's Head, and ftood in the Place call'd *the Ox*. The Holy Martyr *Antipas*, *Cedrenus* fays, was confum'd to afhes there. They alfo us'd to burn Criminals inthe fame Place.

T H E

THE Tower of *Leander* is juft by the Cape of *Scutari*: The Emperor *Manuel* built it on a Rock two hundred Paces from the Tower, and likewife another on *Europe*'s fide, at the Convent of St. *George*, for a Chain to be laid crofs from one to the other, and fo barricade the Canal. 'Tis obferv'd by M. *Gilles*, that formerly there was a Wall built in the Sea, which occupy'd the Paffage now between the Rock whereon is the Tower, and the firm Land of *Afia*: 'Tis likely this was the Work of the fame Emperor; for by this means, the Chain going from one Tower to the other, made it impoffible for Ships to pafs thro the Canal of the *Black Sea*. M. *Gilles* adds, that this Wall was demolifh'd by the *Turks*, on purpofe to employ the Stones elfewhere. They call this Tower the *Virgin*'s *Tower*, but the *Franks* the *Tower of Leander*; tho the Loves of *Hero* and *Leander* were carried on afar off, on the Shore of the Canal of the *Dardanelles*. This Tower is fquare, and has in it fome Pieces of Artillery: it is almoft defencelefs, and inftead of a Garifon, has only a Keeper, who picks up a few Pence among the Janizaries and Merchants of *Conftantinople*, that go thither to folace themfelves.

THO it is not a Cuftom with the *Turks* to rebuild ruin'd Towns, yet has that general Rule fuffer'd an exception in the cafe of *Scutari*, burnt by the *Perfians*. True it is, the *Turks* look on it as a Suburb to *Conftantinople*, or as the firft Baiting-place in *Afia*: 'tis alfo a principal Rendezvous of Merchants and Caravans from *Armenia* and *Perfia*, coming to trade in *Europe*. Formerly the Port of *Scutari* ferv'd as a Retreat to the Gallies of *Chalcedon*; and it was on account of its Situation, that the *Perfians* aiming at the Conqueft of *Greece*, made choice of it, not only for a Place of Arms, but as a Treafury or Bank, for keeping the Gold and Silver they levy'd by way of Tribute from the Towns of *Afia*. Hence it got the Name of *Chryfopolis* or *Gold Town*, as is reported by *Stephens* the Geographer; who however adds, that the moft common Opinion was, that the name of *Chryfopolis* comes from *Chryfes* the Son of *Chryfeis* by *Agamemnon*. *Conftantine Manaffes* fo well defcribes the Situation of *Chryfopolis*, that there's no room to doubt its being the fame as *Scutari*, tho he at the fame time fays that fuch as have taken it for *Uranopolis* are not very wide of the Truth. This was perhaps its name e'er the *Perfians* mafter'd it: the latter name, which fignifies the *Heavenly City*, was no lefs

honoura-

honourable than that of the *Golden City*.   Be it as it will, it was deſtin'd Lett. III.
to be a Harbouring-place for Excifemen; for the *Athenians* erected there-
in, the firſt of any Nation, a Cuſtom-houſe, for the gathering of the Im-
poſts laid on ſuch as uſed the *Black Sea*. *Xenophon* avers they wall'd in
the Town; and yet in *Auguſtus*'s time it made no figure, ſince *Strabo*
calls it but a Village.   At preſent it is a large and beautiful Town, and the
only one upon the *Boſphorus* on the *Aſiatick* Side. *Cedrenus* informs us, that
in the 19th Year of *Conſtantine the Great*, *Licinius*, his Brother-in-law, after being ſeveral times beaten by Sea and Land, was taken Priſoner in
Chryſopolis, and thence carried to *Theſſalonica*, where his Head was
chopt off.

T H E firſt Town of the *Boſphorus* beyond *Scutari*, is *Coſſourgé*, then
Stavros, ſo call'd from a gilt Croſs on the top of a Church built by
Conſtantine.   Next to *Stavros* you diſcover the Village of *Telengel-
cui*, which may have been the Place formerly known by the Name of
Chryſoceramus, or the *Gilded Brick*, on account of a Church cover'd with
Bricks of a golden Colour; for according to M. *Gilles*, who follows *Di-
nyſius Byzantinus* ſtep by ſtep, and has ſet him right in ſome Places,
Chryſoceramus is ſituated after *Stavros*, going up to the old Caſtles of *Aſia*.
Leunclavius makes mention of *Chryſoceramus*, and places between it and
Stavros the Monaſtery of *Akimiti*, or the *Night-watching* Monks.

B E F O R E we arrive at the old Caſtle of *Anatolia*, we meet with two
other Villages, and croſs two Brooks.   The firſt Village is call'd *Coulé*
or *Coulé-bacheſi*, and the other *Candil-bacheſi*. *Coulé-bacheſi* is on the point
which the Antients call'd Cape *Cecrium*, and now *Cecri*, oppoſite to Cape
Eſties, below which is built *Courouchiſmé*. *Candil-bacheſi* is at the Mouth
of the firſt Brook, which empties itſelf into the Gulph of *Napli* ; and
perhaps *Napli* comes from *Nicopolis*, deſcrib'd by *Pliny* to be hereabouts.
M. *Gilles* calls this Brook the Brook of *Napli*, but the *Turks* have given it the
Name of *Ghiock-ſou* or *Green Water*, as well as to that other near the Caſtle ;
ſo that one may almoſt venture to ſay that *Candil-bacheſi* is the antient
Nicopolis of the *Boſphorus*. *Stephanus Byzantinus* ſays no more than
that it is a Town of *Bithynia*; it were to be wiſh'd we could diſcover what
Victory occaſion'd its being ſo call'd.   The ſecond Brook is alſo call'd
Green Water, and is the largeſt Stream of Water that runs into the *Boſ-*

phorus on the *Afiatick* fide. In times paft it went by the Name of *Are-te*, and fome among the *Greeks* ftill call it *Enarete :* but it is proper to obferve that all the Places hereabouts are taken up with the Grand Sig-nior's Gardens, which not only extend from the firft Green Waters to thefe, but to *Sultan Solyman Kiofc;* and from thence they ftretch till they come to the Entrance of the *Black Sea.* All the reft of the Coun-try is fet apart for the Emperor's Diverfion of Hunting; and there are few Places in the World fo fit for it.

I T is certain, as *Leunclavius* obferves, that in the time of the *Greek* Emperors, there were two Caftles on the *Bofphorus;* one on *Afia* fide, the other on that of *Europe;* whereby the Paffage of the Canal in its narroweft part was barr'd. In the Declenfion of the Empire they were let run to ruin, and even before that time they were look'd on rather as Prifons than Citadels. And indeed *Gregoras* affirms they were call'd the Caftles of *Lethe,* or the *Prifons of Forgetfulnefs,* becaufe fuch as were fo unfortunate to be fent thither, were never thought of more. The *Turks* repair'd thefe Caftles at different times, even before they were Mafters of *Conftantinople.* At prefent we fhall only fpeak of that on the Coaft of *Afia.*

T H E old Caftle of *Afia* being fituated on the narroweft part of the Canal, makes it indifputable that it was there where *Darius,* Father of *Xerxes,* caus'd a Bridge to be made in his Expedition againft the *Scy-thians* or *Tartars.* The Execution of this Work was committed to *Man-drocles,* a skilful Engineer of *Samos. Dionyfius Byzantinus* avers that the Engineer had cut a Seat in the Rock for *Darius* to fit and fee the Troops march over the Bridge : 'tis not faid whether this Seat was in *Europe* or *Afia;* nor is there any Poffibility to afcertain it, even fuppofing it were ftill in being, becaufe the *Turks* admit no body to come near their Caftles. They neither know nor care to know whether there were ever fuch Men in the World as *Darius* or *Xerxes :* perhaps they lay their Tails in the very Place which ferv'd as a Throne to the then Mafter of the World.

A F T E R that Prince had feen his Troops march, he order'd two large fquare Stones to be fet up; on one were grav'd in *Affyrian* Cha-racters the Names of the Nations that were in his Pay; the like was done on the other in *Greek* Characters. The Land Army confifted of 700000

Men, and the Fleet of 1600 Ships; but this Army tarried in the *Propon-* *tis*, with Orders to go to the *Bofphorus*, in order to repair to the Mouth of the *Danube*, where another Bridge was built. *Mandrocles*, or *Androcles* as fome call him, was fo well fatisfy'd with the Liberality of *Darius*, that he caus'd a Picture to be drawn, reprefenting the Paffage of the *Per-* *fians* over the Bridge at the *Bofphorus*, in the Prefence of their Prince, who, *Herodotus* fays, was feated on a Throne after the manner of the *Perfians*.   This Piece was plac'd in a Temple of *Juno*, with an Infcrip-tion of four Verfes in *Greek*, which *Herodotus* has recorded.   The old Caftle of *Europe* being over-againft that now under Confideration, *Darius*'s Army muft have paft between the two Caftles, or a little higher up, to avoid the Violence of the Current.

THE place of the old Town of *Ciconium* mention'd by *Dionyfius* *Byzantinus*, is beyond the Caftle of *Afia*; and the Place is ftill call'd *Cormion*, juft by the Gulph of *Manoli*, where there's excellent Fifh.   The Coaft leads to the Village of *Inghircui*, that is the *Fig-Village*.   You crofs a Rivulet at *Inghircui* to enter into the Gulph of *Cartacion* or *Ca-* *tangium*, according to the faid *Dionyfius*.   This Gulph, on the North, is terminated by Cape *Stridia*, or the *Oyfter Cape*, for it affords admirable ones; and the *Greeks* call 'em *Oftridia*.   M. *Gilles* calls this Cape the *Turks* Cape, becaufe it is over againft the Kiofc of *Sultan Solyman*, and parted only by a handfome Rivulet.   This Kiofc has nothing extraordinary; thefe Kiofcs are a fort of Pavilions with large flat Roofing, after the manner of the *Levant*, where they prefer Coolnefs to Magnificence.   The Pavi-lions of the Orientals are open on all fides, and in the middle there are Jets-d'eau's.   That of the Sultan is at the Entrance of a beautiful Gulph, which forms the Elbow of the Canal, where the *Bofphorus* runs inden-ted, tho in the Maps it is fet down to be almoft in a ftrait line.   This is the *Round Gulph* mentioned by *Dionyfius Byzantinus*, or the Gulph of the Sultan, fpoken of by M. *Gilles*, who remark'd on the South of it the Foundation of the famous Monaftery of thofe Monks that fpend the whole Night in Prayer; whereas *Leunclavius* places it between *Stavros* and *Te-* *lengelcui*.   We muft not forget, that the Cape which turns the Gulph *Caftacium* to the South, makes two confiderable Points: the one fhuts in the Gulph on the fide of the *Greater Glari*; the other, which is to-

wards

wards the little *Glari*, forms the Gulph of *Placa*, fhap'd fomewhat like a Table. The two *Glari's* are perhaps the Rocks which *Dionyſius Byzantinus* call'd *Oxyrrhoon* and *Poryrhoon*, for the Waves make a conſidera. ble Noiſe about thoſe Points.

G O I N G up from the Pavilion of Sultan *Solyman*, towards the new Caſtles, we meet *Beicos* or *Becouſſi*, the *Walnut-tree Village*, on which account *Leunclavius* calls it *Megalo-Carya*. The fine Stream that flows into it, and its advantageous Port, give ſome ground to ſuſpect that this is the Place where *Amycus* King of the *Bithynians* kept his Court. There's no other part of this Coaſt that can be ſuppos'd to have ſerv'd for the Abode of ſo formidable a Prince, whom *Valerius Flaccus* calls *the Gyant* ; and *Apollonius* of *Rhodes*, *the moſt daring Man of his Age* ; he was not only a great Wreſtler, but very ſkilful too at Boxing, and at that kind of Exerciſe which was call'd *Pugilation*, in which lay great part of the Merit of the firſt Heroes. Before the Invention of Iron and Arms, ſays *Donatus*, Men exercis'd their Valour in fighting with Hand and Foot, and Tooth. If ſuch Sports were to come in faſhion again, how many Porters would now be reckon'd Heroes? *Amycus* was a brave ſtrapping Fellow, *like thoſe Great Men*, ſays the Poet, *that the Earth brought forth in anger, to oppoſe the Power of* Jupiter ; yet this dreadful Champion met with his Match. According to his uſual Cuſtom, he made an open challenge to the boldeſt *Argonauts* that appear'd on the Coaſts of his Kingdom. *Pollux* the Brother of *Caſtor*, and Son of *Jupiter* and *Leda* ; *Pollux*, I ſay, the greateſt Wreſtler among the *Greeks*, vigorous as a young Lion, overthrew this Coloſſus, tho his Chin was ſcarcely yet cover'd over with the Down of Youth. They firſt gave each other ſtrenuous Puſhes, like Rams that ſtrive to overturn each other to the Ground : after the firſt Heaves they took their *Ceſtus's* in their hands, and nothing was to be heard but *Blows like thoſe of the Hammers that are made uſe of to break the Planks of a Ship*, according to *Apollonius's* Compariſon : And in this manner the Cheeks and Jaws of the Athletes us'd to found in thoſe days ; Each Man drubb'd his Companion without Mercy ; their Teeth were looſen'd, and at laſt beat down their Throats in pieces. Tho the *Ceſtus* was often no more than a Thong of Leather very dry and very hard, yet it would give plaguy Thumps when artfully apply'd. Our Heroes,

weary

weary of this gentle Prelude, after having wip'd their Faces, fell to lufty Fifticuffs: it is very probable they took one another by the Collar, for the Son of *Jupiter* gave him of *Neptune* fuch a Squelch on the Ground, that the Bones of his Ears, tho the hardeft in all the Head, were broken with the Fall. Thus dy'd *Amycus,* who had overcome fo many Strangers, and fo many of his own Subjects. *Apollodorus* and *Valerius Flaccus,* who defcribe his Death in another manner, agree however that he dy'd by the hands of *Pollux.*

AMYCUS was accus'd of laying traps for Strangers, and deftroying them by Treachery; but the *Argonauts* forewarn'd of his Tricks, were too fharp for him: they not only accompanied *Pollux* to the Foreft, which ferv'd for the Field of Battel, but ftaid near him while the Fight lafted. It was a fhame for Coufin Germans, and Sons of Gods and Goddeffes, to ufe one another fo ungenteely. *Pollux* was the Son of *Jupiter* and *Leda,* and *Amycus* the Son of *Neptune* and the Nymph *Melia,* the Daughter of the Ocean, a Hamadryad that prefided among the Afh-Trees. As for the *Ceftus,* it was not always a bare fingle Leathern Thong; there were fometimes feveral of them faftned to a Club, and fome good heavy Knobs of Lead at the end of them.

BEICOS then, to return to our Subject, was in all probability the Capital of the Dominions of *Amycus,* and the fame that was call'd the *Port of Amycus,* and which *Arrian* nam'd *Laurus infana,* or *the Laurel-tree that turn'd Folks Brains.* That Tree which gave its Name to the Place, and which depriv'd the Seamen of their Wits, was perhaps one of thofe kinds of the *Chamærhododendros,* that grow on the Coafts of the *Black Sea,* and which I fhall fpeak of hereafter. That part of *Beicos,* which lies wholly along the Coaft, is ftill call'd *Amya,* as if it were a Corruption of *Amycus:* it may perhaps be the Place where that Prince was buried, for there is mention made of his Tomb in antient Authors. Be that as it will, all this Coaft is fo fruitful, that every Village bears the Name of fome Fruit. The Village which lies above *Beicos,* before you come to the firft Elbow of the Canal, is call'd *Toca,* that is, the Village of *Cherries,* fituated between the Bays of *Monocolos* and *Moucapouris,* parted from each other by a fmall Stream, and by the *Turkifh* Cape, formerly call'd *Aetorhecum.*

A

A LITTLE on this fide the new Caftle of *Anatolia* are the Ruins of an antient Caftle, on one of the Eminences, which, on the fide of *Afia*, form the firft Elbow of the Entrance of the *Bofphorus* : this ruinated Caftle fubfifted in the time of *Dionyfius Byzantinus*. Above the Temple of *Phryxus*, fays that Author, ftands a good ftrong Fort, in a circulary Inclofure, which the *Gauls* deftroyed, as they did many other Places in *Afia*. The *Greek* Emperors maintain'd this Port in repair, till the Decadence of their Empire. It is likely 'twas built by the *Byzantines* after the Retreat of the *Gauls* ; for *Polybius* informs us, that the People of *Byzantium* laid out a great deal of Money to fortify that part of the Country before they went to war with the *Rhodians* and King *Prufias*. This Fortrefs was abfolutely necefTary to their Defign of making themfelves Mafters of the Navigation of the *Pontus*, and of levying Impofts upon all Merchandizes there. The Cape was named *Argyronium*, either by reafon of the great Expence of fortifying it, or becaufe it was purchas'd with a round Sum of the King of *Bithynia* ; for it was fpecify'd in the Articles of Peace, that *Prufias* fhould reftore to the *Byzantines* the Lands, Forts, Slaves, the Materials and Tiles of the Temple that he had demolifh'd during the War : in confequence whereof, the Freedom of Navigation in the *Pontus Euxinus* is entirely reftor'd, to the great Glory of the *Rhodians*. As to the new Caftles beyond thofe Ruins both in *Afia* and *Europe*, they were built not long ago by *Mahomet* IV. to ftop the Incurfions of the *Coffacks*, *Polanders*, and *Mufcovites*, who came very far into the *Bofphorus*.

ALL the Coaft is ftrew'd with old Materials ; for the Antients had fo terrible an Idea of the *Black Sea*, that they durft not venture upon it, till they had rear'd Altars and Temples to all the Gods and Goddeffes of their Acquaintance. All the Strait of the Opening was call'd ' *Hiera*, which fignifies *Sacred Places*. Befides the Temple built on the *Afian* Coaft by *Phryxus* the Son of *Athamantus* and *Nephale*, who carried the Golden Fleece into *Colchis*, the *Argonauts*, who undertook the fame Voyage to fetch back that Treafure into *Greece*, did not fail to implore the Affiftance of the Gods before they trufted themfelves on fo dangerous a Sea. *Apollonius Rhodius*, and his Commentator, who have very well explain'd the Courfe of thofe famous Travellers, let us know, that being

ing

ing detain'd by contrary Winds at the Entrance into the *Pontus,* they crofs'd over from the Court of King *Phineus,* which was in *Europe,* to the Coaft of *Afia,* to raife Altars and Temples to the twelve moft famous Deities of thofe times. According to *Timofthenes* quoted in the Commentary of *Apollonius,* they were the Companions of *Phryxus,* that built the Altars of the twelve Gods; and the *Argonauts* only rais'd one to *Neptune. Ariftides* and *Pliny* make mention of the Temple of that God. *Herodotus,* according to the fame Commentary, pretended that the *Argonauts* facrific'd upon *Phryxus's* Altar. *Polybius* fancied that *Jafon,* in his return from *Colchis,* had built one Temple confecrated to the twelve Deities on the Coaft of *Afia,* oppofite to the Temple of *Serapis,* which was on the *European* fide. Tho thefe Difquifitions are not very ufeful now-a-days, yet nothing can be more agreeable, when a Man is upon the fpot, than to recollect them in his Mind. Upon a cafe of neceffity one might eafily name the Deities fo honour'd. According to the Commentator of *Apollonius Rhodius,* they were *Jupiter, Juno, Neptune, Ceres, Mercury, Vulcan, Apollo, Diana, Vefta, Mars, Venus,* and *Minerva. Jupiter* being the moft potent of the whole Gang, *Jafon* made his Court particularly to him, and endeavour'd to get his Favour above all the reft: Hence it is, that *Arrian, Menippus, Dionyfius Byzantinus,* and *Mela,* make mention of none but the Temple of *Jupiter the Diftributer of good Winds,* notwithftanding thofe of the other Deities were not far off, fince there were as many Temples as Altars. 'Twas probable it was in this Temple of *Jupiter,* that there was plac'd a Statue of that God, fo perfect, that *Cicero* fays, there were but three fuch in the World. It was from the Gate of this Temple, that *Darius* had the pleafure to furvey the *Pontus Euxinus,* or in *Herodotus's* Words, *the Sea moft worthy of Admiration.* We are not to imagine, as fome have done, that this Temple was in one of the *Cyanean* Ifles, for the biggeft of them all can but juft fupport the Column of *Pompey. Herodotus* only fays, that from the Bridge which *Darius* had caus'd to be rais'd over the *Bofphorus* in the Place which we mentioned above, that King went towards the *Cyanean* Ifles, to contemplate the Sea which afforded a wonderful Profpect from the Avenue of the Temple. That Temple muft therefore have been at the Village of *Joro,* a Corruption from *Hieron,* and *Joro* is clofe to the new Caftle of *Afia.*

GOING

GOING along the Coaſt beyond that Caſtle towards the Mouth of the *Black Sea,* you paſs by that Place which *Dionyſius Byzantinus* calls *Pantichium,* and others *Mancipium.* Afterwards you diſcover Cape *Coraca,* or *the Cape of the Crows,* which forms the beginning of the Strait; it is perhaps the Cape of *Bithynia* mention'd by *Ptolomy,* near which was a Temple of *Diana.* Beyond this Cape you find nothing on the *Aſian* Coaſt, that is ſet down in the Authors, except the *Gulph of Vines;* but yet after this you come to the famous Cape *of the Anchor,* ſo call'd becauſe the *Argonauts,* according to *Dionyſius Byzantinus,* were here oblig'd to provide themſelves with an Anchor of Stone. 'Tis likely *Minerva* had forgotten ſo neceſſary a piece of Furniture; ſhe who took care of all the Rigging of the *Argos,* which was the biggeſt and tighteſt Ship that had been known on the Sea before that time. That Veſſel was fit either for Sailing, or Rowing like a Gally; and every Man in her was a Hero. The *Aſian* Lighthouſe is upon this Cape, near which alſo are

[margin] ⁹ *The* Aſian Cyanean Stones.

thoſe ⁹ Rocks accounted ſo dangerous among the Antients, that *Phineus* exhorted *Jaſon* not to go that way, except the Weather was very fair; otherwiſe, ſays he, *your* Argos *will be broken, tho it were made of Iron.* Theſe Rocks are only the Points of an Iſle or Rock ſeparated from the main Land by a narrow Strait which is quite dry, when the Sea is calm, and is fill'd with Water when there is the leaſt Storm: At ſuch a time you can ſee nothing but the higheſt Point of the Rock, the others lying hid under water. This is what makes the Place ſo dangerous, eſpecially to thoſe who are ſo raſh as to paſs thro the Strait, as it ſeems *Phineus* advis'd the *Argonauts* to do. In thoſe firſt Ages, when Navigation was ſcarce in its Infancy, the Seamen never durſt ſtir out of ſight of the Coaſt. As for us, who, I can take my Oath on't, were in no *Argos,* but in a Felucca with four Oars, we affeĉted to keep as far off on't as we could. The *Argonauts* run the hazard; for the Hiſtorians, or rather the Poets inform us, that their Ship ſtuck ſo faſt upon thoſe Rocks, that *Minerva* was forc'd to come down from Heaven to puſh it off with her right Hand, while ſhe ſtrengthen'd herſelf with her left againſt the Points of the Rocks: Topping Mariners, thoſe *Argonauts!* And indeed *Apollonius* very judiciouſly obſerves, that their Hearts were in their Mouths till the Fright was over.

FROM

F R O M the *Afian Cyanean* Iflands, you muft crofs over to thofe of *Eu-rope*, if you would view the other Coaft of the *Bofphorus* to *Conftantino-ple* in order.   Thefe Ifles then, as thofe of *Afia*, are properly nothing but one rough Ifland, the Points of which look like fo many feparate Rocks, when the Sea is much difturb'd.   *Strabo* obferv'd, that towards the Mouth of the *Pontus Euxinus*, there was one little Ifle of each fide, whereas the antient Geographers imagin'd that there were feveral Rocks, as well on the fide of *Europe* as on that of *Afia*, which not only floated on the Wa-ter, but fwam along the Coaft, and joftled one againft the other.   The Foundation of all this Story was nothing but their Points appearing or difappearing, according as the Sea run over them in Tempefts, or left them uncover'd in Calms.   It was never publickly declar'd that they were fix'd till after *Jafon*'s Voyage, becaufe they were then in all probability view'd fo nearly, that it was impoffible to think them moveable: Ne-verthelefs, as moft People are more agreeably entertain'd with Fables than with Truth, they had much ado to throw off their Prepoffeffion. You may fee the whole Rock that is on the fide of *Europe*, when the Sea is gone off: it ftands up in five Points, which look like fo many diftinct Rocks, while the Sea is rough.   This Rock is divided from the Cape of the *European* Light-houfe only by a little Arm of the Sea, which is empty of Water in fair Weather; and it is on the higheft of thefe Points that they fhew a Column, which they have call'd, tho groundlefly, the Column of *Pompey*.   It does not appear by any Paffage in Hiftory, that *Pompey*, after the Defeat of *Mithridates*, rais'd any Monuments here: and befides, the Infcription on the Foot of this Pillar makes mention of *Auguftus*.   When you care-fully examine the Bafis and the Shaft, you muft confefs thofe two Pieces were never made for each other; one would rather imagine the Pillar had been fet upon the Bafis, to ferve as a Guide to fuch Veffels as pafs this way.   The Column, which is about twelve foot high, is adorn'd with a *Corinthian* Chapiter; but 'tis in fo fteep a place, that there is no getting up to it without crawling on all four, and the Bafis is generally under water.   *Dionyfius Byzantinus* fays, the *Romans* fet up an Altar to *Apollo* on this Rock; and this Bafis may be a Remnant of it, for the Feftoons are of Laurel-leaves, which was a Tree facred to that God.   Perhaps, out of Flattery, an Infcription might afterwards be carv'd upon it, in praife of *Auguſtus.* I know not whether the Column be of Marble, or of the Stone of the Country, the Sea would not permit us to examine it cloſely enough ; the Stone of the Country has in its greyiſh Colour ſomething approaching to blue, and this was the Reaſon why theſe Rocks were call'd *Cyanean.*

Phinopolis. I F we may judge by the courſe of the *Argonauts,* the Court of *Phineus,* ſo famous upon account of his Misfortunes and his Predictions, was at the entrance of the *Boſphorus* on the ſide of *Europe.* We read in *Apollonius Rhodius,* that the *Argonauts,* after having work'd thro a violent Tempeſt, after parting from the Dominions of King *Amycus,* caſt Anchor at the Court of *Phineus* to conſult him. That Prince's Court was perhaps at *Mauromolo,* where there is a convenient Port, and a very agreeable Rivulet. May not *Belgrade,* a little Town above *Mauromolo,* be the antient *Salmydeſſa,* where, according to *Apollodorus, Phineus* made his Reſidence? It is certain indeed that the Antients place that City beyond the *Cyanean* Iſles; but as there is no Port on that Coaſt, and *Apollonius* ſaying in ſo many words, that they landed at *Phineus's* Palace, which was on the Sea-ſhore, is it too bold a Conjecture to advance that *Belgrade,* which is naturally a Place compleatly charming, and truly worthy the Abode of a great Prince, is built on the Ruins of *Salmydeſſa,* of which *Mauromolo* was the Haven?

T H E Deſcription *Apollonius* gives of *Phineus,* and the means which that Prince taught the *Argonauts* of paſſing the *Cyanean* Rocks, are extremely ſingular. *Phineus* having notice that this Company of Heroes were arriv'd at his Palace, aroſe from his Bed, (for he remembred *Jupiter* had decreed, that thoſe Demi-Gods ſhould do him Service) and crept half aſleep, leaning with one hand upon a Stick, and ſupporting himſelf with the other againſt the Walls. This good Man ſhook with Old Age and Weakneſs; his Skin, which ſtuck to his Bones, had much ado to hinder them from parting. In this Condition he appear'd like a Spectre at the Entrance of a Hall, where he had no ſooner ſat him down, but he fell aſleep, without being able to utter one ſingle Word. The *Argonauts,* who no doubt expected another kind of a Creature, were ſurpriz'd at ſuch a Figure: at length, *Phineus* more intent upon his own Affairs than upon theirs, recollecting his Spirits a little ; *Heroes,* ſays he, *who are the Glory of* Greece, *for I well know who you are by the Science of*

*

Divina-

Divination which I pofſeſs, leave me not, I conjure you, till you have deliver'd Lett. III. *me from the miſerable Condition I am in.* Can any thing be more terrible than *to die of Hunger in the midſt of Plenty? Thoſe curſed* Harpies *ſnatch the Meat from my Mouth; and if they leave any thing in the Diſhes, they inject it with ſuch an intolerable Stink, that no Mortal can touch it : but it is foretold by the Oracle, that theſe beaſtly Birds ſhall be diſperſed by the Sons of* Boreas.

Ƶ*ETHES* and *Calais*, who were of the Band, were mov'd at the Fate of that wretched Prince, and promis'd him their Aſſiſtance. Supper was immediately brought in; but the moment *Phineus* offer'd to touch the Meat, the *Harpies* iſſuing from certain Clouds, among dreadful Flaſhes of Lightning, fell upon the Table with a ſurprizing Yell, and devour'd every thing there; after which they fled away, leaving behind them a Stink that almoſt poiſon'd the whole Company. The Sons of *Boreas* (who were ſaid to have Wings) did not fail to purſue them, and had ſoon caught them : but *Iris* deſcending from Heaven, told them they muſt not for the world touch their Lives; that they were the Dogs of the Mighty *Jupiter* ; and ſhe ſwore by the River *Styx* they ſhould be ſent ſo far off, that they ſhould never come near *Phineus*'s Houſe any more. This good News was carried to the King, who, that he might be ſure of the Truth, order'd what there was in the Houſe to be brought in; and not hearing the Noiſe of thoſe ugly Beaſts, he laid about him luſtily. By way of Acknowledgment, the good old Man then began to dogmatize, and gave our Heroes ſuch Notices as he thought would help to carry them thro their Voyage without Danger. *Apollodorus* relates theſe Fables with other Circumſtances, whereof a longer Recapitulation would be tedious. I leave it to Men of more Learning to explain the Story of the *Harpies* : Of what conſequence is it to know whether they were Graſhoppers that infected *Phineus*'s Lands, and devour'd his Harveſts, as M. *Bochart*, and the Author of the *Bibliotheque Univerſelle* have imagin'd? whether the Sons of *Boreas* are to be interpreted the North Wind, which drove away thoſe Inſects? whether *Phineus* was ſtript by his Miſtreſſes, who reduc'd him to the laſt Extremity ? whether the *Argonauts*, who, in all antiquity, were accounted Heroes, were not Merchants more daring than the reſt, who went quite into *Colchis* to buy Sheep to ſtock *Greece* with them? All this ſeems to me very obſcure. But I admire the Invention of honeſt *Phi-*

neus,

reus, who not having ever a Compaſs, any more than the *Argonauts,* advis'd them before they ventur'd the Paſſage of the *Cyanean* Iſlands, to let fly a Dove; *If ſhe get ſafe and ſound over thoſe Rocks,* quoth he, *make the beſt of your way with Oars and Sails, and rely more upon the Strength of your Arms than upon the Vows you may make to the Gods : but if the Dove comes back, turn tail, and march home again.* It was impoſſible to have hit upon a cleverer Expedient.

B U T to return to *Phineus's* Court, or rather *Mauromolo.* It is a fine Monaſtery of Caloyers, who pay no other Tribute than one Load of Cherries. The Story goes, that a Sultan having loſt his way in hunting, near that Houſe, and fancying the Monks did not know him, ask'd them for ſomething to eat. The Monks, who knew well enough who he was, preſented him with a piece of Bread and a Plate of Cherries, which were ſo good, that the Sultan exempted the Religious from the Capitation, and only order'd them to bring every a Year a Load of Cherries to the Seraglio.

T H E R E is not at preſent any remarkable Place between *Mauromolo* and the new Caſtle of *Europe,* tho, no doubt, the Antients did not fail to give diſtinguiſhing Names to all this Coaſt, as ſteep and rugged as it is : but you cannot move a Step in any Country where the *Greeks* have had to do, but you diſcover ſome of their Names ſtill in being.

> *Here ev'ry deſart Waſte, and barren Field,*
> *Of beauteous Names will fruitful Harveſts yield.*

W H A T can be a greater Satisfaction to thoſe that we call *Men of Erudition,* than to know that the firſt Nook on the right hand, as you enter the Strait, was formerly call'd *Dios Sacra,* as much as to ſay, *the Sacrifices of Jupiter?* That the next Port was the Port of the *Lycians* in the firſt Ages, and that of the *Myrleans* afterwards? The *Lycians* were a People of *Aſia,* that traded in the *Pontus,* and commonly caſt Anchor in this Port. As for the *Myrleans, Dionyſius Byzantinus* informs us, that ſome ſeditious Folks of *Myrlea* retir'd to this Part of the *Boſphorus*; and *Myrlea* was that Town of *Bithynia,* which *Nicomedes Epiphanes* nam'd *Apamea,* from the Name of his Mother *Apama.* After the Port of the

Lycians, come two other little Ports, which formerly took their Names Lett. III.
from some Altar of *Venus*; for *Aphosiati* seems to be a remnant of *A-*
phrodysium, which *Dionysius Byzantinus* places thereabouts: and as one of
those Ports was frequented by the Merchants of *Ephesus,* it is very probable this is the Port of the *Ephesians* mention'd by the same Author.
But the most remarkable thing here, is a Gut of Water, whose Sand look'd
like Gold, during the time that the Copper Mines which are on this Coast were
wrought: this Water runs close to the Chappel of our Lady *of the Chesnut-*
Trees, at the foot of a Mountain so much higher than the rest thereabouts,
that from the top of it you may see *Constantinople,* the *Black Sea,* and the
Propontis. The Light that was formerly kept in a *Pharos,* built upon the
Point of it, was as serviceable to the Pilots as those of the *European* and
Asian Cyanean Islands; but they have let the Tower run to ruin. They
were very much in the right in setting up Light-houses on the side of
Europe; for the antient *Thracians* were merciless Folks. We read in *Xe-*
nophon, that those who dwelt along the Sea-Coast had mark'd out the
Extent of their Lands very critically: for before this Precaution they
us'd every day to be at Daggers-drawing about the Wrecks that were
thrown upon them, and which every man pretended to lay hands on.
The antient *Thracians* liv'd in those dreadful Caverns that are on the
Strait to the left, as you go from the *European* Castle towards *Pompey's*
Column. Perhaps in these Rocks it might be, that the *Myrleans* had
settled their Abode. As you pass by them, you hear such strong Echoes,
that they are sometimes as loud as the Report of a Cannon, especially towards *Mauromolo.*

A S to the new Castle of *Europe,* it was built by order of *Mahomet* IV.
opposite to that of *Asia*; beyond this Castle are to be seen the Ruins of
an antient Citadel, built by the *Greek* Emperors, or perhaps the *Byzan-*
tines, to guard that important Passage, where they made Exactions upon all the Vessels that went by. *Polybius* says, there was in this Place
a Temple dedicated to *Serapis,* over against that of *Jupiter,* which was
on the *Asian* Shore. The first of those Temples is call'd by *Strabo, the*
Temple of the Byzantines, to distinguish it from that of *Jupiter,* which
he calls *the Temple of the Chalcedonians. Dionysius Byzantinus* gives the
Name of *Amilton* to the Cape, which is at the end of the Strait before you

enter

enter the Gulph of *Saraia*; this is the Cape *Tripition* of the *Greeks*. *Saraia* is a Village over againſt the Gulf of *Scletrine*, whence you croſs the River *Boujoudera*, which waters the fine Country which *Dionyſius* calls the *Lovely Fields*: It is alſo call'd the River *of the deep Gulf*, becauſe beyond *Boujoudera*, the *Boſphorus* winds into that great Elbow, by which it turns to the South Eaſt, making a kind of right Angle with the Mouth of the *Black Sea*. This deep Gulf was alſo call'd *Saronica*, becauſe the Altar of *Saron*, a Hero of *Megara*, or a Sea-God, ſtood on the Banks of it. According to ſome others, the Gulph ends at that famous Rock, entitled, *the Rock of Juſtice*, of which they tell a pretty ridiculous Story, to be found in *Dionyſius Byzantinus*.

TWO Merchants, ſays he, ſailing towards the *Pontus*, depoſited in a Hole of that Rock a Sum of Money, and jointly agreed that neither of them ſhould meddle with it in abſence of the other; but one of them came ſoon afterwards by himſelf, with deſign to ſteal this ſame Money. The Rock would not by any means betray its Truſt, and ſo gain'd the Name of *Equitable*. At a diſtance this Rock appears like a Pine Apple, with the top riſing up, and hollow. This Hole was perhaps what gave occaſion to the Fable of the pretended Treaſure hidden by the Merchants. Sailors are the fitteſt People in the World to invent ſuch Tales, eſpecially in a Calm, when they have nothing elſe to do.

THE Town of *Tarabia* or *Tharapia* is beneath this Rock, upon a little River, at the Mouth of which ſtands the Shelf *Catargo*, which afar off looks like a little Galley. The Mouth of this River forms a tolerable good Port call'd *Pharmacias*, becauſe it was deliver'd down by Tradition, that *Medea* caſting Anchor there, had brought with her out of the Ship her Box of Drugs, by means whereof ſhe perform'd ſo many Miracles. Oppoſite to *Tarabia*, on the other ſide of the River, is the Valley call'd *Lino*, in which is the Gulf *Eudios Calos*, ſpoken of by *Dionyſius Byzantinus*; but lower, as you go down towards *Tenicui*, is the Port of King *Pithecus*, mention'd by the ſame Author. The Coaſt is ſo ſteep from this Place to the Elbow that turns towards the old *European* Caſtle, that the Antients fancy'd theſe Rocks were *Bacchantes*, upon account of the Noiſe made by the beating of the Waves againſt them. The Elbow before you come to *Tenicui*, was formerly cover'd with a

Grove

Grove of Arbute, or Strawberry-Trees, and was call'd *Commarodes* Commaros, which fignifies an *Arbute-Tree*.

AS for *Yenicui*, it is a Village fituated on the Elbow that the Canal makes to run to *Conftantinople. Yenicui* is a *Turkifh* Word, and confequently has no relation to any antient Name, any more than *Neocorion,* which is the Name of the fame Place, and means, in modern *Greek, New Village.* Beyond *Yenicui* ftands *Iftegna*, upon the fartheft part of a little Port: this may be the *Leoftenion* of *Dionyfius*, and *Stephens* of *Byzantium*; fince the *Port of the Women*, which we are going to fpeak of, muft be between the old *European* Caftle and the *Leoftenion.* Now, it is certain the *Port of the Women* of *Dionyfius Byzantinus* is at the Entrance of the River *Ornoufdera*, or of *the Stream of the Hogs,* which runs exactly between the Caftle and *Iftegna.* The Mouth of this River forms the fineft Haven in all the *Bofphorus,* and that Haven has had various Names. The *Greeks* call it *Saranta Copa,* becaufe of its wooden Bridge, which is fuftain'd by forty Beams that ferve inftead of Piles. *Dionyfius Byzantinus* calls it the *Gulf of Lafthenes,* whence it fhould feem, that in *Pliny* we ought to read *Lafthenes,* not *Caftanes*; nay, perhaps *Leofthenes* in *Dionyfius*, to make him agree with *Stephanus Byzantinus.* Be that as it will, the fame Port is *Dionyfius*'s Port of the Women, and *Pliny*'s Port of the Old Men: for as to that which this Author calls by the fame Name, that is in all likelihood the Port of *Iftegna*, fince he makes mention of it juft after the Port of the old Men. The Port of *Sarantacopa* was alfo nam'd the Port of *Phidalia,* the Wife of *Byzas*; fhe who, according to *Stephanus Byzantinus,* having put herfelf at the head of a little Army of Women, overthrew in this place *Strelius*, who endeavour'd to dethrone her Brother *Byzas.*

BALTHALIMANO, or the *Port of the Ax*, with a Village of the fame Name, are fituated between *Ornoufdera* and the old Caftle; but 'tis fo inconfiderable a Haven, that there is no mention of it in antient Authors. All the Coaft quite to the Caftle, is in many places directly perpendicular, and the Waves make fuch a frightful Noife againft it, that the *Greeks* ftill call it *Phonea*, as who fhould fay, *Phonema*, *a Voice repeated.* The Voice tofs'd about by continual Whirlwinds, to ufe the Expreffion of *Stephanus Byzantinus, falls at laft with the Water into a Caldron ftand-*

ing

ing upon the Fire. The Sailors when they go up the Canal, are obliged here to make ufe of ftrong Poles to keep themfelves off the Rocks, and without them they muft unavoidably run upon them, their Oars not being fufficient to prevent their being driven by the South Wind. It is therefore probable that *Darius*'s Bridge was built lower down towards the old *European* Caftle.

THE old Caftle is fituated on the narroweft part of the Canal, upon a Cape oppofite to that where the *Afian* Caftle is built. 'Twas upon thefe Capes that the *Greek* Emperors formerly rais'd Forts, as we faid above: but the *Turks* have fortify'd thefe Places much more ftrongly, which in themfelves are very advantageoufly fituated. *Amurat* or *Mourat* II. having declar'd War againft *Vladiflaus* King of *Poland*, was refolv'd to fecure the Paffage of the *Bofphorus* : and as the *Greek* Caftles were falling to decay, he demolifh'd the Monaftery of *Softhenion*, dedicated to *St. Michael*, and founded by *Conftantine* the Great. The Materials were employ'd in building this Caftle ; and they were excellent, for *Juftinian* and *Bafil* the *Macedonian* had thorowly well repair'd that Convent.

1451. or
10··

Neverthelefs, *Mahomet* II. did not think *Mourat*'s Fortifications prudently laid out ; and to block up *Conftantinople* on all fides, he put them in the condition they are at prefent. This Caftle, as *Chalcondylus* fays, has three great Towers, two on the fide of the Canal, and the third on the brow of the Hill. Thefe Towers are cover'd with Lead, and are thirty foot thick ; the Walls of their Circuit, which is triangular, are about two and twenty foot thick, but they are not terrafs'd. The Portholes for the Cannon are horrible, as they are in the reft of the Caftles of the *Bofphorus* and the *Dardanelles*. The Canons are without Carriages, and require a great deal of time to charge. *Mahomet* II. finifh'd thefe Fortifications in three Months ; he befieg'd *Conftantinople* in the following Spring, and nam'd this Caftle *Chafcefen*, that is, *Cutter off of Heads*. The *Greeks* call it *Neocaftron*, the *New Caftle*, and *Lemocopia*, or the *Caftle of the Strait*. It has been call'd the *Old Caftle*, fince *Mahomet* IV. built thofe at the Entrance of the *Black Sea*. *Mahomet* II. who put 400 Men in Garifon in his Caftle of *Bafcefen*, gave the Command of it to *Pherus Aga*, with Orders to exact Cuftom from all the Veffels, as well *Genoefe* and *Venetian*, as thofe of *Conftantinople*, *Caffa*, *Sinope*, and *Trebifond*, *&c.* that fhould pafs by. The

Gover-

Governour interpreted his Mafter's Orders in a cruel Senfe; for *Erizzo* a *Venetian* Captain neglecting to ftrike fail, had the misfortune to fee his Ship funk by a Stone Bullet of a prodigious fize : and all he could do in this Diforder, was, to make the beft of his way to Shore with about thirty of his Men : but he was impal'd by the Governour's Direction, and the reft beheaded, and their Bodies left unburied upon the Shore.

THE Caftle of *Mahomet* II. is built upon *Polybius*'s Cape of *Mercury*, and that Temple dedicated to the God of Thieves and Merchants, was, according to that Author, built on the narroweft part of the *Bofphorus*, almoft in the middle between *Byzantium* and the Temple of *Jupiter the Diftributer of Winds*. *Dionyfius Byzantinus* calls this Cape the *Red Dog*. Here ended the other foot of the Bridge, over which *Darius* march'd his Army, when he went againft the *Scythians :* The firft foot of that great Work was in *Afia*, at the narroweft part of the *Bofphorus*, oppofite to the other Caftle. As to the Chair that was hollow'd for that Prince to fit in, to fee his Army march, it was in all likelihood on the fide of *Europe* ; and *Dionyfius Byzantinus* agrees that it was the fineft Monument remaining of that antient piece of Hiftory : but this Monument is now loft. The *Mahometans* entirely fubverted the two Coafts of the Canal, for the building not only of the old Caftles, but alfo of that beautiful Village that lies round the *European* one, and which properly receiv'd the Name of *Lemocopia*, when *Mahomet* II. order'd People gather'd from all parts to go and inhabit it.

THE Canal widens from the Caftle to *Courouchifme*, and forms a great Gulph in the fhape of a Bow, on the Banks of which is a Seraglio of the Grand Seignior, then the Village of *Bubec Bachefi*, and next *Arnautcui*, or the Village of the *Albaneze* or *Arnauts*. This Gulph of *Arnautcui* is meant by *Dionyfius Byzantinus* under the Name of the Gulph *of the Ladder*, becaufe in thofe times there was a famous Ladder or Machine compos'd of Beams, which was of great ufe in loading and unloading of Ships, becaufe they went up to it as it were by Steps. Such forts of Machines were call'd *Chelæ*, upon account of the Lord knows what refemblance obferv'd between them and the Claws of a Crab; from *Chelæ* came *Scalæ*, and hence it is that the Ports moft frequented in the *Levant* are call'd *Ladders*. Perhaps the Temple of *Diana* built at *Arnaut-*

cui, and very well known to the Fifhermen by the Name of *Dictynna,* might give occafion to fet Ladders there for the more eafy embarking and landing. Thofe Machines were not rais'd high, but lay almoft flat upon the Sea-fhore, and kept People dry-fhod in their paffing to and fro.

AFTER *Arnautcui* you come to the famous Cape *des Efties,* at the foot whereof ftands *Courouchifme. Efties* is very probably a Remnant of *Eftia,* a Name by which the *Greeks* knew the Goddefs *Verfa,* who perhaps had fome Temple hereabouts. *Courouchifme* was formerly call'd *Afomaton,* from a Church built there by *Conftantine,* in honour of *St. Michael* the Archangel. *Procopius* defcribes the Magnificence of this Church, which was rebuilt by *Juftinian*; but there is no Footftep of it left. We can't fay the fame of the March of the Crabs, which to avoid being borne away with the Current, which is very violent above the Cape, are forced to fcramble along the Rocks, and venture not again into the Canal till they have whetted their Claws to fome purpofe, and as it were carv'd their Steps upon the Rocks.

FROM Cape *Courouchifme* to the point of *Befichtachi,* the Canal runs out into a half Circle, on the fide of which ftand *Ortacui* and *St. Phocas. Ortacui* is a Village built on the Port which the Antients call'd *Clidium* and the *Old Sea Man,* whom fome take to be *Nereus,* others *Proteus,* or fome God of the Waters. The little Port of *St. Phocas* is at the Entrance of a fruitful Valley, known to the Antients upon account of *Archias of Taffos,* who made choice of it to build a City in ; but according to *Stephanus Byzantinus,* the *Chalcedonians* out of jealoufy oppos'd it. Below *St. Phocas* is another Port where the *Rhodians* anchor'd when they came to trade in the *Pontus,* which preferv'd to it the Name of *Rhodacinon.* Thefe *Rhodians* were fo powerful at Sea in thofe days, that they forc'd the *Byzantines* to allow a free Trade upon the *Pontus Euxinus,* that is to fay, to give free Paffage to all Nations that were willing to fail into the *Black Sea,* without exacting any Impofts from them.

THERE now remains only *Befichtachi* or *Befichtas,* before you come to *Fondocli,* the firft of the Suburbs of *Conftantinople,* according to the Route we follow'd. *Befichtachi* formerly bore the Name of *Jafon* the Captain of the *Argonauts.* That Hero, according to *Stephanus Byzantinus,* refted in this Place, where there was nothing but a Foreft of Cyprefs Trees,

and

and a Temple to *Apollo*.   In After-times, or rather many Ages afterwards, Lett. III.
the fame Place took the Name of *Diplocionion*, from two Columns of
Thebaic Stone, which are ftill to be feen near the Tomb of *Barbaroffa*,
who was certainly a much greater Man in Sea-Matters than *Jafon*, tho
born of poor Parents in the Ifland of *Metelin*.   *Barbaroffa* dy'd King of
Algier, and Captain Baffa in 1547.   *Solyman* II. call'd him *Chairadin*, that
is to fay, *a great Captain :* from *Chairadin, Calcondylus* has made it *Chara-
tin*, and *Paulus Jovius Hariadene*.

T O follow exactly the Defcription given us of the *Bofphorus* by *Dionyfius Byzantinus*, we fhould look for the Places, where were formerly
Pentecontarion, Thermaftis, Delphinus, and *Charandas*; the Temple of *Pto-
lemeus Philadelphus, Palinormicon*, and *Aiantium :* but where fhould we
find them? The *Greeks* and *Turks* have turn'd every thing topfy-turvy
fince that time to people *Fondocli* and *Topana*, where lies Cape *Meto-
pon*, which fronts the Point of the Seraglio.

I am, M<small>Y</small> L<small>ORD</small>, *&c.*

LETTER IV.

To *Monseigneur the Count* de Pontchartrain, *Secretary of State,* &c.

My Lord,

Description of the South Coasts of the Black Sea, from the Mouth of it as far as to Sinope.

HATEVER the Antients have said, the *Black Sea* has nothing Black in it, as I may say, beside the Name. The Winds upon it are not more furious, nor Tempests more frequent than in other Seas. We must forgive the Exaggerations of the antient Poets, and particularly the Resentment of *Ovid*: in short, the Sand of the Black Sea is of the same Colour as that of the White Sea, and its Waters are as clear; and if the Coasts of it, which are thought so dangerous, seem dusky at a distance, it is owing to the Woods which overshade it, or to the distance from whence it is view'd. The Weather was so fine and so serene during our Voyage upon it, that we could not forbear giving a sort of Lye to *Valerius Flaccus* the famous *Latin* Poet, who has describ'd the Course of the *Argonauts*, who pass'd for the most celebrated Travellers of Antiquity, but who were notwithstanding mere Children in comparison of *Vincent le Blanc, Tavernier*, and a world of others who have seen the greatest part of the habitable Globe.

THIS Poet assures us that the Sky over the Black Sea is always foul and stormy, and that the Weather is never quiet and settled. For my own part, I do not pretend to affirm this Sea is not subject to great Tempests, having never seen it but in the finest Season of the Year; but I am persuaded that in the Perfection to which Navigation is now brought,

4

one

one might fail there as fafely as in other Seas, if the Veffels were fteer'd Lett. IV.
by good Pilots. The *Greeks* and the *Turks* are not at all more skilful
th an *Typhis* and *Nauplius*, who conducted *Jafon, Hercules, Thefeus,*
and the other Heroes of *Greece*, to the Coafts of *Colchis* or *Mengrelia*. If we may judge by the Route which *Apollonius Rhodius*
fays they took, all their Knowledge reach'd no farther than, according
to the Counfel of *Phineus* the Blind King of *Thrace*, to fhun the Shelves
which are on the South fide of that Sea, without daring to fail out at large;
that is, that they could fail there only in a Calm. The *Greeks* and *Turks*
follow the fame Maxims; they have no Ufe of Sea-Charts, and fcarcely
knowing fo much as that one end of the Needle points to the North, are
out of their wits, if they lofe fight of Land. In fhort, the moft
experienc'd among them, inftead of counting by the Rhomb, pafs for
Men of extraordinary Abilities, if they underftand that to go to *Caffa;*
they muft veer to the left hand as they get out of the *Black Sea* Channel;
and that to go to *Trebifond*, they muft tack to the right.

A S to the Tackling, they know nothing of the matter, and their great
Merit lies in Rowing. *Caftor* and *Pollux, Hercules, Thefeus,* and the o-
ther Demi-Gods, diftinguifh'd themfelves by this Exercife in the Voyage
of the *Argonauts*. Perhaps they were ftronger and more hardy than the
Turks, who often chufe rather to return from whence they came, and to
drive with the Wind, than ftruggle againft it. They may fay, if they
pleafe, that the Waves of the *Black Sea* are fhort, and confequently
ruffled and violent; but it is certain they are more free and open
than thofe of the White Sea, which is broken by a great number
of Channels which lie between the Iflands. The moft troublefome
Circumftance in failing upon the *Black Sea*, is, that there are few good
Ports, and that moft of its Roads are unfhelter'd; but if the Ports were
ever fo good, they would be of no fervice to Pilots, who know not
how to make them in a Storm. In order to make the Navigation of the
Black Sea fafe, any other Nation befide the *Turks* would train up artful
Pilots, repair the Ports, build Moles, and erect good Magazines there;
but the Genius of the *Turks* is not turn'd this way at all. The *Genoefe*
were not wanting to take thefe Precautions in the Declenfion of the *Gre-
cian* Empire, and chiefly in the 13th Century, when they kept all the

Com-

Commerce of the *Black Sea*, after they had feiz'd the beft Places upon it. The Relicks of their Works are yet to be feen there, and efpecially of thofe about the Sea. *Mahomet* II. drove them out entirely; and fince that time the *Turks*, who have let all run to ruin by their Negligence, would never fuffer the *Franks* to navigate there, notwithftanding any Advantages which have been propos'd to them for a Permiffion.

ALL that has been faid concerning this Sea from *Homer*'s time down to the prefent, and all that the *Turks* imagine about it, (who have only tranflated the fame Name into their Tongue) did not make us hefitate one moment as to undertaking the Voyage; but I muft confefs it was upon condition that we fhould go in a Caick, and not in a Saick. The Caicks which fail upon this Sea are Felucca's of four Oars, which hale afhore every Evening, and never put out but in a Calm, or with a fair Wind, to which they hoift a four-corner'd Sail, which they furl very dextroufly when the Gales are over. To avoid the Alarms which happen fometimes upon the Water by night, the Mariners of this Country, who love to fleep at their eafe, hale their Veffel upon the Beach, and make a fort of Tent of the Sail; and this is the Tack they underftand any thing of.

THE Departure of *Numan Cuperli*, the Vifier or Baffa of the three Horfe-Tails, who had been Viceroy of *Erzeron*, feem'd fuch a happy Opportunity, that we ought not to let it flip. He is a Perfon of great Merit, learn'd in the *Arabian* Language, profound in the Knowledge of his Religion, and who at the Age of 36 Years had read over all the Chronicles of the Empire. He is Son of the Grand Vifier *Cuperli*, who dy'd fo glorioufly at the Battel of *Salankemen*, at a time when Fortune feem'd to declare for the *Ottoman* Arms. This *Numan Cuperli* is deftin'd for the greateft Employments of the State. Sultan *Muftapha*, the Brother of *Achmet* now reigning, honour'd him with his Affinity, and gave him one of his Daughters; but fhe was drowned at *Adrianople* in one of the Canals of the Seraglio, before the Marriage was confummated. From being Viceroy of *Erzeron*, he was made Baffa of *Cutaya*, and then Viceroy of *Candia*; and it is not doubted but he will one day be made Grand Vifier. It feems that the *Ottoman* Empire can't be fupported but by the Virtue of the *Cuperli*'s: this Man is belov'd by the People, and univerfally acknowledg'd to be the moft juft and upright Baffa in the Court.   W E

W E determin'd then to follow fo brave and honeft a Perfon.
Ambaffador was fo good as to prefent us to him by M. *le Duc*, his Phyfician in ordinary, who was alfo Phyfician to the Baffa. He affur'd us
of his Protection, in regard to the Emperor of *France*, whofe Forefight
and Care he fhould always admire, in fending abroad, he faid, Perfons
capable of difcovering the Products of Nature in every Country, to learn
upon the fpot the Ufes which may be made of them, with refpect to
Health. Befides, the Baffa was not forry to have Phyficians in his Train,
and he inform'd me that his Father was highly fatisfy'd of the Abilities
of M. *d' Hermange*, whom he had had with him a long time, and in whofe
hands he dy'd at *Salankemen*. Our principal Converfations turn'd upon
the Interefts of the Princes of *Europe*, which he underftood perfectly,
and generally clos'd with a fhort Relation of the chief Curiofities we
had obferv'd. For fear of offending his Houfhold, he afk'd of us in private
the Draughts of the Plants we had obferv'd in our Voyage : I deliver'd
them by his Orders to *Cuperli* Bey, one of his Brothers, who brought
them again after the Baffa had confider'd them alone, and at his leifure.
This Policy is neceffary among the *Turks*, where it is taken ill for good
Muffulmans to take notice of Sciences cultivated by the Chriftians, and
to fhew Marks of their efteeming them. I had occafion to give him a
bit of Phofphorus, and to explain to him the manner of ufing it ; but he
would not let me make the Experiment in his Prefence. Some days after
he acknowledg'd the Chriftians were ingenious People, and that their Sa-
gacity was as much to be commended, as the Idlenefs of the Orientals
was to be blamed. We were fo happy as not to have any of his Family die under our hands. Tho he had M. *de S. Lambert* an able *French*
Phyfician with him, he order'd him to have us to vifit the Sick, which I
confented to only upon condition it was in concert with him : all his
Family were fick upon the Road ; we had the Care of the Mafter firft,
of his Wife, his Mother, his Daughter, and his other Officers ; all this
fucceeded to our Honour, and the Sick recover'd very well.

O U R Equipage was foon ready, tho the Journey was to be very
long ; for in fuch tedious Paffages, I think a Man ought not to load himfelf with any things but what are abfolutely neceffary. We bought
therefore one Tent, four large Leather Sacks to put our Baggage in,

and

and fome Ofier Baskets cover'd with a Skin to preferve our Plants, and the Papers which ferv'd to dry them. The *Levant* Tents are lefs cumberfome than thofe of this Place. They have only one Pole in the middle, which takes off in half when you fold up the Tackle; this fupports a Pavilion of thick clofe-fet Cloth, from which the Water runs off very eafily; the Pavilion is faftned at the Border with Cords, hook'd on to Iron Pins, fix'd in the Ground: near the Top is alfo a Set of Cords which are faftned very firmly by another Row of Pins, at a wider diftance from the middle Pole than the former, and ftrain out the top of the Tent on the Outfide, making a Saliant Angle after the manner of *Manfarde.* We plac'd our three Beds, fo that the Head was next the Pole, and the Feet to the Circumference of the Pavilion, where we alfo put our Sacks and our Baskets. A quarter of an Hour fuffices to erect this mighty Apartment, which has all forts of Conveniences in it. As to our Kitchen Furniture, it confifted of fix Plates, two large Bowls, two Kettles, two Cups, all of Copper tinn'd, two Leather Bottles to carry Water in, one Lanthorn, and fome wooden Ladles; for one can have no other in *Turky*, where the fineft People have no better Veffels than we had.

OUR Clokes were of wonderful Service to us; they were made of a thick Capuchin Cloth, lin'd with a Stuff of equal Subftance to bear the Drudgery: A Cloke is an incomparable Moveable for a Traveller, and ferves in cafe of need for a Bed and a Tent. We furnifh'd our felves in the *Archipelago* with Linen for our Table, and for other Ufes, efpecially with Callicoe Drawers, which ferve inftead of Bed-clothes in this fort of Roads; and we are able to boaft we brought up the Fafhion of it among the *Armenians* of our Caravans. We were oblig'd to quit the *French* Habit at *Conftantinople*, for the Dolyman and the Veft; but as this feem'd to be very troublefome to walk up and down in when we went a fimpling, we got an *Armenian* Habit for Riding, and *Spanifh* leather Boots for walking in the Fields. The *Turkifh* Habit was defign'd for Vifits of Ceremony and Refpect, and the other for Bufinefs.

OUR Friends at *Conftantinople* help'd us to a wonderful Man, who underftood all forts of Trades, and ferv'd us for an Overfeer, a Chamberlain, a Cook, an Interpreter, and a Mafter, if I may fo fay; for generally

we

we were forc'd to let him take his own way.   This dextrous Fellow Lett. IV·
was a *Greek*, as lufty as a *Turk*, and had travell'd all Countries: he drefs'd
Victuals after the *Turkifh* manner, and after the *French*.   Befide the vulgar *Greek*, he fpoke *Turkifh*, *Arabick*, *Italian*, *Ruffian*, and *Provencal*, which
is my natural Tongue.   We were fo well provided in *Janachi* (which
was his Name) that we took no other till we came as far as *Armenia*;
for why fhould we expend his Majefty's Money without occafion?
Befides, a Man fhould make as little clutter as poffible in ftrange Countries, where he is fent only for the fake of Obfervations.   *Janachi* had alfo
an excellent Quality for a Traveller; he was fuch a Coward as a Man
of Senfe fhould be: for who the duce would ramble about the World to
find People to quarrel with, unlefs he were of the Character of a *Don
Quixot?* But upon the whole, a Man may go a great way with a little
Cowardice, and a good deal of Sobriety.   Our Officer had the firft of
thefe Qualifications in a fublime degree; but as he knew nothing of the
fecond, as robuft as he was, he could not refift the Power of Wine, and was
every now and then under a kind of Eclipfe: however, to do him juftice,
he chofe his times fo well, that his Liquor operated only when he was
on horfeback, and then he dos'd quietly; and our Affairs fuffer'd no Damage.

O U R Embaffador was fo good, as to procure us a Commandment
of the Port *gratis*; that is, he would needs pay all the Fees, out of refpect to your Lordfhip, and we are fenfible we owe all the Civilities he
heap'd upon us to you.   I have tranflated this Paffport literally, to fhew
the Form the *Turks* ufe on fuch an occafion.

COMMANDMENT,

*To the Baffa's, Beglerbegs, Sangiack-begs, Cadi's, and other Commanders
upon the Road from* Conftantinople *to* Trebifond, Erzeron, Aleppo, Da-
mafcus, *&c. as well by Sea as by Land.*

"KNOW ye, at the Arrival of this fublime Commandment, that
" the Pattern of the Great-Ones of the Religion of the Meffiah,
" M. *de Ferriol*, Embaffador of the Emperor of *France*, refiding at my fu-

Vol. II.   S   " preme

" preme Port (whofe End be happy) has fent a Requeft to my Imperial
" Camp, by which he gives me to underftand, that one of the Doctors
" of *France*, named *Tournefort*, particularly experienc'd in the Knowledge
" of Plants, is fent out from *France*, with four other Perfons, to fearch after
" Plants which are not in their Kingdom ; and having defired my Com-
" mandment, that in the Places he is to pafs thro, be it by Sea
" or by Land, no one may give him Lett or Hindrance, nor do any
" damage to his Goods nor his Baggage, he employing himfelf only in
" things belonging to his Art, not intermeddling in the Affairs of our
" tributary Subjects, nor exceeding the Bounds of his Condition, but
" behaving himfelf as he ought to do: this my Commandment has
" been given for this time only, that he may meet with no Oppofition
" in his Paffage. And I ordain that upon his arriving with this noble
" Commandment, ye comport yourfelves conformably to the Orders
" contain'd in it upon this Subject; and that the faid Doctor, with four
" Perfons of his Retinue only, not intermeddling in the Affairs of our
" tributary Subjects, and keeping within the Bounds of his Duty in e-
" very Place of our Jurifdiction where he arrives, for this time only, ye
" make no oppofition to his Paffage, nor do any prejudice to thofe of
" his Retinue, nor to his Baggage: and that doing nothing on your part
" contrary to the Conftitutions Imperial, ye caufe to be deliver'd to him
" for his Money, at the current Price, the things he hath need of, by
" them who fell them; and that ye execute all this which is contain'd
" in my noble Commandment, when it fhall be prefented to you.
" Which after ye have read, return it back to the hands of him who
" bears it, and yield Credence to the noble Sign with which it is mark'd.
" Written at the beginning of the Moon *Zilcadeh* of the *Egira* 1112.
" Ordain'd in the Plain of *Daout Baffa*.

W E took leave of the Embaffador *April* the 13th, and lay that Night
at *Ortacui*, upon the Canal of the *Black Sea*, in the Seraglio of *Mahomet Bey*, Page to the Grand Signior. *Mahomet* had given the ufe of
this Apartment to *M. Chabert*, Apothecary of *Provence*, who was eftablifh'd long fince at *Conftantinople*, where he was in a world of Practice:
This poor Man, a little after our Departure, had the lot of moft who

come.

come to feek their Fortune in this mighty City, that is, to die of the Lett. IV.
Peftilence, with which he was taken off when he leaft expected it. His
Son, who was Apothecary to the Baffa, and was of great affiftance to
us upon the Road, by the Knowledge he had of the Languages of
the Country, went with us to wait upon his Lord in *Mahomet*'s
Houfe, which pafs'd for one of the fineft upon the Channel.

T H E next day we took a view of the Country round about: it confifted of fmall Hills delightfully green, but which produc'd only common
Plants. As to the Seraglio, it made no great appearance, no more than
the other Houfes of the *Levant*, tho the Apartments were handfome,
and there had been a great deal of Expence upon it. All the Cielings
are painted with Hiftory, and gilded after the *Turkifh* Tafte, that is, with
Ornaments fo fmall and trifling, that they were more proper for a
piece of Embroidery than for a Hall. Thefe Halls are wainfcoted
neatly enough, and inftead of Pictures are fet round with *Arabick* Sen-
tences taken out of the Alcoran. But whatever care is taken of the
Decorations of thefe Places, the Cielings are too low, which is the common fault of the Buildings in the *Levant*, where Proportion is never
obferv'd. This fault appears on the Outfide; for the Roofs are
fo low, that one would think they muft fall in upon the Houfes,
and indeed they deprive them of half their Light. Tho the Rooms
have two Rows of Windows, they are ne'er the lighter: thofe Windows
are ufually fquare, with another fmaller Window which is arch'd over
each. The Baths are what chiefly diftinguifh the Houfes of the Great,
from thofe of the Vulgar. Tho the *Turks* erect Baths only for Conveniency, yet they often fet them off with fome Ornament; thofe of the
Bey's Houfe are pav'd and lin'd with Marble: they temper the Water
in them by means of a leaden Pipe, out of which they draw as much
hot as they think fit. The Galleries and Coridors, which are of painted
Wood, run quite round the Houfe; but the Stair-cafe is a Scandal to the reft;
but they know not how to build better in *Turky*, where the Architects
only fet up a kind of wooden Ladder cover'd with a Shed: 'tis ftill worfe
among the *Greeks*, where even this Ladder is expos'd to the Rain and
Sun. The Court-yard of the Houfe I am now fpeaking of, would be tolerably handfome, were it not cramp'd by a Bafon, where they fet up

:heir

their Caiques; for thefe Caiques on the *Black Sea* ferve the purpofes of Coaches, Carts, and Waggons: they are put to all manner of Ufes, of which Fifhing is none of the leaft advantageous. From the Court you go into the Gardens, which would be very fine, were they not too much ftraiten'd by the Hills that furround them; but the Park is well feated, and of a confiderable extent. This is the Model of a *Turkifh* Country-houfe, and tho they are not comparable to thofe about *Paris* they are not without their Beauties, and have particularly fomething of Magnificence in them. We pafs'd our time not at all unpleafantly in that of *Mahomet Bey*.

AT length the Baffa appear'd on the Canal the 26th of *April*, with eight great Caiques or Feluccas, in which were part of his Family, the reft being gone before in Saiques, in order to wait for him at *Trebifond*. The Felucca which carried the Women was fo cover'd in with wooden Lattices made Net-wife, that they could fcarce breathe in them. The Baffa had only his Mother, his Wife, one of his Daughters, fix Slaves of the fame Sex to wait on them, and fome Eunuchs. Our Felucca was the ninth Veffel in this little Fleet, and brought up the Rear. Whether it be that the *Turks* are not over-fond of joining Company with Chriftians, or that they fancy'd 'twould be a Difrefpect to the Baffa, for us to be drawn up in the fame line with the Caiques of his Houfhold, his Intendant had given Orders that a certain Space fhould be left between our Felucca and the reft. 'Twas in vain for me to bid our Sailors go forward: they would have been hang'd before they would have gone nearer, or have landed before the reft. Tho we had hir'd our Veffel at the fame Price with the Baffa, namely at 400 Livres, for our Voyage from *Conftantinople* to *Trebifond*; yet we had but four Sailors and one Steerfman, whereas in the reft there were fpare Sailors to relieve the others: but 'tis no wonder the Natives, and efpecially great Men, fhould be better ferv'd than Strangers. One day I could not help finding fault with their burdening our Felucca with fome Sheep that incumber'd the Baffa's Kitchen: But I thought 'twould even be my beft way to hold my peace, when I heard that they began to call us Dogs and Infidels; fo, that we might have a quiet Voyage, we were forc'd to learn to bear the *Turkifh* Civility.

WE

W E therefore fet out in the tail of the Fleet, after having embrac'd Lett. IV.
our Friends who came to take their leaves of us at *Ortacui*, and pafs'd
the firft Caftles wholly by rowing, for there was no Wind ftirring.   We
arriv'd at the laft Caftles in the fame Calm, and had the pleafure to
enter the *Black Sea* with all the Safety that could be.   Tho this Sea appear'd to us then to be as pacifick as that of *America*, yet we could
not help feeling a little Palpitation at Heart at fight of that immenfe
Quantity of Water.   We landed about *Quindi*, that is to fay, about four
o' Clock, at the entrance of the River *Riva*, eighteen Miles from *Ortacui*.   We encamp'd along the Shore in Meadows not wholly free from
Marfhes : and as we were inform'd of the Manners of the Country, we
pitch'd our Tent at a good diftance from thofe of the Muffulmans, to
fhew our refpect, and to allow them full freedom in making their Ablutions ; for which purpofe they fet up little Clofets of Cloth, in which
one Perfon might have as much room as was neceffary for wafhing himfelf at his eafe.   The Baffa's Tent was upon a Carpet-fpot of Ground,
on the brow of a little Hill in a thin Wood : the Womens Apartment
was not far from it ; it confifted of two Pavilions furrounded by Ditches,
round which they might walk without being feen, behind a great Inclofure of Cloth painted grey and green.   Here the Baffa, and his Brother the Bey, fpent the Night and part of the Day.   The Guard of the
Ladies was entrufted to Eunuchs as black as Jet, whofe Vifages I did
not like in the leaft ; for they made horrible Grimaces, and roll'd their
Eyes in a frightful manner, when I went in or out of the Inclofure, to
fee the Baffa's Daughter, who was troubled with a fad Cough.

R I V A, which we juft now call'd a River, is really no better than
a Brook about as broad as that *des Gobelins*, all flimy, and hardly
deep enough at the Mouth to be a Retreat for Boats : yet the Antients
have made it very famous, under the Name of *Rhebas*. *Dionyfius* the
Geographer, who made three Verfes in its favour, calls it an amiable
River. *Apollonius Rhodius*, on the contrary, mentions it as a rapid
Torrent : And at prefent it is neither amiable nor rapid, nor does it
look as if it had ever been either.   Its Sources are towards the *Bofphorus*, on the fide of Sultan *Solyman* Kiofc, in a pretty flat Country, from
whence it runs into marfhy Meadows among Rufhes.   It is no great

*

wonder

wonder that *Phineus* fhould give the *Argonauts* fo terrible an Idea of this River, when he look'd upon the *Cyanean* Ifles to be the moft dangerous Rocks in the World. *Arrian* reckon'd it 11 Miles 250 Paces from the Temple of *Jupiter* to the River *Rhebas*, that is, from the new Caftle of *Afia* to the *Riva*; this Author is of admirable Exactnefs, and no body was fo well acquainted as he was with the *Black Sea*, all the Coafts of which he has defcrib'd, after having furvey'd them in quality of one of the Generals of the Emperor *Adrian*, to whom he dedicates the Defcription of it, under the title of the *Periple of the Pontus Euxinus.*

I DON'T know how they manag'd it in the days of that Emperor, as to the landing of their Women; but I know that at prefent, among the *Turks*, whenever they want to get them out of the Boats, they make every body retire without much Compliment: the very Sailors hide themfelves when they have laid the Planks for their Paffage; and if the Shore happen to be fuch, that the Caiques cannot come near enough, they cover the Ladies, or rather wrap them up in five or fix Bales, and the Sailors take them on their Backs like Bundles of Goods. When they are fet down, the Slaves take them out; and the Eunuchs bawl and threaten every body, to make them get further off, tho at a Mile diftance already. The Baffa's Footmen then fled into the Woods, and were fo far from waiting upon the Ladies, that they would fooner have let them drown, than but turn their Heads that way.

FOR fear we fhould be ignorant of this laudable Cuftom, the Baffa's Lieutenant took care to inform us of it at our firft Vifit. *As you come from a far Country,* fays he, *'tis fit I give you notice of certain things which among us it is abfolutely neceffary you fhould know: Always to get as far from the Quarter of the Women as you can; never to walk upon Heights from whence you may difcover their Tents; never to make any diforder in fow'd Ground, when you fearch for your Plants; and particularly, not to give Wine to the Baffa's People.* We return'd him our humble Thanks for his Goodnefs to us. As for the Ladies, 'tis certain we never dreamt of them; the Love of Plants entirely poffefs'd us. As to the Wine-matter, the Baffa's Footmen came and begg'd it fo earneftly, that fometimes we could not refufe them; for which reafon I begg'd the Steward to forbid them abfolutely from having any thing to fay to us.

*

THAT

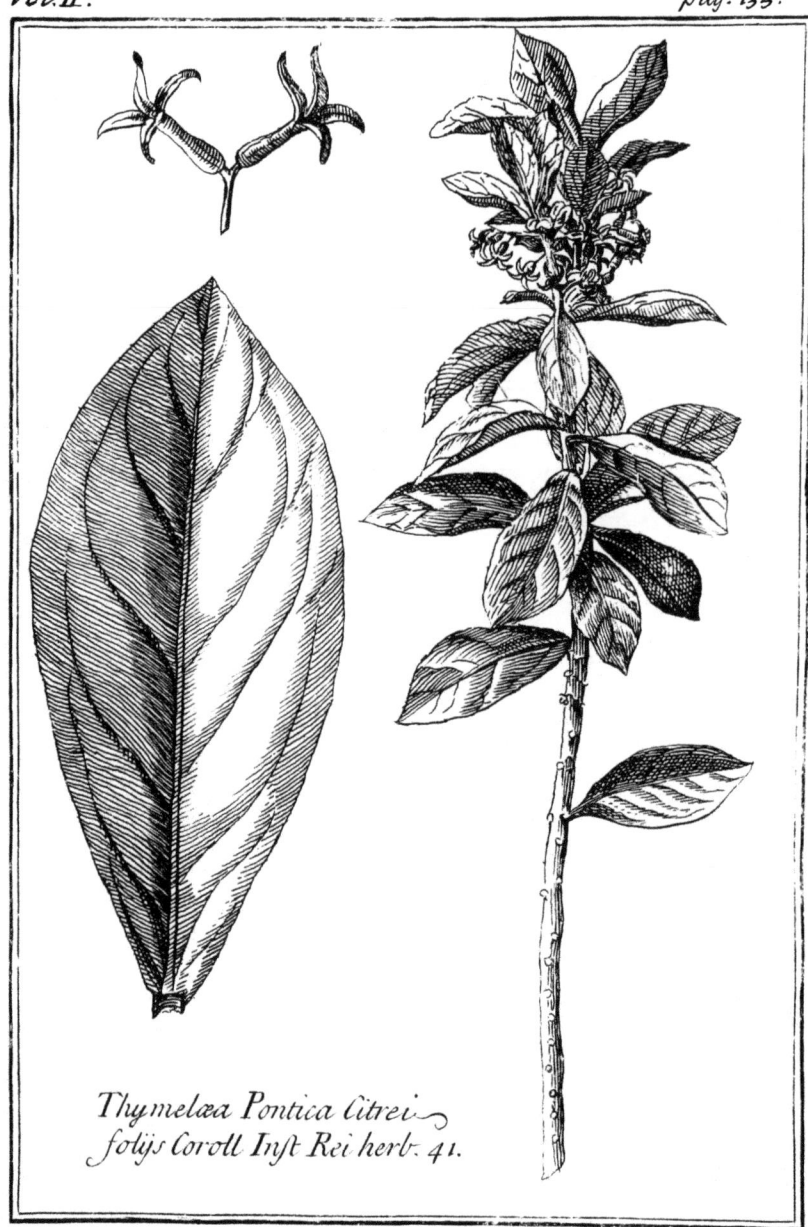

Thymelæa Pontica Citrei folijs Coroll Inst Rei herb. 41.

THAT Steward feem'd to be a good fort of Man, and very well
belov'd in his Mafter's Houfe, tho not chofen by him; for the Grand
Vifier, that he may have a thorow Infight into the Baffa's, and be fully
inform'd of all their Actions, generally names fuch Officers to them.
The Man we are now fpeaking of, told us we fhould lie by every Evening, about the Quindi, let the Weather be what it wou'd; that the Baffa would take fome days Reft on the way; that whenever we pleas'd, we
might have fome of his People to accompany us in our Walks; and in a
word, that he would favour our Searches as much as he could. He offer'd us his Arm, that we might feel his Pulfe, and then order'd Coffee
and Tobacco to be brought in. We, in return, offer'd him any thing
that depended upon our Function; and he efcaped with only two Bleedings
and one Purgation during the whole Voyage.

WE foon found the Difference that there is between the *Black Sea* and
the *Archipelago*. Tho it was but the 17th of *April*, it never gave over
raining, whereas in the *Archipelago* it feldom rains after *March*. We were
therefore oblig'd to cut a Trench round our Tent to drain the Water
from it; befides, the North Wind, which began to blow, did not at all
help to warm our Lodging, and the Rain continu'd to fall in Sheets:
yet for all this we travers'd with pleafure, fometimes the Coafts, fometimes the Fields, and efpecially the Banks of the Stream; which grew
fo marfhy, that we were every moment forc'd to come back again for
fear of fticking: we were at laft conftrain'd to keep to the higher
Grounds, but we exhaufted thofe in five or fix days; and then we began
to be really vex'd with the North Wind and Rain. It was thought
convenient to go higher up the River inftead of putting out to Sea;
and we were frightned when we faw they thought of nothing but making
Provifions for a long Stay. The Baffa's People offer'd us Meat very civilly, but we fent for it as they did, two days Journey from the Camp.
Nothing fo much alleviated our Uneafinefs, as two admirable Plants, of
which here is the Defcription.

THYMELOEA Pontica, *Citrei foliis.* Corol. Inft. Rei Herb. 41.
Its Root, which is half a foot long, about the neck is as big as the little
Finger, ligneous, hard, divided into fome Fibres cover'd with a Bark of
an Orange Colour. This Root produces a Stalk of about two foot high,

branchy

branchy fometimes from its very beginning, about three lines thick, firm, but fo pliant that there's no breaking it, cloth'd with a grey Bark, accompany'd towards the top with Leaves plac'd without order, in Figure and Confiftence like thofe of the Orange-tree; the biggeft are about four inches long, and two broad, pointed at each end, fleek, bright green, and fhining, and the under parts rifing in a pretty large Rib, which diftributes Veffels to the Rims. From the extremity of the Stalk and Branches come forth about the end of *April* young Sprigs terminated by new Leaves, among which grow Flowers faftned ufually two to two on a tail nine or ten lines long. Each Flower is a Pipe of a greenifh yellow, approaching fomewhat to an Orange-Colour, a line broad and above half an Inch long, divided into four parts oppofite to each other like a Crofs, almoft five lines long to one broad, a little ftreak'd in Gutters, and growing fmaller and fmaller to the point. Four very fhort Filaments appear at the entrance of the Pipe, laden with tops whitifh and flender, furmounted by four other Filaments of the like form. The Piftile which is at the bottom of the Pipe, is an oval Button a line long, bright green, fleek, terminated by a little white head. The Fruit was as yet only a young green Berry, in which the young Seeds were diftinguifhable. The whole Plant is pretty bufhy. The Leaves being bruifed, fmell like thofe of the Elder-tree, and are of a mucilaginous tafte, which leaves a pretty confiderable Impreffion of Fire, as does all the reft of the Plant. The Smell of the Flower is fweet, but is foon gone. This Plant grows on Hills and in thin Woods. Of all the known Species of this Genus, this has the biggeft Leaves.

THE following Plant is no lefs confiderable for the Singularity of its Flower, I nam'd it

BLATTARIA Orientalis, Bugulæ folio, flore maximo virefcente, Lituris luteis in femicirculum ftriato. Coroll. Inft. Rei Herb. 8.

THE Root confifts of three or four flefhy Knobs, from one to three inches long, from two lines to half an inch thick, white, brittle, cover'd with a chapt brown Skin, garnifh'd with fome pretty thin Fibres faftned to a Neck as big as a Man's little Finger. The firft Leaves that this Root puts forth, are almoft oval, like thofe of the Bugle, bunchy, wavy towards the Rims, an inch and half or two inches long, fifteen lines broad, fupported by a Stalk of two lines long, flat at top, rounded beneath, purple, and

and running to the extremity of the Leaves in feveral Veffels of the fame Lett. IV.
Colour. The Stalk is commonly but about nine or ten inches high, and
one line thick, flightly hair'd, accompany'd with Leaves feven or eight
lines long, to four or five lines broad. Thofe below are fleek, the o_
thers interfpers'd with fome Hairs like the Stalks. From their Bafis, towards the top, grow Flowers pretty compact and difpos'd in manner of a
great Ear of Corn. Each Flower is a Bafon of near fifteen lines diameter,
cut in five rounded parts, whereof the two uppermoft are fomewhat lefs
than the others. The bottom of this Flower is Sea-green, as are
alfo the Rims, which draw a little nearer to yellow; but the rounded
points before mention'd are ftrip'd in a Semi-circle of a bright yellow,
which goes quite thro. From the hole in the center of this Flower run
two fillets, purplifh, mix'd with white, which end at the yellowifh Semi-
circle of the two upper parts; and from the fame rim of that hole rife
two whitifh Stamina terminated by crooked Summits fill'd with yellow Duft. Befides thefe Stamina, there appear on the rims of the fame
hole, fome Locks, purplifh, hairy, cottony, and filky. The Cup is a
Bafon, pale-green, four lines long, cut in five parts almoft to the Center, whereof three are much narrower than the others. The Piftile which
is juft in the middle, is rounded, hairy, a line long, terminated by a Fil-
let much longer. We were convinc'd by the cods which remain'd of the
Fruit of the preceding Year, that this Plant is a true Species of the
Herbe aux Mites, varying not only in the height of its Stalk, but alfo
in the colour and largenefs of its Flowers.

WHILE we were agreeably amus'd in obferving of Plants, we were
threatned with fpending the reft of *April* in this Marfh; but by good luck
the North-wind ceas'd the 26th. The Sea continued difturb'd with it two
days longer; but by Oars and Ropes we at length came out of the Mouth
of the *Riva,* the 28th of *April.* Our Fleet kept along the Shore, and ftopt
at *Kilia,* a Village thirty Miles from *Riva.* The *Turks* landed to fay their
Prayers; but afterwards we took the advantage of the South-weft Wind,
to go as far as the River *Ava* or *Ayala,* twenty-four Miles from *Kilia.*
All this Country, or to fpeak more properly, all the Coafts of the *Black
Sea,* quite to *Trebifond,* are admirable for their Verdure; and moft of
the Woods extend fo far into the Land, that you lofe fight of them.

Vol. II.   T   'Tis

'Tis a wonder the *Turks* have retain'd the antient Name of the River *A-va*, for they call it *Sagari* or *Sacari*; and this Name is certainly deriv'd from *Sangarios*, a River famous in antient Authors, and which serv'd as a Limit of *Bithynia*. *Strabo* tells us it was made navigable, and that its Sources came from a Village call'd *Sangias*, near *Peſtinuntum*, a Town of *Phrygia*, well known by the Temple of the Mother of the Gods. *Lucullus* was encamped on its Banks, when he learnt the Loſs of the Battel of *Chalcedon*, where *Mithridates* defeated *Cotta* who commanded part of the *Roman* Army. *Lucullus* advanc'd as far as *Cizicus*, which *Mithridates* intended to beſiege, fell upon his Army, and cut it in pieces. As for the other Rivulets, which, according to *Strabo* and *Arrian*, ran between *Chalcedon* and *Heraclea Pontica*, they muſt either be dried up, or reduc'd almoſt to nothing; for our Sailors aſſured us they knew of none between *Riva* and *Ava*.

THE 29th of *April*, tho there was a great Calm, we made forty Miles only by rowing, and encamp'd about noon on the Shore of *Dichilites*. Our Sailors being in for it, row'd us next day as far as the Mouth of the little River *Anaplia*, full 60 Miles. The firſt of *May* we came to *Penderachi*. The River *Anaplia*, according to *Arrian*'s Deſcription, muſt be that which he calls *Hypius*, ſince there is no other quite to *Heraclea*, which is now called *Eregri* or *Penderachi*. As ſmall as the River *Anaplia* is, it was of great ſervice to *Mithridates*; he retired into its Mouth with his Fleet, after having loſt ſome Gallies in the Storm. As the bad Weather oblig'd him to ſtay there, he corrupted *Lamachus* the moſt powerful Nobleman in *Heraclea*, who by his Brigues got the King of *Pontus* and his Troops receiv'd there.

Eregri

PENDERACHI is a little Town built on the Ruins of the antient *Heraclea*: this latter muſt have been one of the fineſt Cities in all the Eaſt, if we may judge by its Ruins, eſpecially by the old Walls built of huge Stones that are ſtill on the Sea-ſhore. As to the compaſs of the City, which is fortify'd from diſtance to diſtance by ſquare Towers, that indeed ſeems to be no older than the *Greek* Emperor.s On every hand you diſcover Columns, Architraves, and Inſcriptions very much defac'd. Near a Moſque is the Door of a *Turk*'s Houſe, the Mounters whereof are pieces of Marble, on which is legible on one ſide P. B. A. TPAIAN, and

on

ELEGRI.

on the other ΤΟΚΡΑΤΩΡΙ, which are the Remains of an Inſcription Lett. IV.
of the Emperor *Trajan*.   This City was built on a high Coaſt which go-
verns the Sea, and ſeems to have been deſign'd to command the whole
Country: landward there ſtill remains an antient Gate, perfectly ſimple, built of great pieces of Marble.   They aſſured us that further off
there were other Remains of Antiquity; but Night coming on, and the
Tents of the Women being ſet up near thoſe Ruins, we durſt not go to
view them.   And which was a further Misfortune, that we did not expect,
there was no getting a Guide: the *Greeks* were celebrating their *Eaſter*,
and were reſolv'd not to loſe the Fruit of the Money they had given
the Cadi for leave to drink and dance heartily that day.   We therefore
walk'd out at a venture Eaſtward, as far as the Marſhes below the City,
where probably the Waters of the *Lycus* ſubſide.

WE could not poſſibly get over thoſe Marſhes; and in returning towards the Ruins of the Town, we found an admirable Species of *Sphon-*
dylium, which at firſt we took for *Dioſcorides*'s *Heraclean Panacea*; but the
Flowers of this are white, whereas thoſe of *Dioſcorides*'s Plant muſt be yellow.   'Twas the Name of the *Heraclean* that miſled us, for according
to that Author it was call'd *Heraclean Panacea*, upon account of its *Her-*
culean Efficacy.   *Dioſcorides*'s Plant grew naturally in *Bœotia*, *Phocis*,
Macedon, on the Coaſts of *Africa*, and yielded the Juice which they
call *Opopanax*, which probably differ'd from that which is call'd ſo now.
Be this as it will, the Plant that grows in the Ruins of *Heraclea* is a
very fine one, and the biggeſt of all the known kinds of Plants with
Umbrello Flowers: 'twas for this reaſon I gave it the name of

SPHONDYLIUM Orientale maximum, Cor. Inſt. Rei Herb. 22.
The Stalk is about five foot high, an inch and a half thick, hollow from
one joint to the other, channel'd, pale green, hairy, accompany'd with
Leaves two foot and a half long, and two foot broad, cut quite to their
Ridges in three great parts, the middlemoſt of which is again cut in
three pieces, and the middlemoſt of thoſe two cut in the ſame manner.
The upper part of all theſe Leaves are ſmooth, and the under white and
hairy, and are ſuſtain'd by a Stalk thicker than a Man's Thumb, ſolid,
fleſhy, embracing the Stalk by two great Wings, which form a kind of

T 2

ſheath

ſheath of nine or ten inches long. From the junctures of theſe Leaves riſe great Branches as high as the Stalk, and ſometimes higher, laden with white Flowers, exactly like thoſe of the common *Sphondylium* : but the Umbellas that ſupport them are a foot and a half diameter ; the Seeds, tho green and very backward, were much bigger than thoſe of the other Species of this Kind. This Plant grows in the Ruins of thoſe fine Walls that are upon the Port, and that to us ſeem'd to be of the remoteſt Antiquity.

IT is doubted whether *Strabo* meant that this City had a good Port, or whether we are to let that Word ſtand in him which ſays that it had none at all. For my part, I believe that the old Mole which is entirely ruin'd, and which is ſuppos'd to have been the Work of the *Geno-eſe*, was formerly built upon the Foundation of ſome other more antient Mole, which defended the Veſſels of the *Heracleans* againſt the North-wind : for the Road which forms the Cape or Peninſula of *Acheruſia*, is too open, and of no great ſervice even to Saiques, ſo far is it from being a Port fit for Ships of War. Yet *Arrian* ſays poſitively that the Port of *Heraclea* was good for ſuch Veſſels. *Xenophon* informs us, that the *Heracleans* had very many of them, and that they furniſh'd ſome to favour the Retreat of the ten thouſand, who look'd upon this to be a *Greek* City, either as founded by the *Megareans,* the *Bœotians,* the *Miletians,* or by *Hercules* himſelf. The beautiful Medal of *Julia Domna,* which is in the King's Collection, and whereof the Reverſe repreſents a *Nep-tune* holding a Dolphin in his right hand, and a Trident in his left, plain-ly denotes the Power this City had at Sea : but nothing is a greater Honour to its antient Navigation, than the Fleet it ſent to the Aſſiſtance of *Ptolemy,* after the Death of *Lyſimachus,* one of the Succeſſors of *Alex-ander.* 'Twas by means of this Succour that *Ptolemy* beat *Antigonus* ; and *Memnon* obſerves, that there was among the reſt one Ship call'd the *Lion,* of ſurprizing Beauty, and ſo prodigiouſly big, that its Complement was above three thouſand Men. The *Heracleans* ſent *Antigonus* the Son of *Demetrius* thirteen Galleys againſt *Antiochus,* and forty to the *Byzan-tines* who were attack'd by the ſame Prince. We alſo know that the City of *Heraclea* maintain'd for eleven Years in the Service of the *Romans* two cover'd Gallies, which were of great uſe to them againſt their Neigh-bours,

bours, and even againſt thoſe People of *Africa* call'd *Marrucini*, whence Lett. IV.
perhaps is deriv'd the Name of the People of *Morocco*. Hiſtory is full
of Inſtances of the Naval Power of the *Heracleans*, and conſequently
of the Goodneſs of their Port. After *Mithridates* had caus'd *Scio* to be
plunder'd by *Dorylaus*, upon pretence that it favour'd the *Rhodians*;
they put the moſt illuſtrious of the Inhabitants on board a few Ships,
by that Prince's Order, to diſperſe them throughout the Kingdom of *Pontus* : but the *Heracleans* were ſo generous as to ſtop them, to carry them
into their Port, and to ſend back thoſe unfortunate Men laden with
Preſents. Laſtly, the *Heracleans* had ſome Years afterwards the misfortune to be beaten themſelves by *Triarius*, General of the *Roman* Fleet,
conſiſting of 43 Ships, which ſurpriz'd that of *Heraclea*, which had but
30, and thoſe equipt in haſte. Where ſhould this great number of Veſſels be ſhelter'd, but in the Mole we are ſpeaking of, ſince there is no
Port near that Place? If *Lamachus*, the *Athenian* General ſent to raiſe
Contributions upon the *Heracleans*, had been Maſter of the Entrance of
this Mole, he had not loſt his Fleet by Tempeſt, while he was ravaging
the Country with the Troops he had landed. Not being in a condition to return to *Athens*, either by Land or Sea, he was ſent home, ſays
Juſtin, by the People of *Heraclea*, who thought themſelves recompens'd
for the Miſchiefs the *Athenians* had done their Lands, by having an opportunity of winning their Friendſhip by Civilities.

T H E Cavern by which *Hercules* was feign'd to have deſcended into
Hell, and to have brought out *Cerberus*, and which was ſhewn in *Xenophon*'s
time in the Peninſula *Acheruſia*, is much harder to find than the antient
Port of *Heraclea*, tho it was two *Stadia* deep. It muſt have been clos'd
up ſince that time ; for it is certain that there was a Cavern of that
Name, which gave occaſion to the Fable of *Cerberus*. It was not whol-
ly without grounds that a Medal was ſtruck with the Head of the third
Gordian, whereof the Reverſe is a *Hercules* knocking down the *Cerberus*,
after having dragg'd him out of the Cave. M. *Foucaut* Counſellor of
State has one of *Macrinus*, wherein that Dog is ſtanding at the feet of
Hercules, who holds a Club in his right hand. If *Hercules* was not the
Founder of *Heraclea*, he was certainly held in great Veneration there.
Pauſanias informs us, that they celebrated all that Hero's Labours.

There.

There is a Medal of *Severus*, in which *Hercules* holds a Club in one hand, and in the other three golden Apples of the *Hesperian* Garden. Upon a Medal of *Caracalla*, *Hercules* is reprefented overcoming *Achelous*, in the fhape of a Bull. The Fight of that Demi-God with *Hippolita* the *Amazon* is exprefs'd upon a Medal of *Macrinus*; the Combat with the *Erymanthian* Boar upon one of *Heliogabalus*: and the Legends of all thefe Medals are in the name of the *Heracleans*. When *Cotta* took the City of *Heraclea*, he found in the Market-place a Statue of *Hercules*, all the Attributes whereof were of pure Gold. To fhew the Fruitfulnefs of their Fields, the *Heracleans* caus'd Medals to be ftruck with Ears of Wheat and *Cornucopias*; and to exprefs the goodnefs of Medicinal Plants that grew about their City, they reprefented upon a Medal of *Diadumenus*, an *Æfculapius* leaning on a Stick, round which a Serpent was twifted.

WE have no Medal remaining, that I know of, of the Kings, or rather Tyrants of this City. The Extract of *Memnon* preferv'd to us by *Photius* muft comfort us for the lofs of the Hiftory which *Nymphis* of *Heraclea* had wrote of his Country. That Author made his Name illuftrious, not only by his Writings, but alfo by that famous Embaffy wherein he obliged the *Galatians* to retire, at the time when they were wafting with Fire and Sword the whole Country round *Heraclea*.

THIS City in the firft times was not only free, but alfo famous for its Colonies *Clearchus*, one of its Citizens, who during his Exile had ftudied *Plato*'s Philofophy at *Athens*, was recall'd to appeafe the People who demanded new Laws, and a new Partition of Lands: the Senate oppos'd it vigoroufly, but *Clearchus* who was animated with no very *Platonick* Spirit, made himfelf Mafter of Affairs by means of the People: he committed a thoufand Cruelties in the City; and *Diodorus Siculus* tells us that he made *Dionyfius* of *Syracufe* his Model in the Art of Government. *Theopompus* a famous Hiftorian of *Scio* relates, that the Citizens of *Heraclea* durft not go to make their court to *Clearchus*, till they had firft breakfafted upon fome Rue, very well knowing he would prefent them with a Glafs of Hemlock, to fend them to the other World.

CLEARCHUS was kill'd in the twelfth Year of his Reign, while the Bacchanals were celebrating in the City. *Diodorus* tells us that his

Son

Son *Timotheus* was elected in his ftead, and that he reign'd 15 Years; Lett. IV. but *Juftin* makes his Brother *Satyrus* the Succeffor of *Clearchus.* *Suidas* informs us too, that *Clearchus* was not the firft Tyrant of *Heraclea,* fince he faw in a Dream, *Evopius* another Tyrant of his Country : and *Memnon,* who is the fitteft Man to be confulted, fince he fpent twelve Books of his Hiftory in handling that of *Heraclea,* is of *Juftin's* Opinion. *Memnon,* in giving the Character of *Satyrus,* fays, he not only exceeded his Brother in Cruelty, but all the other Tyrants in the World. Being taken with a Canker that eat away all his lower Belly, quite to the Entrails, after having fuffer'd as much as he deferv'd, he threw up the Care of the Government to his Nephew *Timotheus,* in the 65th Year of his Age, and 7th of his Reign.

TIMOTHEVS perfectly well deferv'd his Name, and was an accomplifh'd Prince both in Peace and War; and accordingly he obtain'd the Title of *Benefactor* and *Saviour of his Country.* Before he died, he gave a fhare in the Government to his Brother *Dionyfius,* who taking advantage of the Retreat of the *Perfians,* whom *Alexander* had juft then beaten at the Battel of the *Granicus,* extended the Limits of the Kingdom of *Heraclea* a great way. After the Death of *Alexander* and *Perdiccas, Dionyfius* married *Amaftris* the Daughter of *Oxathris,* Brother of *Darius,* and Coufin of that beautiful *Statyra* who was worthy of having *Alexander* for her Husband. *Alexander* himfelf, before his death, had taken care to marry *Amaftris,* to *Craterus* one of his Favourites; who being afterwards enamour'd of *Philas* the Daughter of *Antipater,* was not difpleas'd that *Amaftris,* or *Ameftris* according to *Diodorus Siculus,* fhould marry *Dionyfius.* That Prince was a Man of Honour, and quitted the Name of Tyrant for that of King, which he maintain'd with great Dignity : and it was certainly this King that *Strabo* had in view, when he fays, there were Tyrants and Kings of *Heraclea.* King *Dionyfius* grew fo big and fat amidft all thefe Felicities, that he fell into a kind of Lethargy, which they could fcarce recover him from, even by running Needles deep into his Flefh. *Nymphis* afcrib'd this Diftemper to *Clearchus,* Son of the firft Tyrant of *Heraclea*; he fays, that Prince fhut himfelf up in a Box, out of which he peep'd only with his Head to give Audience. We may believe what we pleafe of this Story : good King *Dionyfius,* as fat as he

was,

was, made a fhift to have three Children by *Amaftris* : *Clearchus, Oxa-thris*, and a Daughter of the fame Name. He left the care of his Children and the Adminiftration of the Kingdom to his Wife, and dy'd 55 Years old, after having reign'd thirty Years, and deferv'd the Name of a very merciful Prince. *Antigonus*, one of *Alexander*'s Succeffors, took care of the Education of *Dionyfius*'s Children, and of the Affairs of *Heraclea*. But *Lyfimachus* having married *Amaftris*, was Mafter of the City, even long after having deferted that Princefs; for being retired to *Sardis*, he married *Arfinoe* the Daughter of *Ptolomeus Philadelphus*.

NEVERTHELESS *Clearchus*, the fecond of the Name, afcended the Throne of *Heraclea* with his Brother *Oxathris*; but thofe Princes render'd themfelves odious by a horrible Affaffination of their own Mother, whom they caus'd to be fmother'd in a Ship, in which fhe was probably going from *Heraclea* to *Amaftris*, a Town fhe had lately founded, and call'd by her own Name. *Lyfimachus* who then reign'd in *Macedon*, fhock'd at fo black an Action, and out of a juft return of Tendernefs for *Amaftris* his firft Wife, came to *Heraclea*, and put to death the two Parricide Princes; fo that it is not likely they reigned 17 Years, as *Diodorus Siculus* will have it, who calls the younger *Zathras*, inftead of *Oxathris*. *Lyfimachus*, according to *Memnon*, reftor'd the City to full liberty, but it did not long enjoy it; for *Arfinoe*, who had a great power over that Prince, having obtain'd the poffeffion of it, gave the Government of it to *Heraclitus*, who was its feventh Tyrant.

THE *Heracleans*, after the death of *Lyfimachus*, having a mind to fhake off the Yoke of Tyranny, beneath which they had groan'd for 75 Years, made a Propofal to *Heraclitus* that he fhould withdraw with his Riches; but the Tyrant was fo enraged at their Prefumption, that he prepared to punifh the chief Men of the City : however he happen'd not to be ftrong enough for them; for they threw him into Chains, razed the Walls of the Citadel even with the ground, and after having fent an Embaffy to *Seleucus*, another of *Alexander*'s Succeffors, proclaim'd *Phocrites* Adminiftrator of the City. *Seleucus* having given their Embaffadors a very fcurvy Reception, they made a League with *Mithridates* King of *Pontus*, with the *Byzantines*, with the *Chalcedonians*, and even recall'd all their Exiles.

THE

T H E Republick of *Heraclea* maintain'd itfelf honourably till the time when the *Romans* became formidable in *Afia.* To make fure of the Senate, that Republick fent a Deputation to *Paulus Emilius,* and to the two *Scipio's*; and it was no fault of the *Heracleans,* that *Antiochus* did not make his peace with the *Romans.* At length, fo good an Intelligence was fix'd between *Rome* and *Heraclea,* that thofe two Cities made League offenfive and defenfive, the Articles whereof were wrote upon Tables of Brafs at *Rome,* in the Temple of *Jupiter Capitolinus,* and at *Heraclea* in that of the fame God. Yet *Heraclea* was ftrenuoufly befieg'd by *Prufias* King of *Bithynia,* who had certainly carried it, but for a Stone from a Sling, which broke his Thigh, and oblig'd him to retire juft as he was mounting to the Affault. After this the *Galatians* very much difturb'd this City, but they were forc'd to retire. Notwithftanding her Alliance with the *Romans,* fhe thought it her true Intereft to obferve a Neutrality, during the War that the *Romans* waged with *Mithridates,* under the Command of *Murena.* Terrify'd on the one hand at this formidable Power, and alarm'd with the nearnefs of the King of *Pontus,* *Heraclea* at firft refus'd that Prince's Fleet entrance into her Port, and furnifh'd him only with Provifions. Afterwards, by the Perfuafion of *Archelaus* General of the Fleet, the *Heracleans* gave him five Gallies, and cut the Throats of all the *Romans* that were in their City to exact the Tribute, with fuch Secrecy, that it was never known. At length, *Mithridates* himfelf was receiv'd in the Place by means of his old Friend *Lamachus,* whom he corrupted with Money.

T H A T Prince left *Cannacorix* there in Garifon with four thoufand Men; but *Lucullus,* having beaten *Mithridates,* caus'd the City to be befieg'd by *Cotta,* who having taken it by treachery, and totally pillag'd it, reduc'd it to afhes. He receiv'd the Sirname of *Ponticus* at *Rome*, but the immenfe Riches he brought from *Heraclea,* occafion'd him violent Troubles. He was accus'd in open Senate by one of the moft illuftrious Citizens, who painted in fuch lively Colours the Conflagration of a powerful City which had been deficient in her Alliance with the *Romans,* only thro the Fraud of her Magiftrates, and Treachery of her Enemies, that a Senator could not forbear faying to *Cotta, We gave you orders to take* Heraclea, *but not to deftroy it.* All the Captives were fent

Vol. II.   U   home

home by the Senate's Direction, and the Inhabitants again settled in the possession of their Goods. They were allow'd the Use of their Port, and Freedom of Commerce. *Britagoras* spar'd for nothing that might re-people it; and made his court a long while to *Julius Cæsar*, tho in vain, to obtain the primitive Liberty of its Citizens. It was probably about this time that the *Romans* sent the Colony thither, spoken of by *Strabo*, and of which one part was receiv'd in the City, and the other in the Country. Before the Battel of *Actium*, M. *Anthony* gave that quarter of *Heraclea* to *Adiatorix* Son of *Demenecelius* King of the *Galatians*; and this latter, as he said, by *Anthony*'s Permission, cut the Throats of all the *Romans* in it: but after the Defeat of that General, he was carry'd along in Triumph, and put to death with his Son. After this Expedition, *Heraclea* was made part of the Province of the *Pontus*, which was join'd to *Bithynia*. Thus was this City incorporated into the *Roman* Empire, under which it still flourish'd, as appears from the Remains of the Inscription of *Trajan*, mention'd above.

HERACLEA afterwards fell into the hands of the *Greek* Emperors; and 'twas in the Decadence of that Empire, that it receiv'd the Name of *Penderachi*; which, according to the *Greek* Pronunciation, seems to be a Corruption of *Heraclea Pontica*. It was possess'd by the Emperors of *Trebisond* after the *French* enjoy'd the Empire of *Constantinople*; but *Theodore Lascaris* won it from *David Comnenus* Emperor of *Trebisond*. The *Genoese* seiz'd *Penderachi* among their Eastern Conquests, and kept it till *Mahomet* II. the greatest Captain of his Age, drove them from it. Since that time it has continued to the *Turks*; they call it *Eregri*, and that Name too seems to retain something of *Heraclea*. At present they know nothing in the world of Tyrants, *Romans*, or *Genoese*. One single Cadi administers Justice, a Waivode collects the Land-Tax and Capitation of the *Greeks*: the *Turks* pay only the Prince's Dues; happy that they can smoke at their ease among those fine Ruins, without knowing or caring what pass'd there heretofore.

WE were not long enough in *Penderachi* to disentangle its History, for we only lay there, and departed the second of *May*, in such fine Weather, that we made 80 Miles with all the pleasure imaginable. About four in the Afternoon we enter'd the River *Partheni*, whose Name the

Greeks

Greeks have retain'd ftill; but the *Turks* call it *Dolap*. The River is not Lett. IV.
a very great one, tho it was one of thofe which the ten thoufand were
afraid to pafs. *Strabo* and *Arrian* tell us it feparated *Paphlagonia* from *Bi-*
thynia. If that Author were to come to life again, he would find it ftill
as beautiful as he defcrib'd it. It flows among thofe flowry Meadows
which obtain'd it the Name of *Virgin*. *Dionyfius Byzantinus* had been
more exact, had he made it run thro the Country of *Amaftris*, inftead
of thro the middle of the City: and he imagines too that the Name of
Virgin was given it upon account of *Diana*, who was ador'd on its Banks.
The Citizens of *Amaftris* reprefented it upon a Medal of *Marcus Aurelius*;
the River appears like a young Man lying down, holding a Reed in his
right hand, with one Elbow leaning upon fome Rocks, out of which gufhes
his Stream. *Pliny* was not well acquainted with the difpofition of
thefe Coafts, for he has plac'd the River *Partheni* a great way beyond
Amaftris, and even further than *Stephane*, whom we fhall fpeak of by
and by. Yet we difcover'd *Amaftris* the next day, which was the third of
May, about nine in the morning; and we lay by that day in the River of *Sita*,
after having gone 70 Miles, partly by Rowing, and partly by Sailing.

AMASTRIS, which they now call *Amaftro*, and not *Famaftro*,
as our Maps write it, is a pitiful Village built on the Ruins of the
antient City *Amaftris*, by the Queen we before fpoke of, who united in
it four Villages, *Sefame*, *Cytore*, *Cromne*, and *Tios*: but the Inhabitants of
Tios foon afterwards left that Society; and *Sefame*, which was as it were
the Citadel of the Town, is what properly took the Name of *Amaftris*.
We muft read *Arrian* before we can well underftand *Strabo*: for *Arrian*
reckoning 90 *Stadia* from the River *Parthenius* to *Amaftris*, 60 *Stadia*
from *Amaftris* to *Erythine*, as many thence to *Cromna*, and from *Cromna*
to *Cytore*, where was a Port, 90 *Stadia*; we muft infallibly conclude that
the aforefaid Queen *Amaftris*, to people her new Town, fetch'd thither
the Inhabitants of all thofe Villages. Befides, *Memnon* fays it in fo many
words, and informs us further that this Alteration happen'd after the Retreat of *Amaftris*, who was provok'd at *Lyfimachus* her Husband's having
married *Arfinoe* at *Sardis*. Now fince, according to *Strabo*, the Citadel
which was before call'd *Sefame*, took the Name of *Amaftris*, it is out of
all doubt that the antient City of *Sefame* mention'd by *Stephanus Byzanti-*

U 2

nus,

nus, where he fays *Phineus* fix'd his firſt Abode, was fituated where *A-maſtro* now ſtands. *Pliny* agrees that heretofore *Amaſtris* was call'd *Se-fame*, and that Mount *Cytore*, ſo famous for its Box-trees, with which all the Coaſts of the *Black Sea* are cover'd, was diſtant from *Tios* 63 Miles. *Cytore* was a Port dependant upon *Sinope*, but *Amaſtris* follow'd the Fate of *Heraclea*. The Situation of *Amaſtris* is advantageous, for it ſtands upon the Iſthmus of a Peninſula, whoſe two Bendings form ſo many Ports: in *Arrian*'s time there was one very convenient for Ships of War; both are now fill'd up with Sand. That Author ſpeaks of *Amaſtris* as of a *Greek* City, becauſe its Foundreſs, tho a *Perſian*, was Queen of *Heraclea*, and it was firſt begun by a Colony of *Greeks*. The Goodneſs of the Ports of *Amaſtris* gave occaſion to the Senate and People of that City to ſtrike Medals: there are ſome of *Nerva*, of *M. Aurelius*, of the younger *Fau-ſtina*, of *Lucius Verus*, the Reverſes whereof repreſent a Fortune ſtanding, holding in her right hand a Rudder, and in her left a *Cornucopia*. Neither did they fail to ſtrike others in honour of *Neptune*, as that of *Antoninus Pius* in the King's Cloſet, on which that God holds with his right hand a Dolphin, and with his left a Trident. It is ſomewhat wonderful there ſhould be ſo many Medals of a City, which never made much noiſe in Hiſtory: they ſtruck them in honour of almoſt all the Deities. The *Diana* of *Epheſus* was not forgot. The King has a Medal of *Domitia* Wife of *Do-mitian*, on the Reverſe whereof that *Diana* is repreſented. There are Medals of *Amaſtris* ſtampt with the Head of *Antoninus Pius*, with Re-verſes of *Jupiter*, of *Juno*, of the Mother of the Gods, of *Mercury*, of *Caſtor* and *Pollux*. There is particularly one with the Head of *M. Aurelius*, and a Reverſe of *Homer*, as if the Town of *Amaſtris* expected Glo-ry upon account of the Birth of that great Man. There is not any Medal of this Town more beautiful than that which the King has with the Head of *Julia Mæſa*: the Reverſe repreſents *Bacchus* ſtanding, dreſt like a Wo-man holding a Pot in his right hand; *Jupiter* is on his left hand ſtanding too, but with very different Attributes, for he has a Spear in his right, and a Bolt in his left hand. The Medal of *M. Aurelius* plainly ſhews this City muſt have gain'd ſome conſiderable Advantages over its Neigh-bours, ſince its Reverſe is a Woman with Trophies on her left hand. That of *Fauſtina* the younger, and of *Gordian Pius*, are remarkable for

*   their

their Reverſes, whereon is a Victory which in her right hand holds a Lett. IV.
Crown, and a Palm in her left.   That of *Lucius Verus* is no leſs valuable; it is a winged Victory with the ſame Attributes.   The King has a
fine one, with the Head of the ſame Emperor : *Mars* quite naked is on
the Reverſe, his Helmet on his Head, in the poſture of a Man marching along with a Spear in his right hand, and a Buckler in his left.   With
relation to Phyſick, I have a ſort of kindneſs for the Citizens of *A-
maſtris*, for their having ſtruck ſeveral Medals in its Honour : We ſee a
great many *Eſculapius's* of *Amaſtris* with Sticks, round which a Serpent
is winded.   The Goddeſs *Salus* is repreſented upon ſome others, ſtill not
forgetting the Serpents; moſt of the Heads are of *Adrian, Antoninus Pius,
M. Aurelius,* and *Fauſtina* the younger.

T H E R E remains no Medal of the Foundreſs *Amaſtris,* who was ſtifled
at Sea by order of her Brothers.   After her Death *Lyſimachus* gave the Towns
of *Amaſtris, Heraclea,* and *Tios* to his Wife *Arſinoe,* who deliver'd them
to *Hercules* the ſeventh Tyrant or King of *Heraclea.*   His Reign was not
long, for *Lyſimachus* dying ſome time afterwards, *Heraclea* and *Amaſtris*
ſhook off the Yoke.   *Amaſtris* was even diſmember'd from the Kingdom of the *Heracleans :* and when *Antiochus* the Son of *Seleucus* declar'd
War againſt *Nicomedes* King of *Bithynia,* this *Nicomedes* who ſtood in
need of the Aſſiſtance of the *Heracleans,* was never able to put them a-
gain in poſſeſſion of *Amaſtris,* becauſe it was enjoy'd by *Eumenes,* who
rather choſe to make a Preſent of it to *Ariobarzanes,* Son of *Mithridates,*
than to reſtore it to thoſe of *Heraclea.*

A F T E R the taking of *Heraclea* by *Cotta, Triarius,* by that General's
Order, ſeiz'd *Amaſtris,* where *Cannacorix* was retir'd ; and from that time
the City remain'd under the Dominion of the *Romans* and their Emperors, till the Eſtabliſhment of the *Greek* Emperors.   It was part of the
Empire of *Trebiſond* founded by the *Comnenii,* after the *French* were
ſettled at *Conſtantinople :* but *Theodore Laſcaris* having defeated *Iathin* Sultan of *Iconium,* took *Amaſtris* in 1210, with *Heraclea* and ſome other
Places.   *Amaſtris* was in the poſſeſſion of the *Genoeſe,* when *Mahomet* II.
took *Conſtantinople* and *Pera.*   They thought fit to declare War againſt
him, upon his refuſing to reſtore them *Pera.*   *Mahomet* went in Perſon to
Amaſtris, with a numerous Artillery, which made ſo ſtrong an Impreſ-
ſion,

fion, not upon the Walls of the Town, but upon the Minds of the Inhabitants, that they open'd him the Gates. He left there only the third part of the Inhabitants, and caus'd the reft to be tranfported to Conftantinople.

WE will leave the City of *Amaftro* in the hands of the *Turks*, and purfue our Journey. The 4th of *May* we left the River *Sita*, which I find neither in Maps nor Books: we went but 30 Miles further, and the North-wind obliged us to encamp on a wretched flat Shore, where we had much ado to defend ourfelves from the Wind. The 5th of *May* we doubled Cape *Pifello*, which the Antients knew under the Name of *Carambis*, and which they make to be juft oppofite to the Ram's Head of the *Cherfonefus Taurica*, now call'd the *Little* or *Crim Tartary*. The Antients, as *Strabo* obferves, compar'd the *Black Sea* to a Bow bent, the String being reprefented by the South Coaft, which would be almoft in a ftrait line, but for Cape *Pifello*.

THAT day, the 5th, we made but 50 Miles, and encamp'd on the Banks of the Sea at *Abono*, where are nothing but wretched Cazerns for a great number of Workmen that are employ'd in making Cordage for the Grand Signior's Ships and Gallies. I forgot to mention that the Coafts of the *Black Sea* furnifh in abundance every thing neceffary for ftocking the Arfenals, Magazines, and Ports of that Emperor. As they are cover'd with Forefts and Villages, the Inhabitants are oblig'd to cut and faw the Wood for the Navy. Some make the Nails, others the Sails, the Cables, and all the neceffary Rigging. There are Janizaries fet to overfee thefe Workmen, and Commiffioners to prefs Seamen. From hence the Sultans have had their ftrongeft Fleets in the time of their Conquefts, and nothing would be eafier than to reftore their Navy. The Country is excellent; it abounds in Provifions, as Corn, Rice, Meat, Butter, Cheefe; and the People live very foberly.

ABONO feems to be the Remnant of the Name of an antient Town call'd *The Walls of Abonos*. If I wrote to a Man of mufty Erudition, I would value myfelf highly upon this pretended Difcovery; but as I have the honour to write to a Minifter that knows the juft Value of things, I hardly dare propofe this Conjecture. Be it as it will, thofe Walls of *A-*
bono

bono were never any other than a paultry Village, whofe Name *Strabo*, Lett. IV.
Arrian, *Ptolemy*, and *Stephanus Byzantinus* have preferv'd.

I AM much fonder of an admirable Species of *Chamærhododendros*,
with yellow Flowers, which we difcover'd: it may not only ferve to explain a Paffage of *Pliny*, as may alfo another fine Species of that Genus,
with purple Flowers, which we faw beyond *Penderachi*; but alfo help us
to account for the fad Accident that happen'd to the ten thoufand, who after the Defeat of the young *Cyrus*, retired into their own Country along
the Coafts of the *Black Sea*. I fhall do myfelf the honour, my Lord, to
fend you the Defcription of thofe two Plants, when the Fruits of it are
grown compleat.

W E left *Abono* the 16th of *May*, with defign to go to *Sinope*; but the
Rain oblig'd us to ftop half way, and to encamp along the Shore 40
Miles from that City. You fee fome very pretty Villages on the Coaft
at the Entrance of Woods of a furprizing Beauty. *Stephanio* is none of
the leaft of them: that Name has fuch a refemblance with that of *Ste-*
phane, which we find in *Pliny*, *Arrian*, *Marcian* of *Heraclea*, and *Stephanus*
Byzantinus, that there's no room to doubt that it is deriv'd from it, and
that confequently the antient City was not far from this Village.

T H E Sea ran fo high the next day, the 17th, that we were forc'd to
land in a Creek eight Miles from *Sinope*, whither we went the fame day
on foot, fimpling all the way: we ftay'd there two days.

I am, MY LORD, *&c,*

LETTER V.

To *Monseigneur* the *Count* de Pontchartrain, *Secretary of State, &c.*

My Lord,

Description of the Coasts of the Black Sea, *from* Sinope *to* Trebisond.

IT were to be wish'd that among the many Regulations that have been made in *France* for the Promotion of the Sciences and polite Arts, there had been one relating directly to the improving of Geography: for the Faults committed by the Geographers are very essential, and often occasion Travellers, Pilots, and sometimes Generals themselves, to take false Measures. I would have some Token of Capacity requir'd of Geographers, before they were allowed to publish Maps; and they should be obliged to travel themselves for a certain time, since they undertake to be Guides to others in their Travels.

I THINK nothing is so difficult as to make an exact Geographical Map. It is necessary for that purpose to have been upon the spot, of which one pretends to give a Plan, to take the Measures of it with good Instruments, and to make the necessary Observations as to the Heavens. Our most famous Geographers usually do their Work in their own Country, without knowing the Places they represent; they copy the Maps that are already extant; they rely upon imperfect Relations of Travels, and fancy themselves wondrous Artists when they have grav'd on the Margins of their Performances a few silly Ornaments, that generally have nothing in the world to do with the Countries they describe. The Sea-Charts are more exact than the others, frequent Shipwrecks having at length made

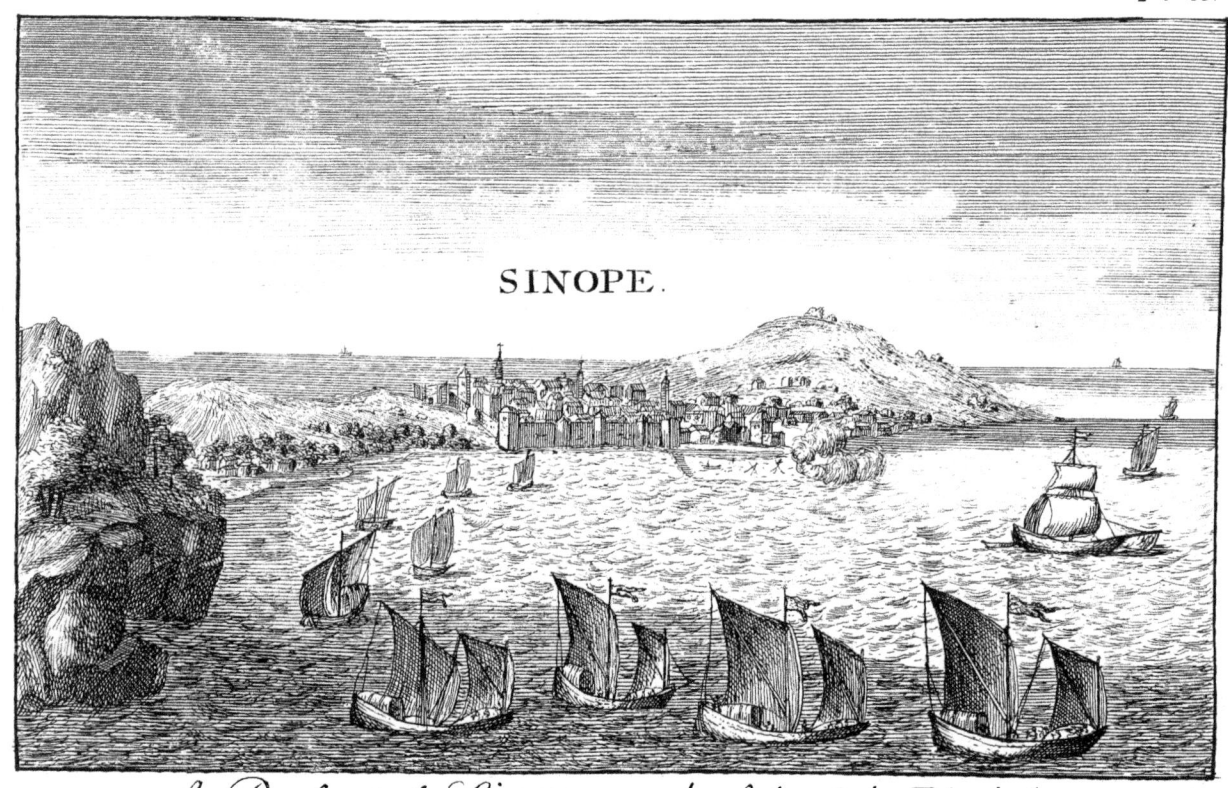

SINOPE.

A Prospect of Sinope on the side of the Black sea. 94.

made them feel the Neceffity there is of knowing the Coafts, yet the Windings of thefe Coafts are generally ill drawn. In fhort, if we have any certain Informations with refpect to Geography, as no doubt we have, we are oblig'd for them to the Aftronomers, who by repeated Obfervations have determin'd the Pofition of an infinite number of Places. How much do we owe to the Difcoveries of *Galileo*, and of thofe who follow his Steps? M. *Caffini* not only deferves the Name of the greateft Aftronomer of this Age, but alfo that of the greateft Geographer that ever was. If we have excellent Maps of Meff. *de Lifle*, the reafon is, becaufe they are fkilful Cofmographers, and keep a Correfpondence with the moft learned Aftronomers and moft experienc'd Travellers. How many Geographers in *France*, *Holland*, and *Italy*, where moft new Charts are made, whether of Land or Sea; how many, I fay, apply themfelves to Aftronomy? Moft of them build Kingdoms, Provinces, nay, Maps of the whole World, by their Fire-fide, Rule and Compafs in hand, without having ever been beyond the Smoke of their own Chimneys, or confulting thofe that have.

THE Pofition of *Sinope* is what put me out of humour with our Geographers. It is fo well defcrib'd in *Polybius* and *Strabo*, that 'tis unpardonable not to know that this City ftands upon the Ifthmus of a Peninfula about fix Miles in circuit, ending in a confiderable Cape. Yet *Sinope* is fet down in our Maps upon a ftrait open Shore, without the leaft appearance of any Port, tho it has two very good ones, and very well defcrib'd by *Strabo*. This advantageous Situation was no doubt what invited the *Milefians* to build a Town on it, or at leaft to fend a Colony thither; for *Autolicus*, one of the *Argonauts*, was reckon'd the Founder of it. *Plutarch* and the Scholiaft of *Apollonius Rhodius* go further back to look for the Origin of this City, but no body now is affected with fuch Difquifitions. The Inhabitants of *Sinope* undertook to fortify all the Avenues of their Cape, that they might be able to refift the Undertakings of that *Mithridates*, who, according to *Polybius*, defcended from one of the feven *Perfians* that put the Magi to death, and govern'd the Country which *Darius* had given as a Recompence to his Anceftors, upon the Coaft of the *Pontus Euxinus*: it was perhaps the fame *Mithridates* that was Founder of the Kingdom of the *Pontus*.

WE are not to miftake this Founde r for the great *Mithridates Eupator,* Son of *Mithridates Evergetes.* *Eupator* was born at *Sinope,* he was bred there, he honour'd it with Benefits, fortify'd it, and put it in a condition to refift *Murena,* General of the *Roman* Army, after *Sylla* was withdrawn out of *Afia.* At laft *Mithridates* made *Sinope* the Capital of his Dominions, and *Pompey* would have him buried there. *Pharnaces* was the firft that depriv'd this City of its Liberty. This *Pharnaces* was not the Son of the great *Mithridates,* but his Grandfather; for according to the Genealogy of the Kings of the *Pontus,* drawn up by *Tollius,* there was a *Pharnaces* who was Father of *Mithridates Evergetes.* *Lucullus* added *Sinope* to the *Roman* Conquefts, in delivering that place from the Yoke of the *Cilicians,* who had got poffeffion of it under pretence of holding it for *Mithridates.* The *Cilicians,* at the approach of the *Roman* Troops, fet fire to the Town, and made their efcapes in the night; but *Lucullus,* whom the true Citizens look'd upon to be their Deliverer, enter'd *Sinope,* and put to the fword eight thoufand *Cilicians,* who were not as quick as the reft. He reftor'd the Inhabitants to the Poffeffion of their Effects, and did them all manner of good Offices, ftruck with having feen in a Dream the Founder of their City the day he made his Entry. The *Romans* fent a Colony thither, which enjoy'd part of the City and of the Country. This Country is fuch as *Strabo* defcrib'd it, that is to fay, all the Ground between the Town and the Cape is wholly thrown into Fields and Gardens. *Appian* relates the taking of *Sinope* in another manner, however he agrees in the Dream and Clemency of *Lucullus.* That General, according to *Plutarch,* in his purfuit of the Fugitives, found upon the Sea-fide the Statue of the aforefaid *Autolicus,* which they had not time to carry off with them, fo he brought it away. Twas a fine piece of Workmanfhip; they paid it divine Honours, and fancy'd that it gave Oracles.

'TWAS probably about this time that they ftruck at *Sinope* the Medal I have brought from thence, or at leaft it might be ftruck upon *Lucullus*'s account. On one fide 'tis a Head naked, after the *Roman* manner, which feems to me to be that General's; on the Reverfe is a *Cornucopia,* which fhews the Riches that the Ports of *Sinope* drew thither. It is plac'd between the two Caps of *Caftor* and *Pollux*; and thefe Caps,

over

over which are a couple of Stars, inform us that thofe Sons of *Jupiter* Lett. V.
and *Leda* favour'd the Navigation of the *Sinopians.* The Colonies they
fettled, fhew that their Naval Power was very extenfive; but nothing
reflects more Glory upon this City, than the Succour it gave to the Remnant of the Army of the ten thoufand *Lacedemonians,* whofe Retreat is
one of the fineft Pieces of the *Greek* Hiftory.

T H E *Sinopians* even affected under the *Roman* Emperors to preferve
to their City the Name of a *Roman* Colony. *Patinus* has given us the
Type of two Medals, whofe Legends make mention of it; one is of the
Head of *Caracalla,* and the other of that of *Geta*: the Reverfe of this
latter is a Fifh, which puts me in mind of the great Fifh-trade that they
ftill drive in this Town. Except the Cordage they fend to *Conftantinople,* they deal in nothing but Salt-fifh and Train-oil. Their chief Salt-
fifh are Mackrels, and Pilchers or young Tunnies: the Oil is drawn from
Dolphins and Sea-Calves. As to the Medal of *Caracalla,* it reprefents *Pluto* leaning on a Bed; his Head is laden with a Bufhel, an Eagle perch'd
upon his left fift, and he holds with his right a *Hafta pura,* that is to
fay, a Spear without an Iron Head. *Tacitus,* after fpeaking of the pretended Miracles of *Vefpafian,* who they tell us reftor'd a blind Man to
fight, and made a Cripple walk in the City of *Alexandria,* relates in what
manner the Statue of *Pluto,* or the *Jupiter* of *Sinope,* was tranfported to
Alexandria, by order of *Ptolemy* the firft King of *Egypt.* That Prince fent
a famous Embaffy to the King of *Sinope,* call'd *Scydrothemis,* who being
prevail'd upon by Prefents of a great Value, after having amus'd the Deputies three Years with various Put-offs, at length confented that the
God fhould be gone, but it muft not be done without a Miracle. To fatisfy the People, who grudg'd *Europe* fo great a Felicity,
and were very apprehenfive of the fatal Confequences of that Deity's departure, they gave out that the Temple was fallen, and that the Statue
went on board all alone, and of its own free Motion. *What Stuff will
not People broach, when they have a mind to make a Miracle of a thing?*
The Report went, that in three days it pafs'd from *Sinope* to *Alexandria,*
where they rais'd it a magnificent Temple, upon the fame fpot where
formerly had been one confecrated to *Serapis* and *Ifis*; and it is likely

it retain'd the Name of *Serapis* for the fame reafon; for *Euftathius* obferves, that the *Serapis* of the *Egyptians* is the fame as the *Jupiter* of *Sinope*.

PHARNACES having by his Revolt oblig'd the great *Mithridates* his Father to kill himfelf, pretended to be a Friend to the *Romans*, and was contented with the *Cimmerian Bofphorus*, which *Pompey* granted him: but fome time afterwards flattering himfelf with hopes that he might be able to recover the other Kingdoms of his Father, while *Pompey* and *Julius Cæfar* kept the whole *Roman* Empire in a Combuftion, he pull'd off the Mask, and took feveral Towns on the Coafts of the *Pontus Euxinus*; *Sinope* was not one of the laft. He was afterwards beaten by *Cæfar*, and oblig'd to yield *Sinope* to *Domitius Calvinus*, who had Orders to continue the War againft *Pharnaces*. It is not known what ill Treatment the City might have then; but it is certain the Walls were very fine in *Strabo*'s time, who liv'd under *Auguftus*; the prefent were built under the laft *Greek* Emperors. The Walls have double Ramparts, defended by Towers moftly triangular and pentagonal, which prefent but one Angle. The Town is commanded landward, and would require two Fleets to befiege it by Sea. The Caftle is very much neglected now. There are but few Janizaries in the Town, and they will admit of no *Jews*. The *Turks*, who miftruft the *Greeks*, oblige them to lodge in a great Suburb, that is without any defence. We found no Infcription either in the City or Parts adjacent; but to make us amends, befides the Fragments of Marble Pillars that are fet in the Walls, we faw a prodigious quantity in the Burying-place of the *Turks* amidft feveral Chapiters, Bafes, and Pedeftals of the fame kind: they are the Remains of the Ruins of the magnificent *Gymnafium*, *Forum*, and Porticoes fpoken of by *Strabo*, not to mention the antient Temples of the Town. The Baffa encamp'd with all his Houfe at the foot of the Walls between the Town and the Suburb. As for us who were look'd upon as Mifcreants, tho we were treated at the Baffa's with all the Complaifance in the world, we lodg'd in the Suburb, at the Houfe of a *Greek*, who fold very good Wine of high Growth, for they have no low Vines. The Water here is excellent, and they cultivate Olive-trees of a reafonable fize: but as fine as this Country is, it produces none but common Plants, except one Species of Wormwood that grows in the Sand along the Sea-

fhore,

shore, and which in all probability must be the *Absynthium Ponticum* of the Antients, which I believe is known to no modern Author. Perhaps it is more common towards the Mouths of the *Danube*, for *Ovid* says the Fields there produce nothing so common as Wormwood. Perhaps too he speaks poetically, and uses the Word *Absynthium* only to express in a lively manner the Bitterness of his Banishment.

THE Plant we are speaking of is an Under-shrub, two foot high, hard, bushy, and branchy from the very bottom, where it is as big as a Man's little finger, and reddish. The rest, as well as the Branches, is cottony and white. The whole Plant is garnish'd with Leaves of the same colour, pretty soft, almost round, two inches broad, but more slenderly cut than that kind which is cultivated in the Gardens by the name of the *Little Absynthium,* or the *Absynthium* of *Galen.* From the junctures of the Leaves of our *Absynthium Ponticum,* arise Branches and Sprigs laden with Leaves less round, and yet more slenderly cut; the last that grow towards the extremity of the Branches, which are close enough to one another, are not above half an inch long and half a line broad, and are usually quite plain, or have at most but one or two Divisions. The Flowers grow in abundance all along the Branches and Sprigs, which are more cottony and whiter than the rest of the Plant. Each Flower is a Button two lines long, consisting of very slender Leaves posited like Scales, and cover'd with a pretty thick Down, which said leaves enwrap seven or eight Fleurons of a pale yellow, very slender, divided into five points in the place where they open; they let out a little Sheath of a deeper colour, a-cross which juts out a greenish Thred. Each Fleuron bears upon an Embryo of Seed, which is not ripen'd till the latter Season; it is very small and brown. This kind of Wormwood has been cultivated in the King's Garden above twenty Years, and I can't find how it came thither. Perhaps some Missionary might bring the Seed of it from the Coast of the *Black Sea.* The Root of this kind of Wormwood is hard, ligneous, reddish, divided into Fibres, wavy and hairy. The Leaves and Flowers are extremely bitter: their Smell is not so strong as that of the common Wormwood that grows naturally in the *Alps,* and which thrives in all the Gardens of *Europe.*

CHA-

CHARATICE, a *Mahometan* Captain, furpriz'd *Sinope,* and pillag'd it, with defign to carry off the Treafures which the Emperors had depofited there; but he was oblig'd to leave the Place, without meddling with the Riches, by order from the Sultan his Mafter, who courted the Friendfhip of *Alexis Comnenes,* and had fent him an Embaffador. The Government of the Town was given to *Conftantine Dalafthenes,* a Kinfman of the Emperor's, and the greateft Captain of that Age. When the *French* and *Venetians* made themfelves Mafters of *Conftantinople, Sinope* fell into the power of the *Comnenes,* and was one of the chief Cities of the Empire of *Trebifond. Sinope* afterwards became a Principality independent of *Trebifond;* and it was probably fome Sultan that made a Conqueft of it at the time when they fpread themfelves over *Afia Minor:* for *Ducas* relates, that *Mahomet* II. being at *Angora* in 1461. was faluted there, and receiv'd the Prefents of *Ifmael* Prince of *Sinope,* by the hands of his Son. *Mahomet* bid him give his Father to underftand that he muft deliver him up his Dominions; 'twas a Compliment of no very eafy digeftion, but the *Turkifh* Fleet appearing before the Town, made *Ifmael* refolve to obey. *Chalcondylus* fays, he exchang'd his Principality for the City of *Philipopolis* in *Thrace,* tho there were 400 pieces of Artillery on the Ramparts of *Sinope.* By the fame Treaty *Mahomet* acquir'd *Caftamene,* a very ftrong Town depending upon the fame Principality. Thofe *Turks* that upbraid the Chriftians with carrying on bloody Wars againft one another, are poorly acquainted with the Hiftory of their own Empire; for the firft Sultans made no fcruple to defpoil the firft *Mahometans,* whofe Lands lay, as we call it, convenient for them. 'Tis univerfally known that they conquer'd *Afia Minor* only from Princes of their own Religion, who had erected themfelves into petty Sovereigns at the coft of the *Greeks.*

ONE cannot pafs by *Sinope* without calling to mind the famous *Cynick* Philofopher *Diogenes;* that *Diogenes,* whofe fharp Sayings *Alexander* fo much admir'd, was a Native of this Place. You know, my Lord, *Alexander* told his Courtiers one day, that were he not *Alexander,* he could wifh to be *Diogenes,* which he faid upon occafion of an Anfwer that Philofopher made him; for that Prince honouring him with a Vifit at *Corinth,* ask'd him, *If he had need of any thing? Diogenes* anfwer'd, *He had*

need

need of nothing but the Warmth of the Sun, and that therefore he wifh'd he *would ftand a little on one fide, and not take that from him.* His Epitaph is to be feen on an antient Marble at *Venice,* in the Court of the Houfe of *Erizzo*; it is grav'd beneath the figure of a Dog, fitting upon his Breech, and may be thus tranflated:

Qu. *S P E A K, Dog, whofe Tomb do you watch fo carefully?* Anf. *The Dog's.* Qu. *Who is it you call Dog?* Anf. *Diogenes.* Qu. *Of what Country was he?* Anf. *Of* Sinope, *the fame that formerly liv'd in a Tub, and that now has the Stars for his abode.*

T H E *Terra Sinopiana,* which *Strabo, Diofcorides, Pliny,* and *Vitruvius* mention'd, is not green, as many believe, imagining that the green Colour which in Heraldry is call'd *Sinople,* took its Name from it. The *Terra Sinopiana* is a kind of Bolus more or lefs deep, which was formerly found about this City, and which they brought to it to diftribute it. What proves that 'twas really nothing but Bolus, is, that the Authors above quoted affirm, that 'twas as fine as that of *Spain:* every body knows that there is very fine Bolus found in many parts of that Kingdom, where they call it *Almagra*; and this Bolus, in all likelihood, is a natural *Saffron of Mars.* Yet it is poffible there may be fome fort of green Earth in the Country of *Sinope,* for *Chalcondylus* fays there is excellent Copper near it; and, I believe, the green Earth, which the Antients call'd *Theodotion,* to have been, properly fpeaking, nothing but natural *Verdigreafe,* juft as it is found in the Copper Mines. The Antients had an efteem for the green Earth of *Scio,* but the People there know nothing of it now, or at leaft no body could give us any Information about it.

W E departed from *Sinope* the 10th of *May,* and got but 18 Miles, becaufe the ill Weather carried us to *Carfa,* as the Natives pronounce it. This Village is call'd *Carofa* in our Maps, and this Name has yet more fimilitude to that given it by the Antients; for *Arrian* calls it *Caroufa,* and fays with good reafon, 'tis a pitiful Port, a hundred and fifty ftadia diftant from *Sinope,* which is juft eighteen Miles and a half. 'Tis furprizing that the Meafures of the Antients fhould fometimes anfwer fo exactly to the modern Computation.

T H E 11th of *May* we encamp'd upon the Shore of the Ifland form'd by the Branches of the River *Halys,* 30 Miles from *Carfa.* Here is another

other Blunder in our Geographers, who make this River run from the South, whereas it comes from the Eaſt. They have no other Excuſe, but that *Herodotus* committed the ſame Miſtake; yet 'tis a long while ago ſince *Arrian* correct̃ed it, who review'd the Places in perſon, by order of the Emperor *Adrian*. *Strabo*, who was of that Country, perfect̃ly well deſcribes the courſe of the *Halys*. Its Sources, ſays he, are in the greater *Cappadocia*, whence it flows towards the Weſt, and then winds towards the North thro *Galilea* and *Paphlagonia*. It took its Name from the ſalt Grounds thro which it paſſes. Indeed all thoſe Parts are full of a foſſile Salt; it is found even in the great Roads and arable Lands: its Saltneſs approaches a little to Bitter. *Strabo*, who omitted nothing in his Deſcriptions, juſtly obſerves that the Coaſts from *Sinope* quite to *Bithynia* are cover'd with Timber proper for building of Ships, that the Fields are full of Olive-trees, and that the Joiners of *Sinope* made beautiful Tables of Walnut and Maple Wood. All this is ſtill practis'd, except that inſtead of Tables which are not us'd in *Turky*, they uſe the Maple and Walnut-tree Wood, in making of Sophas, and wainſcoting Rooms: ſo that 'twas not this part of the *Black Sea* that *Ovid* declaim'd ſo vehemently againſt, in his third Letter written to *Rufinus* from the *Pontus*.

T H E next day we perform'd no more than twenty Miles, the North Wind forcing us in ſpite of our teeths to caſt Anchor at the Mouth of the *Caſalmac*, in the Port which the Antients nam'd *Ancon*. The *Caſalmac*, which is the biggeſt River upon all this Coaſt, was heretofore known by the Name of *Iris*. *Strabo* did not forget to tell us that it ran thro *Amaſia* his own Country, and that it receiv'd the River *Themiſcyra* before it falls into the *Pontus Euxinus*.

W E left behind us upon the Sea-ſhore a Village built on the Ruins of *Amyſus*, an antient Colony of the *Athenians*, according to *Arrian*. *Theopompus*, who in *Strabo* aſcribes the Foundation of it to the *Mileſians*, agrees with him; and thereby he informs us of the reaſon why the Town was call'd *Pireum*, which was the Name of one of the Ports of *Athens*. The Town of *Amiſus* was a long while free, nay, and appear'd ſo jealous of its Liberty, that mention was almoſt conſtantly made of it in its Medals. There are Medals of that Legend, with the Heads of *Ælius, Antoninus Pius, Caracalla, Diadumenus, Maximin, Tranquillin*. *Alexander* the Great being

being in *Asia*, restor'd the Liberty of *Amisus*; the Siege and taking of that
City by *Lucullus* are describ'd very copiously in *Plutarch*.   That *Roman*
Captain not thinking fit to press it too closely, left *Murena* before
it; but return'd thither after the Defeat of *Mithridates*, and had easily
carried it, but for the Engineer *Callimachus*, who after having heartily fatigued the *Roman* Troops, and finding he could no longer defend the Town,
set it on fire.   *Lucullus*, with all his Authority, could not extinguish the
Flame; and he began to be very uneasy that he should be less happy upon
such an occasion than *Sylla*, who had sav'd the City of *Athens* from being
consum'd.   But Heaven back'd his Wishes, and the Rain fell time enough to
save part of *Amisus*: *Lucullus* caus'd the rest to be rebuilt, and affected to
shew the Citizens as much Clemency as *Alexander* had shewn the *Athenians*: in short, *Amisus* was restor'd to its former Liberty.   As to the Town
of *Eupatoria*, which *Mithridates* had built, and call'd by his Name, near
to *Amisus*, it was taken by Storm, and level'd with the Ground,
during the Siege of *Amisus*.   It was afterwards rebuilt, and but one
Town made of these two, which was call'd *Pompeiopolis*, or the *Town of
Pompey*; but it did not long enjoy its Liberty, *Pharnaces* the Son of *Mithridates* besieg'd it during the Wars of *Cæsar* and *Pompey*, and won it with
such mighty Difficulties, that to be reveng'd upon the Inhabitants, he cut
all their Throats with the utmost Cruelty.   *Cæsar*, now Master of the
World, beat *Pharnaces*, and oblig'd him to submit.   He thought, says
Dion Cassius, he made the Citizens of *Amisus* sufficient Amends for all
the Misfortunes they had undergone, by granting them that Liberty which
was so dear to them.   *Mark Anthony*, according to *Strabo*, put the Town
again into the hands of its Kings; and which was whimsical enough, the
Tyrant *Strato* having given it very ill Usage, *Augustus*, after the Battel
of *Actium*, allow'd it its antient Liberty.

I T was perhaps upon this occasion that the beautiful Medal which
is in the King's Closet, might be struck with the Head of *Ælius Cæsar*.
The Reverse is Justice standing, holding a pair of Scales in her hand; for
the Epoch ΡΞΘ agrees with that of *Augustus*.   The Peasants that work'd
at making of Cordage brought us some Medals which are pretty common,
among which was one of the Town of *Amisus* which was not so common; on one side is the Head of *Minerva*, on the other *Perseus* having

juft cut off the Head of *Medufa*. We obferv'd above that *Amifus* was a Colony of *Athens* : no doubt they ftill ador'd that *Minerva*, and as fhe had a great fhare in *Perfeus*'s Expedition, they reprefented upon the Reverfe one of that Hero's greateft Actions.

ONE cannot pafs by thefe Coafts without calling to mind that the *Cafalmac* water'd part of that beautiful Plain of *Themifcyra*, where the famous *Amazons* had their little Empire, if we may venture to fay thus much of Women, who are ufually counted imaginary : yet *Strabo*, who places them in thefe parts, informs us, that the *Thermodon* water'd the reft of their Country. This River agreeably recalls the Idea of thofe Heroines, of whom it is certain many Fables have been invented. But be that as it will, the Sight of this Coaft gave us a great deal of Delight. 'Tis a flat Country, divided into Woods and Lawns, which begin from *Sinope* ; whereas from *Sinope* to *Conftantinople* the Country rifes in little Hills of admirable Verdure.

THE 13th of *May* we again encamp'd upon the Coafts of the *Amazons*, very ill fatisfy'd with our Searches, for we could not find any rare Plants ; and thofe ran more in our heads than any thing we are told of thofe illuftrious Women. Our Journey was no more fuccefsful the next day, for the Rain made us lofe all our time. They would fain perfuade us on the 15th, that we had travell'd 50 Miles, but we thought 'em very fhort ones, and we entred very early in the River of *Tetradi*, which the *Turks* call *Cherfanbaderefi*. The next day we drew up into that of *Argyropotami*, in *Turkifh Chairguelu*, which is but forty Miles from *Tetradi*.

WE receiv'd a vaft deal of Pleafure this day, even much more than if we had met with the *Amazons* ; and yet 'twas nothing but a kind of Elephant-plant, of a foot and half high, which all the Hedges were full of. We muft range this Plant under the Genus of Elephants with *Fabius Columna*, the moft exact of all the Botanifts of the laft Age. The Flower of this kind of Plant is fo like the Head of an Elephant by its Probofcis, that every body muft agree in the thought of that learned Man. Give me leave, my Lord, to fend you the Defcription of it ; for the Species of Elephant that grows on the Coaft of the *Black Sea* is not exactly the fame as *Columna* found in the Kingdom of *Naples*.

FROM

FROM a hairy, reddifh Root, rife feveral Stalks a foot and a half or two foot high, about, a line and a half thick, fquare, pale green, thick-fet with little Hairs, hollow from joint to joint, towards the bottom rifing into fome Tubercula, whitifh, pretty flat, wrinkled, flefhy, two or three lines long, and plac'd almoft like Scales. The Leaves grow two by two oppofite crofs-wife to thofe below and thofe above, from one to two inches long, and nine or ten lines broad, travers'd by a Rib, accompanied with pretty big Nerves almoft parallel to each other, and which grow crooked and fubdivided as they come towards the Rims. Otherwife thefe Leaves are of the fame texture with thofe of the *Yellow-flower'd Pedicolary*, brown-green, rough beneath, ftrew'd with little Hairs on each fide, moderately indented, and fupported by a flender Pedicule two lines long. From the junctures of thefe Leaves, which grow fmaller and fmaller to the top, rife Branches oppofite crofs-wife like the Leaves; and along thefe Branches grow Flowers, fometimes alone, fometimes oppofite two and two, yellow, and fix or feven lines long. Each Flower begins by a Pipe of about two lines long, which opening divides into two lips, the undermoft whereof is an inch long, and fometimes more broad, flafh'd in three pieces pretty well rounded, falling down like a Ruff, and mark'd at the beginning of its Divifion with the fpot of a deep Fillemot Colour. The upper Lip is a little longer than the lower, and begins with a kind of Head-piece, flat at top like the head of a Dog, about three lines broad and four long to the Orbits, which are mark'd by two great Points of a deep red, a third part of a line diameter. From thefe Orbits the Head-piece turns up a little, and lengthens out like the Trunk of an Elephant. It is hollow, four or five lines long, obtufe or blunt at the end, and lets out the thred of the Piftile. At the Birth of this Trunk, before it folds it felf gutterwife, you fee two little Hooks half a line long, crooked inways; the Stamina are hidden in the Head-piece, and garnifh'd with yellowifh Summits: the Piftile is an oval Button, a line long terminated by a Thred: the Cup is four or five lines long, pale green, flafh'd deep into three parts hairy, ray'd, the middlemoft whereof, which is the biggeft, is hollow like a Gutter. The Piftile comes to be a Fruit, flat, membranous, blackifh, almoft fquare, but rounded at the Corners, divided into two Apartments length-ways, and full of Seeds, a little crooked, a line

Y 2   and

and a half long, blackifh, channel'd length-ways. The whole Plant has a graffy tafte, and no flavour; its Flowers fmell like thofe of the Lillies of the Vallies; a fat Soil and fhady Place.

THE 14th of *May*, after going twenty eight Miles, we anchor'd in the Mouth of the little River *Vatiza*, clofe to a Village of the fame Name, whither we went to get Refrefhments: the Wind was North, and the Sea a little high, fo a Council was fummon'd; and as Opinions were divided, the Baffa was in fufpence whether he fhould go forward or no. I had the honour to induce him to ftay not only that day, but the next too, affuring him upon the word of a Phyfician that the fick Folks of his Family had need of Reft, and efpecially his Preacher, whom he honour'd with his Efteem. And indeed, this Intermiffion did the Patients both good and pleafure: the Sailors were the only People that grumbled; for being paid for the Voyage in the whole, they were for making the beft of their way. For my part, I was overjoy'd at having it in my power to fearch fo fine a Country, fo that I gave very little heed to any thing they faid. The Hills of *Vatiza* are cover'd with a *Laurel Cherry-tree*, and a *Guaiacum* of *Padua*, higher than our Oaks; we were never weary of admiring them. There is alfo a Species of *Micocoulier*, with large Leaves, the Fruit of which is half an inch diameter. We obferv'd an infinite number of other fine Plants; but we were forc'd to decamp the next day. The Sea ftill feem'd turbulent to the Baffa's Attendants; and tho the Sailors affur'd us it was as fmooth as Oil, which is a comparifon they make every where at Sea, we got but 20 Miles before dinner. We moor'd at the foot of an old demolifh'd Caftle, whofe Name we could not learn; but we were not very uneafy about it, for the Ruins had no appearances of any great Antiquity. You muft not, my Lord, form a difadvantageous Idea of the *Black Sea* upon this Relation: we never ftirr'd but in perfect Calms; the North Winds which they were in fuch dread of, and the Sea which always feem'd rough to thefe good Muffulmans, gave our Boats but very moderate Shakes, and did not hinder the Saiques from going to and fro. Our March put me in mind of thofe luxurious Times fo well defcrib'd by *Boileau* in his *Lutrin:*

All Night they refted, and all Day they fnor'd.

THIS

Hypericum Orientale, Ptarmicæ folijs Coroll Rei herb. 18

THIS was exactly the Life of our Court. They wak'd only to
smoke, drink Coffee, eat Rice, and drink Water; not a word either of
Hunting or Fishing. We travell'd but twelve Miles this day, and those
by Rowing, and landed on a flat Shore, in a delicious place, abounding
with fine Plants.

THE 26th of *May* somebody took it in his head to report (one
would think he did it only to make the Sailors give themselves to the
Devil) that 'twas an unlucky Day: this one word hinder'd us from setting out till after dinner; so that the Hour of Prayer being come, we
anchor'd two Miles from *Cerasonte*, which the *Greeks* call *Kirisontho*.
The desire we had to see that Town, made me pretend that we wanted
Honey for our sick Folks, and that we must go thither to buy some.
They answer'd, 'twas an unlucky Day, and God would take care of the
sick People. We were comforted for this Disappointment by the Discovery of an admirable Species of *St. John's-wort*; and indeed nothing less
than so fine a Plant could have softned our Discontents; for whom had we
to tell them to, in a Country where we saw neither Man nor Beast? When
we found no beautiful Plants, Reading supply'd the place of all other Di-
versions.

THE old Stocks of this kind of *St. John's wort* have a Root two or
three lines thick, hard, ligneous, lying sloping, and above half a foot
long. That of the young ones is a Tuft of yellowish curl'd Fibres, three
or four inches long. The Stalks are from half a foot to a foot high,
some strait, others horizontal, and then standing up again, pale-green, a
line thick, garnish'd with a little Thred, which descends from one
leaf to another. These leaves, which grow two by two, are an inch or
fifteen lines long, and two lines broad, pale-green also, of the same contexture as those of our *St. John's-wort*, close, without any appearance of transl.
parent points, indented about the Rims almost like those of the *Sneeze-
wort* that grows in our Meadows, fastned to the Stalk without any Pedicule, and terminated at the bottom by two very pointed Ears two lines
long, but slash'd deeper than the rest of the Leaf. From their junctures
rise Branches garnish'd with the like Leaves, tho shorter and broader.
Those Branches form a Cluster like that of the common *St. John's-wort*.
The Flowers of the Species I am describing, consist of five yellow Leaves,
eight

eight or nine lines long, three lines broad, rounded at the point, but narrower at the Bafis. From the midft of thefe Leaves rifes a Tuft of yellow Stamina, fhorter than the Leaves, garnifh'd with little Summits. They furround a Piftile two lines and a half long, greenifh, terminating in two horns. The Cup is three lines long, flafh'd in five indented parts as neatly as the Leaves. The Piftile comes to be a Fruit of a deep red, three lines high, divided into five Apartments, full of very fmall brown Seeds, which fall out of the point of the Fruit when thorowly ripe. The whole Plant has a refinous Smell. It varies confiderably as to bignefs; you may find fome with very fhort Stalks, and whofe Leaves are extremely flender. The Flower varies alfo, for there are fome whofe Leaves are even ten lines long. The Leaves are bitter, a little gluy, and fmell refinous.

THE 21ft of *May* we pafs'd by *Cerafonte*, a pretty large Town built at the foot of a little Hill upon the Sea-fhore, between two very fteep Rocks. The ruinated Caftle, which was the work of the Emperors of *Trebifond*, is upon the Summit of a Rock to the right as you enter the Port; and this Port is proper enough for Saiques. There were feveral then there, that only ftaid for a fair Wind to proceed to *Conftantinople.* The Country of *Cerafonte* feem'd to us to be very good for fimpling. It confifts of little Hills cover'd with Woods, wherein Cherry-trees grow naturally. St. *Jerom* believ'd thefe Trees took their Name from this Town; and *Ammianus Marcellinus* tells us that *Lucullus* was the firft that from hence carried Cherry-trees to *Rome.* Cherry-trees, fays *Pliny*, were not known before the Battel which *Lucullus* fought with *Mithridates*, and 'twas a hundred Years longer before they pafs'd into *England.* *Cerafonte*, according to *Arrian*, was afterwards nam'd *Pharnacia*; 'twas a Colony of *Sinope*, to which it paid Tribute, as *Xenophon* obferves; yet *Strabo* and *Ptolemy* diftinguifh *Pharnacea* from *Cerafonte.* 'Twas at *Cerafonte* that the ten thoufand *Greeks* who had been at the Battel of *Babylon*, in the Army of the young *Cyrus*, pafs'd in Review before their Generals. They continued there ten days, and after all their Fatigues their Army was diminifh'd only fourteen hundred Men. In thofe times a Diftinction was made between the *Greek* Cities, that is to fay, Colonies of the *Greeks*, upon the Coafts of the *Pontus Euxinus*, and the other Towns, built by the Natives, whom the *Greeks* look'd upon as *Barbarians* and declar'd Enemies.

CERASONTE.

TRIPOLI

A Prospect of Tripoli on the side of the Black sea.   *97.*

Vitis Idæa Orientalis maxima
Cerasi folio flore variegato Coroll.
Inst. Rei herb. 42.

nemies. The Remains of the ten thousand carefully avoided such Towns, and sought the *Greek* Colonies; but they were generally forc'd to cut thro with sword in hand. Tho *Cerasonte* was never any very considerable place, we neverthelefs have Medals left of it. There are some with the Head of *Marcus Aurelius*, on the Reverse whereof i a Satyr standing upright, in his right hand holding a Flambeau, and a Crook in his left. By this it appears that it was not a Town of Naval Commerce; it rather valued it self upon its Woods and Flocks.

WE put in that day 36 Miles from *Cerasonte* to fetch some Provifions from *Tripoli*, a Village mention'd by *Arrian* and *Pliny*, and which you will here find a Draught of. Afterwards our little Fleet came to anchor three Miles below it, at the entrance of a River that probably bore the same Name as the Town in *Pliny*'s time. Some Mines of Copper were formerly wrought along this River, for you still find there Recrements of that Metal, cover'd with Vitrifications enamel'd white and green. All thefe Coafts are agreeable, and Nature has here preferv'd it felf in its Beauty, becaufe there have not been this long while Inhabitants enough to exhauft it. We obferv'd a Shrub, which in all appearance muft be the *Uva Urfina* or Bearsberry of *Galen*.

THIS Shrub grows up to the height of a Man. The Stalk is as thick as one's Arm, the Wood whitifh, the Bark flender mix'd with brown, chapt, and the firft Rind eafily comes off. This Stalk puts forth feveral Branches from the very bottom, as thick as a Man's Thumb, fometimes more, fubdivided into Boughs clothed in a Bark pale-green. All thefe Boughs are laden with new Shoots, cover'd with a clean fhining Bark, garnifh'd with Leaves like thofe of the Cherry-tree, two inches and a half long, and one and a half broad, moderately indented about the edges, pointed at each end, bright-green, fometimes reddifh, fleek, rifing into a Rib beneath, and ftrew'd with very fhort Hairs. The Flowers grow amidft thefe Leaves upon Stalks an inch and a half long, inclining downwards, rang'd upon a line in the junctures of the Leaves, which as yet are but half an inch long, and their Pedicule is but three or four lines long. Each Flower is like a Bell, about four lines diameter, and five lines high, of a dirty white, beautify'd with large purple ftreaks on that fide which is expos'd to the Sun, flafh'd into five points, fometimes more, and thofe points are

a.

a little bending outwards. This Flower varies. Upon some Stocks it is quite white, and upon others it has a little of the Purple without being striped. Of whatever Colour it be, it has always a hole in the bottom, and is articulated with the Cup. Round the hole of the Flower rise ten Stamina a line and a half long, whitish, a little crooked, each laden with a Summit of the same length, deep, yellow, approaching to Fillemot. The Cup is a greenish Button, flat before, and as it were pyramidal behind, a line and a half long, flash'd into five parts, which form a little Bason, heightned with a kind of Wod hollow in the middle, as in the other sorts of this kind. From the Center of this Bason runs a slender Thred 4 or 5 lines long. The Leaves of this Plant have a tartish, graffy taste: the Flowers have no smell. I only saw the Fruit of it when it was green, and about three lines long, acrid, and hollow before like a Navel. This is the biggest known Species of the *Vitis Idæa.* 'Tis probably the same that *Galen* call'd ΑρκτοϛαφυλΘ, or *Bear-berry*: that Author says it grows in the Kingdom of the *Pontus,* and that its Leaves are like those of the *Arbute-tree*; which is true, if you compare these Leaves with those of the *Adrachne Arbute-tree,* which is as common in *Greece,* and more common in *Asia,* which was the Country of *Galen,* than our common *Arbute.*

WE got but 35 Miles the 22d of *May,* and our Tents were pitch'd near a Water-mill, within sight of *Trebisond,* which the *Turks* call *Tarabosan,* where we arriv'd the next day in four hours, by Sailing and Rowing. This Town is famous in History for nothing but the retreat of the *Comnenes,* who after the taking of *Constantinople* by the *French* and by the *Venetians,* made it the Seat of their Empire. Antiently *Trebisond* was look'd upon to be a Colony of *Sinope,* to which it even paid Tribute, as we are inform'd by *Xenophon,* who pass'd by *Trebisond* when he led back the Remains of the ten thousand. *Xenophon* relates the melancholy Accident that happen'd to them upon eating too much Honey. Here, my Lord, is a Description of the Plants from which the Bees suck it.

C H A M Æ R H O D O D E N D R O S Pontica maxima, Mespili folio, flore luteo. Coroll. Inst. Rei Herb. 42.

THIS Shrub grows to seven or eight feet in height, and produces Trunk almost as big as a Man's Leg, accompanied with several smal-

ler

ler Stems divided into unequal Branches, weak, brittle, white, but cover'd with a sleek greyish Bark, except at the extremities, where they are hairy, and garnish'd with Clusters of Leaves pretty like those of the wild *Medlar-tree*, 4 inches long, and a foot and a half broad, pointed at each end, bright green, haired slightly, except at the edges, where the Hairs form a kind of Eyebrow. The Rib of these Leaves is pretty strong, and distributes itself into Nerves all over the Surface. This Rib is only a continuation of the tail of the Leaves, which commonly is 3 or 4 lines long, and one thick. The Flowers grow in Clusters, 18 or 20 together, at the extremity of the Branches, sustain'd by Pedicules an inch long, hairy, and which rise from the bosoms of little Leaves, membranous, whitish, 7 or 8 lines long, and 3 broad. Each Flower is a Pipe two lines and a half diameter, superficially gutter'd, hairy, of a greenish yellow. It opens above an inch wide, and divides into five parts, the middlemost whereof is above an inch long, almost as broad, turning backwards as well as the rest, and terminated like a *Gothick* Arch, pale yellow, tho of a gold-colour towards the middle. The other parts are a little narrower and shorter, pale yellow also. This Flower which is pierced behind, articulates with the Pistile, which is pyramidical, channel'd, two lines long, whitish green, thinly hairy, terminating in a crooked Thred two inches long, rounded at the end like a Button, pale green. Round the hole of the Flower grow five Stamina shorter than the Pistile, unequal, crooked, laden with Summits a line and a half long, full of yellowish Dust. The Stamina are of the same colour, hairy from the beginning almost to the middle, and all the Flowers lean on their sides like those of the *Bastard Dittany*. The Pistile in time comes to be a Fruit of about 15 lines long, and 6 or 7 diameter, hard, brown, pointed, rising into 5 Ribs. It opens from the point to the basis into 7 or 8 parts, hollow'd gutterwise, which joining with the Axis that runs thro the middle of it, form so many Apartments full of Seeds. The Leaves of this Plant are stiptick. The Smell of the Flower is something like that of the *Honey-Suckle*, but stronger, and hurtful to the Brain.

C H A M Æ R H O D O D E N D R O S Pontica maxima, folio Lauroce-rasi, flore cæruleo purpurascente. Coroll. Instit. Rei Herb. 42.

THIS Species generally grows the height of a Man. Its chief Stock is almoſt as big as a Man's Leg. Its Root runs to five or ſix foot long, at firſt divided into ſome other Roots as big as a Man's Arm, diſtributed into Subdiviſions one inch thick. Theſe laſt diminiſh inſenſibly, accompany'd with abundance of Hairs. They are hard, ligneous, cover'd with a brown Bark, and produce ſeveral Stalks of different ſizes, which ſurround the Trunk. The Wood of it is white, brittle, cloth'd with a greyiſh Bark, deeper in ſome parts than in others. The Branches are pretty buſhy, and grow from the very bottom, ill form'd, unequal, garniſh'd with Leaves only towards the Extremities. Theſe Leaves, tho rang'd without order, are exceeding beautiful, and are exactly like thoſe of the *Laurel Cherry-tree.* The biggeſt are ſeven or eight inches long, and about two or three broad, and terminate in a point at each end, bright green, ſleek, almoſt ſhining, firm and ſolid. The back, which is only a continuation of the tail, which is almoſt two inches long, riſes out into a great Rib ridg'd before, the chief Subdiviſions whereof are as it were alternate. The Leaves diminiſh in proportion as they approach the Summits, tho often even there you ſhall ſee ſome that are larger than the under ones. From the end of *April* to the end of *June,* theſe Summits are laden with Cluſters 4 or 5 inches diameter, conſiſting each of twenty or thirty Flowers, at the bottom of which is a Leaf but an inch and a half long, membranous, whitiſh, 4 or 5 lines broad, hollow and pointed: the Pedicule of the Flowers is from an inch to 15 lines long, but it is only about half a line thick. Each Flower is of one ſingle piece, an inch and a half or two inches long, ſtraitned at bottom, open'd and ſlaſh'd into five or ſix parts. The uppermoſt, which is ſometimes the biggeſt, is about ſeven or eight lines broad, rounded at the end, as are alſo the reſt, a little curl'd, adorn'd towards the middle with ſome yellow points ſtanding cloſe together like a great ſpot. The under parts are a little ſmaller, and ſlaſh'd deeper than the others. As to the Colour of this Flower, it is uſually of a Violet-colour, approaching a little to gridelin. Some of theſe Stocks have white Flowers, and others purple more or leſs deep, but all theſe Flowers are mark'd with the ſame yellow points, which I juſt now mentioned; and their Stamina, which grow in a tuft, are more or leſs ting'd with Purple, tho white and cottony at their firſt Birth. Theſe

Stamina

Stamina are unequal, crooked, and furround the Piftile. Their Summits lie fideways, and are two lines long, and one broad, divided into two purfes full of a yellowifh Duft. The Cup is but about a line and a half long, flightly channel'd into 5, 6 or 7 purple Ribs. The Piftile is a kind of Cone two lines high, heightned at its Bafis with a Hem greenifh, and as it were curled. A purple Thred crooked, and 15 or 18 lines long, terminates this young Fruit, and ends in a Button pale green. The Cluf-ters of Flowers are very clammy before they blow. When they are gone, the Piftile becomes a cylindrical Fruit, from an inch to 15 lines long, a-bout 4 lines thick, gutter'd, rounded at each end. It opens at top into 5 or 6 parts, and fhews as many Apartments which divide it lengthways, feparated from each other by the wings of an Axis that runs thro the mid-dle. It is this Axis that is terminated by the Thred of the Piftile; and far from drying, it becomes longer while the Fruit is green, and does not fall when it is ripe. The Seeds are extremely fmall, bright brown, al-moft a line long. The Leaves of this Plant are ftiptick: the Flowers have an agreeable Smell, but it is foon gone.

THIS Plant loves a fat moift Soil, and grows on the Coafts of the *Black Sea* by the fide of Streams from the River ' *Ava* to *Trebifond*. This ' Species is reckon'd unwholefom. The Cattel never eat it but when they can find no better Nourifhment. As beautiful as the Flower is, I did not judge it convenient to prefent it to the Baffa *Numan Cuperli Beglerbey* of *Erzeron*, when I had the honour to accompany him upon the *Black Sea*; but as to the Flower of the preceding Species, I thought it fo very fine, that I made up great Nofegays of it to put in his Tent: but I was told by his Chiaia that this Flower caus'd Vapours and Dizzinefs. I thought he rally'd very pleafantly, for the Baffa complain'd of thofe Diftempers. The Chiaia gave me to underftand that he was in earneft, and affur'd me he had lately been inform'd by the Natives that this Flower was prejudi-cial to the Brain. Thofe good People, from a very antient Tradition, grounded perhaps upon feveral Obfervations, maintain alfo that the Honey which the Bees make after fucking that Flower, ftupifies thofe who eat of it, and caufes Loathings.

DIOSCORIDES mentions this Honey almoft in the fame Terms. *About* Heraclea Pontica, *fays he, in certain Seafons of the Year the Honey*

†   Z 2   *makes*

makes thoſe mad who eat of it ; and this certainly proceeds from the quality of the Flowers from which it is diſtill'd. They ſweat abundantly, but they are eas'd by giving them Rue, Salt-meats, and Metheglin, in proportion as they vomit. This Honey, adds the ſame Author, *is very acid, and cauſes Sneezing. It takes away Redneſs from the Face, if pounded with Coſtus. Mixed with Salt or Aloes, it diſperſes the black Spots that remain after Bruiſes : If Dogs or Swine ſwallow the Excrement of Perſons who have eaten of that Honey, they fall into the ſame Accidents.*

P L I N Y has diſtinguiſh'd the Hiſtory of the two Shrubs before mention'd better than either *Dioſcorides* or *Ariſtotle :* this latter imagined *that the Bees gather'd this Honey from the Box-trees ; that it depriv'd thoſe of their Senſes who eat of it, and were in health before ; and that on the contrary, it cured thoſe who were already mad.* Pliny ſpeaks of it thus : *In ſome Years,* ſays he, *the Honey is very dangerous about* Heraclea Pontica ; *Authors know not what Flowers the Bees extract it from. Here is what we have learnt of the Matter : There is a Plant in thoſe parts call'd Ægolethron, whoſe Flowers in a wet Spring acquire a very dangerous Quality when they fade. The Honey which the Bees make of them is more liquid than uſual, more heavy and redder ; its Smell cauſes Sneezing : Thoſe who have eaten of it, ſweat horribly, lie upon the Ground, and call for nothing but Coolers.* He then adds the ſame things that are ſpoken of by *Dioſcorides,* whoſe Words he ſeems to have only tranſlated : but beſides the Name of *Ægolethron* which is not in that Author, here follows an excellent Remark that we owe entirely to *Pliny.*

T H E R E *is found,* continued he, *upon the ſame Coaſt of the* Pontus *another ſort of Honey, which is call'd* Mœnomenon, *becauſe it makes thoſe mad that eat of it. 'Tis thought the Bees collect it from the Flower of the* Rhododendros, *which is frequent among the Foreſts. The People of thoſe parts, tho they pay the* Romans *a part of their Tribute in Wax, are very cautious how they offer them their Honey.*

I T H I N K one may from theſe Words of *Pliny* determine the Names of our two Species of *Chamærhododendros.* The firſt in all probability is the *Ægolethron* of that Author ; for the ſecond which produces the purple Flowers, comes much nearer to the *Rhododendros,* and may be call'd *Rhododendros Pontica Plinii,* to diſtinguiſh it from the *common Rhododendros,*

dendros, which is our *Rofe-Laurel*, known to *Pliny* by the Name of *Rho-dodaphne* and *Nerium*. It is certain the Rofe-Laurel grows not upon the Coafts of the *Pontus Euxinus*. That Plant loves warm Climates. You find few or none of them after paffing the *Dardanelles*, but it is very common by the fide of Streams in the Iflands of the *Archipelago*; fo that the *Rhododendros* of the *Pontus* cannot be our *Rofe-Laurel*. It is therefore very probable that the *Chamærhododendros* with purple Flowers is the *Rhododendros* of *Pliny*.

WHEN the Army of the ten thoufand came near to *Trebifond*, a very ftrange Accident befel it, which caus'd a great Confternation among the Troops, according to *Xenophon*, who was one of the principal Leaders of it. *As there were a great many Bee-hives*, fays that Author, *the Soldiers did not fpare the Honey: they were taken with a voiding upwards and downwards, attended with Deliriums ; fo that the leaft affected feem'd like Men drunk, and the others like mad Men, or People on the point of death. The Earth was ftrew'd with Bodies as after a Battel ; no body however died of it, and the Diftemper ceas'd the next day about the fame hour that it began ; fo that the Soldiers rofe the third and fourth days, but in the condition People are in after taking a ftrong Potion.*

DIODORUS SICULUS relates the fame Fact in the fame Circumftances. There is all the likelihood in the world that this Honey was fuck'd from the Flowers of fome of our Species of *Chamærhododendros*. All the Country about *Trebifond* is full of them; and Father *Lambert* a Theatin Miffionary agrees that the Honey which the Bees extract from a certain Shrub in *Colchis* or *Mengrelia*, is dangerous and caufes Vomitings. He calls this Shrub *Oleandro Giallo*, that is to fay, *yellow Rofe-Laurel*, which without difpute is our *Chamærhododendros Pontica maxima, Mefpili folio, flore luteo*. The Flower, fays that Father, *is in a medium between the Smell of Musk and that of yellow Wax*. To us the Smell feem'd to be like that of the Honey-Suckle, but incomparably ftronger.

THE ten thoufand were receiv'd at *Trebifond* with all the tokens of Kindnefs that Men ufually fhew to their Countrymen when they return from a far Country ; for *Diodorus Siculus* obferves, that *Trebifond* was a *Greek* City founded by thofe of *Sinope*, who defcended from the *Milefians*.

The

The fame Author tells us that the ten thoufand fojourn'd a Month in *Tre-bifond*, facrific'd there to *Jupiter* and *Hercules*, and celebrated Games.

TREBISOND in all probability fell into the hands of the *Romans*, when *Mithridates* found himfelf incapable of refifting them. It would be impertinent to relate in what manner it was taken under *Valerian* by the *Scythians*, known to us by the Name of *Tartars*, were it not that the Hiftorian who fpeaks of it, defcribes alfo the ftate of the Place. *Zozimus* obferves then that 'twas a great City, well peopled, fortify'd with a double Wall. The neighbouring Inhabitants were fled thither with their Wealth, as to a place where they fhould be fafe from all Dangers. Befides the common Garifon, ten thoufand additional Men were thrown into the Town; but thefe Soldiers fleeping upon truft, and fancying themfelves entirely fecure, were furpriz'd in the Night by the *Barbarians*, who having heap'd up Fafcines againft the Walls, got into the place by that means, flew a part of the Troops, demolifh'd the Temple and all the fineft Edifices; after which, laden with immenfe Riches, they carried away a great number of Captives.

THE *Greek* Emperors were Mafters of *Trebifond* in their turn. In the time of *John Comnenes* Emperor of *Conftantinople*, *Conftantine Gabras* had fet himfelf up there for a petty Tyrant. The Emperor would willingly have driven him from it, but the defire he had to take *Antioch* from the Chriftians, diverted him. Laftly, *Trebifond* was the Capital of a Dutchy or Principality in the Difpofal of the Emperors of *Conftantinople*; for *Alexis Comnenes*, firnam'd the *Great*, took poffeffion of it in 1204. with the Title of Duke, when the *French* and *Venetians* made themfelves Mafters of *Conftantinople*, under *Baldwin* Earl of *Flanders*.

THE diftance of *Conftantinople* from *Trebifond*, and the new Troubles that arofe to difturb the *Latins*, favour'd the Eftablifhment of *Comnenes*; but *Nicætas* obferves, that he was only allow'd the Title of Duke, and that *John Comnenes* was the Man that permitted the *Greeks* to call him *Emperor* of *Trebifond*, as if they thereby meant that 'twas *Comnenes* who was their true Emperor, fince *Michael Paleologus*, who made his Refidence at *Conftantinople*, had quitted the *Greek* Rite, to embrace that of *Rome*. It is very certain that *Vincent de Beauvais* calls *Alexis Comnenes* barely *Lord* of *Trebifond*. Be this as it will, the *Sovereignty* of this Town,

not

not to ufe the word *Empire,* began in the Year 1204. under *Alexis* Lett. V.
Comnenes, and ended in 146 . when *Mahomet* II. ftript *David Comnenes.*
That unfortunate Prince had married *Irene* Daughter of the Emperor *John
Cantacuzene* ; but he in vain implor'd the Affiftance of the Chriftians, to
fave the Wrecks of his Empire. He was forc'd to yield to the Conqueror,
who carried him to *Conftantinople* with all his Family, which was maffacred
fometime afterwards. *Phranzez* even fays, that *Comnenes* dy'd of a Blow
with the Fift which he receiv'd of the Sultan. Thus ended the Empire of
Trebifond, after having lafted above two Centuries and a half.

THE Town of *Trebifond* is built on the Sea-fide, at the foot of a
little Hill pretty fteep ; its Walls are almoft fquare, high, embattel'd, and
tho they are not of the firft Ages, yet it is very probable they ftand upon the Foundations of the antient Inclofure, which got this Town the
Name of *Trapezium.* Every one knows *Trapezion* in *Greek* fignifies *a Table*; and the Plan of this Town is a long Square, very much refembling
a Table. The Walls are not the fame as thofe defcrib'd by *Zozimus*; the
prefent are built of the Ruins of antient Edifices, as appears by old
pieces of Marble fet in feveral parts, and whofe Infcriptions are not legible, becaufe they are too high. The Town is big, and not well peopled: There are more Woods and Gardens in it than Houfes ; and thofe
Houfes that are there, tho well built, are but one Story high. The Caftle,
which is pretty large, but very much neglected, is fituated upon a flat
Rock that is commandable ; but its Ditches are very fine, being generally
cut in the Rock. The Infcription that is on the Gate of this Caftle, the
Arch whereof is a Semi-circle, fhews that *the Emperor* Juftinian *repair'd
the Edifices of the Town.* It is a wonder *Procopius* fhould not mention this,
when he fpends three whole Books in defcribing even the moft inconfiderable Buildings erected by that Prince in every corner of his Empire. That
Hiftorian barely tells us, that *Juftinian* built an Aqueduct at *Trebifond,* and
call'd it *the Aqueduct of St.* Eugenius *the Martyr.* To return to our
Infcription, the Characters of it are good and frefh ; but the Stone being fix'd in the Wall, and almoft a foot and a half deep beyond the reft,
there is no reading the laft line becaufe of the Shade. Here is what we
could read of it, after having to the beft of our power clear'd away the
Cobwebs with a Pole, round which we had wrapt a Handkerchief.

EN

ΕΝ ΩΝΟΜΑΤΙ ΤΟΥ ΔΕΣΠΟΤΟΥ ΗΜΩΝ ΙΗΣΟΥ ΧΡΙΣ-
ΤΟΥ ΘΕΟΥ ΗΜΩΝ ΑΥΤΩΚΡΑΤΟΡ ΚΑΙΣΑΡ ΦΑ
ΙΟΥΣΤΙΝΙΑΝΟC ΑΛΑΜΑΝΙΚΟC ΓΟΘΙΚΟC ΦΡΑΝΓΙΚΟC
ΓΕΡΜΑΝΙΚΟC ΠΑΡΤΙΚΟC ΑΛΑΝΙΚΟC ΟΥΑΝΔΑΛΙΚΟC.
ΑΦΡΙΚΟC ΕΥCΕΒΗC ΕΥΤΙΧΗC ΕΝΔΟΞΟC ΝΙΚΗΤΗC
ΠΡΟΠΕΟΥΧΟC ΑΕΙ CΕΒΑCΤΟC ΑΥΤΟΥC ΑΝΕΝΕΩCΕΝ
ΦΙΛΟΤΙΜΙΛ ΤΑΔΗΜΟC ΚΤΙCΜΑΤΑ ΤΗC ΠΟΛΕΟC
ΕΠΟΥΔΗΚΑ ΕΠΙΜΕΛΙΑ ΟΥΡΑΝΙΟΥ ΤΟΝ ΘΕΟΦΙΛΕΟ.....
ΧC ΥΠ Γ

IN the Veſtibulum of a Convent of *Greek* Nuns, there is a Chriſt very ill painted, with two Figures beſide him: we there read the following Words, painted in wretched Characters and corrupted *Greek.*

ΑΛΕΞΙΟC ΕΝ ΧΩ ΤΟ ΘΟΠΙϛΟC ΒΑCΙΛΕΥ ΚΕ ΑΥΤΟ-
ΚΡΑΤΟΡΩΚ ΠΑCΙC ΑΝΑΤΟΛΗC Ο ΜΕΓΑC ΚΟΝΜΗΝΟC
ΘΕΟΔΩΡΑ ΧΥ ΧΑΡΗΤΙ ΕΥCΕΒΕϛΑΤΗ ΔΕCΠΗΤΑ
ΚΕ ΑΥΤΟΚΡΑΤΟΡΗΚΑ ΠΑCΙC ΑΝΑΤΟΛΗC
ΗΡΙΝΗ ΧΥ ΜΗΤΗΡ ΑΕΤΟΥ ΕΥCΕΒΕϛΑΤΟΥ ΒΑCΙ-
ΛΕΟC ΚΥΡΙΟΥ ΑΛΕΞΙΟΥ ΤΟΥ ΜΕΓΑΛΟΥ ΚΟΜΝΗΝΟΥ.

ACCORDING to the Obſervations of the Gentlemen of the Academy Royal of Sciences, the Height of the Pole at *Trebiſond* is 40 *d.* 45 *m.* and the Longitude 63.

THE Port of *Trebiſond* call'd *Platana,* is to the Eaſt of the Town. The Emperor *Adrian* caus'd it to be repair'd, as we are inform'd by *Arrian.* It appears by the Medals of this Town, that the Port got it a very great Trade; *Goltzius* gives us two with the Head of *Apollo.* We know that God was adored in *Cappadocia,* whereof *Trebiſond* was not the leaſt City. On the Reverſe of one of thoſe Medals is an Anchor, and on the Reverſe of the other the Prow of a Ship: This Port is now proper for nothing but Saiques: The Mole which the *Genoeſe* are ſaid to have built there, is almoſt deſtroy'd, and the *Turks* give themſelves very little trouble about repairing ſuch Works. Perhaps what remains is the Ruins of *Adrian*'s Port; for according to *Arrian,* that Emperor had

made

TREBISONDE.

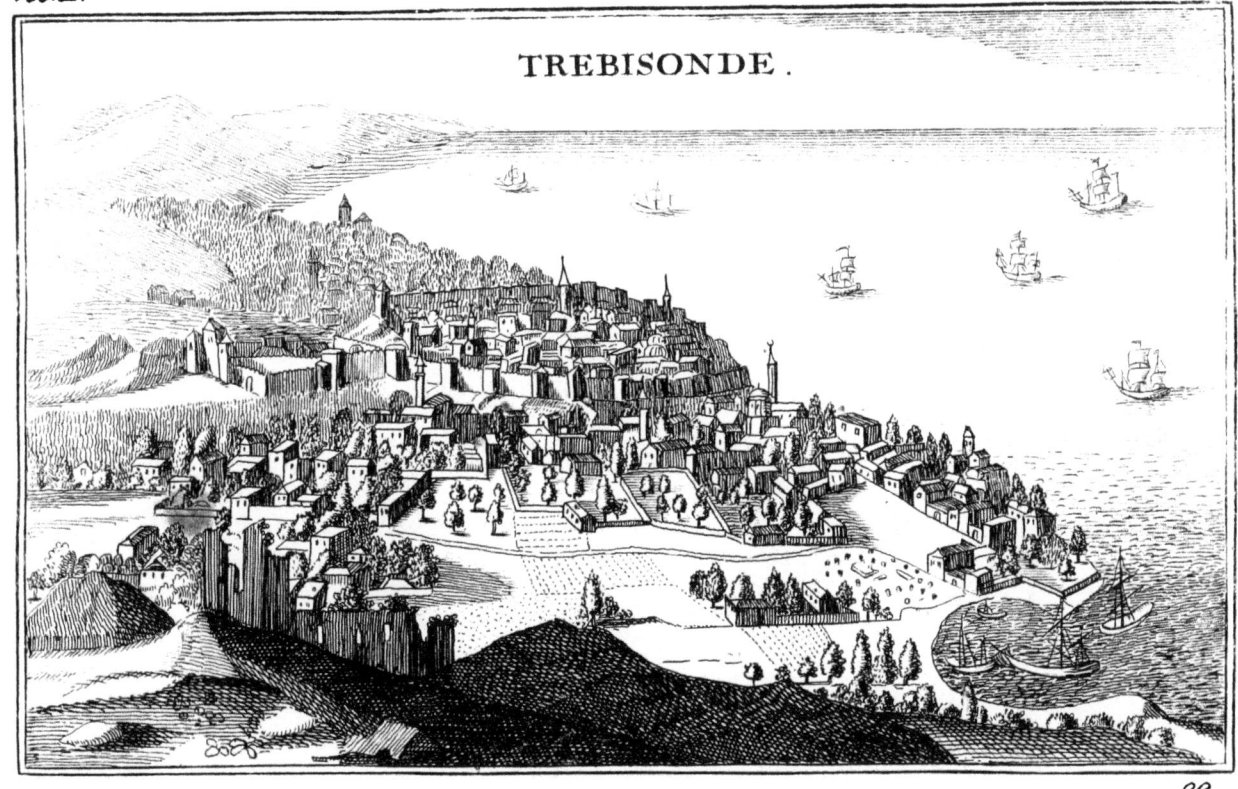

made a confiderable Jettee there, to defend the Ships which before could come to an Anchor there only at fome certain times of the Year, and even then too they lay upon the Sands.

WE fimpled the 24th and 25th of *May* about the Town. Here are very fine Plants. The 26th we went to fee the *Sancta Sophia*, an antient *Greek* Church, two Miles from the Town, near the Sea-fide. Part of this Building is turn'd into a Mofque, the reft is ruinate. We found but four Columns there, which were of an Afh-colour'd Marble. I know not whether this Church was built by *Juftinian*, as was that of *Sancta Sophia* at *Conftantinople*; 'tis indeed the Tradition of the Country, but they cannot prove it by any Infcription. *Procopius* himfelf does not mention it. The Ruins of this Church put me in mind of two great Men that this City has produc'd, *George* of *Trebifond*, and Cardinal *Beffarion*. 'Tis indeed confefs'd that *George* was only originally of *Trebifond*, but born in *Candia*. Be this as it will, he flourifh'd in the fifteenth Century, under the Pontificate of *Nicholas* V. to whom he was Secretary. *George* had before taught Rhetorick and Philofophy in *Rome*; but his Fondnefs for *Ariftotle*, bred mortal Quarrels between him and *Beffarion*, who never fwore but by *Plato*. *Beffarion* was a learned Man too, but his Embaffies hinder'd him too much: However, he wrote feveral Treatifes, and particularly collected a noble Library, which by his Will he bequeathed to the Senate of *Venice*. They preferve it with fo much Care, that they will communicate the Manufcripts to no Soul; fo that it is to be reckon'd a buried Treafure.

THO the Country of *Trebifond* is fruitful in fine Plants, yet it is not comparable in that point to thofe charming Hills whereon is built the great Convent of *St. John*, 20 Miles from the City to the South-Eaft. Finer Forefts are not among the *Alps*. The Mountains round this Convent produce Beech-trees, Oaks, Yoke-Elms, Guaiacs, Afh and Fir-trees of a prodigious Height. The Houfe of the Religious is built of nothing but Wood, clofe againft a very fteep Rock, at the bottom of the fineft Solitude in the World. The View of this Convent is bounded by nothing but the moft charming Profpects; and I could gladly here have fpent the reft of my days. Thofe that dwell here are a few folitary People, wholly employ'd about their Affairs temporal and fpiritual, without Cookery, Learning, Politenefs, or Books.

Vol. II. A a can

Who can live without all thefe ? They go up to theHoufe by a very rough kind of Stair-cafe, and of a very fingular Structure. It confifts of two Trunks of Afh, as big as the Mafts of a Ship, reclin'd againft the Wall, and plac'd upon the fame line like the Mounters of a Ladder; inftead of Steps or Rounds, they have only cut in them a few large Notches from fpace to fpace with Axes, and on each fide they have very wifely fet a Pole, to keep People from breaking their Necks; for without the Affiftance of thofe, I would defy the beft Rope-dancers in *Europe* to clamber up it. Our Heads fometimes turn'd as we came down it, and we had certainly tumbled Head-foremoft without thofe Supports. The very firft Inhabitants of the Earth could not make a plainer Ladder; the bare fight of it gives an Idea of the Infancy of the World. All the parts round this Convent are a perfect I-mage of mere Nature; a vaft number of Springs form a lovely Stream, full of excellent Trouts, and which runs thro verdant Meads and fhady Groves, that one would think muft infpire the nobleft Sentiments; but not one of thefe Monks is in the leaft affected with all this, tho there are about forty of them. We look'd upon their Houfe to be a fort of Cave, to which thefe good Folks are retir'd to avoid the Infults of the *Turks,* and to pray at their eafe. Thefe Hermits poffefs all the Country for above fix Miles about. They have feveral Farms among thefe Moun-tains, and a good many Houfes even in *Trebifond* : we lodg'd there in a large Convent that belong'd to them. What fignifies all this Wealth to thofe who muft not enjoy it? They dare not build a handfome Church or Convent, for fear the *Turks* fhould exact from them the Sums fet apart for thofe Structures when they are once begun.

AFTER having vifited the Country round the Convent, wherein are Plants that furnifh the moft agreeable Amufement in the World, we afcended to the higheft Places thereabouts, which were but very lately clear'd from the Snow, and from whence we could fee others ftill co-ver'd with it. The Natives give the Name of Πεύκ☺ to the common Firs, which differ in nothing from thofe that grow upon the *Alps* and *Pyrenean* Mountains; but they have retain'd the Name of Ελάτη for an-other fine Species of Fir which I never before had feen. Its Fruit, which is all fcaly, and in a manner cylindrical, tho a little more fwelling, is but two inches and a half long, and eight or nine lines thick, ending in a

point,

point, hanging downwards, confifting of Scales, foft, brown, fmall, rounded, which cover Seeds extremely little and oily.   The Trunk and Branches of this Tree are of the bignefs of thofe of the common *Picea.*
Its Leaves are but 4 or 5 lines long, they are fhining, deep green, firm,
ftiff, but half a line broad, with 4 little corners, and difpos'd like thofe
of our Firs, that is to fay, like a flatted Branch.

   W E were forc'd to quit this fine Country to go to *Trebifond* for our
Baggage.   We had very critical notice that the Baffa was juft gone, and
we found it no falfe alarm, for we met him upon the way.   We need not
fay we beftirr'd ourfelves to follow him : Woe had been to us, had we
loft fo rare an Opportunity.   We were forc'd to flave all night to get
our things pack'd up, and to provide Bifcuit and Rice, the things moft
neceffary here in a March, for Water is to be found eafily.   As good
Luck would have it, the Baffa encamp'd that day, the 2d of *June*, but
about 4 Hours Journey from the Town.   The next day we came up with
him with much ado, and found him fourteen Miles off of his firft Camp.

I am, My L o r d, *&c.*

LETTER VI.

To Monſeigneur the Count de Pontchartrain, *Secretary of State,* &c.

MY LORD,

Journey to Armenia and Georgia, **T**HE Towns of this Country are very well govern'd, and you hear of no Thieves in any of them ; they all keep to the Country, and plague none but Travellers; and 'tis pretended too that they are leſs cruel than our Highwaymen. For my part, I believe the contrary, and that a Man who ſhould expoſe himſelf alone, upon a great Road here, would ſoon be at his Journey's end. If theſe Rogues murder no body, 'tis for want of Opportunity, for People always travel a good many together. Theſe Companies, which they call *Caravans,* are Meetings or Aſſemblies of Travellers, more or leſs numerous, in proportion to the Danger. Every Man is arm'd his own way, and upon occaſion defends himſelf as well as he can. When the Caravans are conſiderable, they have a Leader that directs their Marches. The Center is leſs expos'd than the Rear ; and 'tis not always the wiſeſt courſe to ſtay for the moſt numerous Caravans, as moſt Travellers imagine ; the beſt way is to catch at thoſe wherein there are moſt *Turks* and *Franks,* that is to ſay, People fit to defend themſelves. The *Greeks* and *Armenians* have no ſtomach for fighting, and ſo are often made to pay off Scores (as they call it there) for the Blood of a Thief they never kill'd. Travellers are not expos'd to theſe Misfortunes in *America* ; thoſe *Indians* whom we look upon as Savages, thoſe *Iroquois,* whoſe very Name is a Bugbear to Children, kill none but thoſe with whom they are at war.

If they eat Chriftians, they do it not in time of Peace. I don't think Lett. VI.
'tis lefs cruel to ftab a Man to get his Purfe, than 'tis to kill him to
eat. What matter is't to the Wretch whether he is eaten or ftript, after
his Death?

PEOPLE therefore are forc'd to go in Caravans in the *Levant*,
the Robbers do the fame, that they may be able to make themfelves
Lords of the others by Club-Law. We join'd the Caravan of the Baffa
of *Erzeron* on the 3d of *June*, a day's Journey from *Trebifond*, and by the
way we met with I know not how many Merchants coming from the
neighbouring Provinces to improve fo favourable an Opportunity. The
Thieves fled from us with more diligence than they follow'd other Caravans, becaufe when a Baffa is in march, fo many Robbers taken, fo
many Heads off in an inftant: They do them this honour after having
call'd them *Jaours*, that is to fay, *Infidels*. Befides that we were very
much at eafe as to that Article, we were alfo overjoy'd at the Baffa's
travelling but twelve or fifteen Miles a day, which allow'd us full time to
view the Country as much as we pleas'd.

OUR Caravan confifted of above fix hundred People, but not above
three hundred of them belong'd to the Baffa; the reft were Merchants
and Paffengers: this made a very good Shew. 'Twas a Novelty to us
to fee Horfes and Mules mingled with a great number of Camels. The
Women were in Litters terminating like a Cradle, the Top cover'd with
Oil-cloth; the reft was lattic'd on all fides more carefully than the Par-
lours of the aufiereft Nuns. Some of thofe Litters look'd like Cages
plac'd on the back of a Horfe, and they were cover'd with a painted
Cloth, which was fupported by Hoops; a Stranger could not eafily
have guefs'd whether they had Apes in 'em or reafonable Creatures.

THE *Chiaia* was the firft Officer of the Houfhold. We have among
us no Place anfwerable to this; for he is above a Steward, and, as it
were, the Subftitute of his Mafter: Nay, often he is his Mafter's Mafter.
The *Divan Effendi*, or *Head of the Council*, was the fecond Officer. The
Baffa had his *Cotja* or *Chaplain*, whom they alfo call *Mouphti*, feveral
Secretaries, threefcore and ten *Boffinois* for his Guard, a vaft Number of
Chaoux, Muficians or Players upon Inftruments, a terrible Rout of Foot-
men

men or *Chiodars,* without reckoning Pages. His Phyſician was of *Bur-gundy,* and his Apothecary of *Provence*: In what part of the World are there not *Frenchmen?*

THE *Chaoux Bachi,* or Chief of the Chaoux, march'd a day's Journey before, bearing a Horſe's Tail, to mark out the *Conac,* that is to ſay, the Place where the Baſſa was to encamp. The Maſter Chaoux receiv'd Orders about it every night, like our Quarter-maſters. He was attended by a good many Officers to prepare the Camp, and *Arabians* to ſet up the Tents. All theſe march'd on horſeback with Lances and Staves tipt with Iron. The Baſſa's Muſick was diſagreeable in nothing but their repeating conſtantly the ſame Tune, as if they had never learnt above one Leſſon. Tho their Inſtruments were different from ours, yet they began to grow familiar to our Ears. One day the Baſſa did me the honour to ask me *how I lik'd his Muſick*; I anſwer'd, *It was excellent, but a little too uniform*: he reply'd, *That in Uniformity conſiſted the Beauty of every thing.* Tis true, Uniformity is one of the Chief of that Nobleman's Virtues, for he ſeems to be of the moſt unchangeable Temper in the world. The firſt Chamade uſually began an hour before our March; this was to call every body up. The ſecond was beat about half an hour afterwards, which was the Signal for filing off. The third began at the Moving of the Baſſa, who always kept in the Rear of the Caravan, at about 4 or 500 Paces diſtance. The Muſick ſtruck up or ceas'd during the March, according to the Caprice of the Muſicians, who redoubled their Conſort when we arriv'd at the Conac, where before the Baſſa's Tent they ſtuck up the two other Horſe-Tails that had been us'd in the March. The Chaoux Bachi having receiv'd his Orders, took the third Tail, and went his ways to mark out the next day's Camp.

WE were ſoon broke to this Regimen. We roſe at the firſt Chamade, and mounted our Horſes at the ſecond; the Baſſa's Officers drove the People away like ſo many Sheep, crying *Aideder, Aideder,* that is to ſay, *March, March.* They will allow no body whatſoever to mingle with the Houſhold, and he that ſhould be ſurpriz'd among them, would expoſe himſelf to a few Baſtinades. The *Turks* are Men of Order in every thing they do, and eſpecially in their Marches. The *Catergis* or Carriers roſe an Hour before the Signal, and every thing was laden be-

fore

fore notice was beat for the March. I often admir'd their Exactnefs; all was done in filence, and commonly we had not fo much as known that they were loading, but for the Lights that fhone about the Camp.

THIS Day, the 4th of *June*, we pafs'd along very high Mountains, ftill advancing towards the South-Eaft. We did not take the fhorteft Cut to *Erzeron*, the Baffa's Defign being to follow the moft convenient and the eveneft Road he could find; moft of the Merchants were out of humour at this, but we were extremely glad of it, knowing we fhould fee more of the Country, and that a fafer Caravan could never be wifh'd for. We obferv'd this Day the fame Plants that we had feen about *Trebifond*; but what gave us moft pleafure was, that we knew by the March of the Caravan that we fhould have time enough in confcience to find out Plants, both upon the Road and upon the neighbouring Hills. For this purpofe, in the Morning we got to the Head of the Caravan, and each of us taking a Bag, detach'd ourfelves fome Paces from it, now to the right, now to the left, to gather what we could find. The Merchants laught heartily at feeing us mount and remount every moment, only to pick a few Herbs, which they defpis'd becaufe they knew nothing of them. Sometimes we led our Horfes by the Bridles ourfelves, and fometimes gave 'em to our Carriers, that we might get in our Harveft more at eafe. At the next Lodging we defcribed our Plants while our Meat was in our Mouths, and M. *Aubriet* drew all he could.

I FEAR, my Lord, the detail of our March by days Journeys will be tedious, but 'twill not be unferviceable to Geography and the knowledge of the Country. I am even fatisfy'd that this long Relation will be much lefs unpleafant to you than to others, becaufe you know how to make fuch good ufe of the minuteft Circumftances that you have an account of. Men more fkilful than me may alfo perhaps improve by this Journal; a Mountain, a great Plain, a narrow Pafs, a River often help to determine the Places in which the greateft Actions formerly happen'd.

THE 5th of *June* we travell'd from four in the Morning till Noon acrofs great Mountains cover'd with Oaks, Beech-trees, common Firs, and others with very fmall Fruit, the like to which we had feen in the

Mountains of the Monaſtery of *St. John* of *Trebiſond.* We obſerv'd beſides the *common Yoke-Elm,* another Species much ſmaller in all its parts.
Its Leaves are but an inch long, and its Fruits are very ſhort. This
Yoke-Elm has ſeeded in the King's Garden, and is not alter'd.
The Sorts of *Chamærhododendros,* both with purple, and yellow Flowers,
frequently appear'd by the ſide of Streams. We encamp'd that day in
a Plain which was cover'd with Snow, and had as yet produc'd nothing
at all. Tho theſe Mountains are lower than the *Alps* and *Pyrenees,* they
are full as backward, for the Snow here melts not till the end of *Auguſt.*
Among many rare Plants, we obſerv'd a fine Species of *Crow-foot,* with
great Cluſters of White Flowers.

ITS Leaves are 3 or 4 inches broad, by their Slaſhes reſembling *Wolf's-
Bane,* bright green, ſleek, neatly vein'd, ſtrew'd with Hairs about the
Rims, and beneath ſuſtain'd by a Pedicule 4 or 5 inches long, pale-green,
hairy, 2 lines thick, pretty round, fiſtulous, 4 lines broad at the Baſis, where
it is hollow gutterwiſe. The Stalk is about a foot high, hollow alſo,
pale-green and hairy, about 2 lines thick, quite bare except towards the
top, where it ſupports a Cluſter of 7 or 8 Flowers, ſurrounded with 4
or 5 Leaves, no more than two inches or two and a half long, and one
inch broad, flaſh'd into three principal parts, and re-flaſh'd again almoſt
like the other Leaves. Tho the Cluſter is pretty cloſe, each Flower is
neverthelefs ſuſtain'd by a Pedicule about 15 lines long. The Flowers
are two inches diameter, conſiſting of 5 or 6 white Leaves, an inch long,
and 8 or 9 lines broad, rounded at their point, but pointed at their firſt
Growth. In the middle of theſe Leaves is a Piſtile or Button with ſeveral Seeds, terminated by a crooked Thred, and cover'd with a Tuft of
white Stamina half an inch long, laden with Apices greeniſh-yellow a
line long. Theſe Flowers are without Cup, have no Smell, no Acridity,
any more than the reſt of the Plant. Upon ſome Stocks the Flowers
have a touch of the Purple. We had not time enough to pull up the
Root of it.

THE 6th of *June* we ſet out at three in the Morning, and till Noon
croſs'd over great Mountains quite bald, which afforded very diſagreeable Proſpects, for we could ſee neither Tree nor Shrub, but only a ſorry Down blaſted by the Snow, which was but new melted. There
was

Ranunculus Orientalis, Aconiti lycoctoni folio flore
magno albo Coroll. Inst. Rei herb. 20.

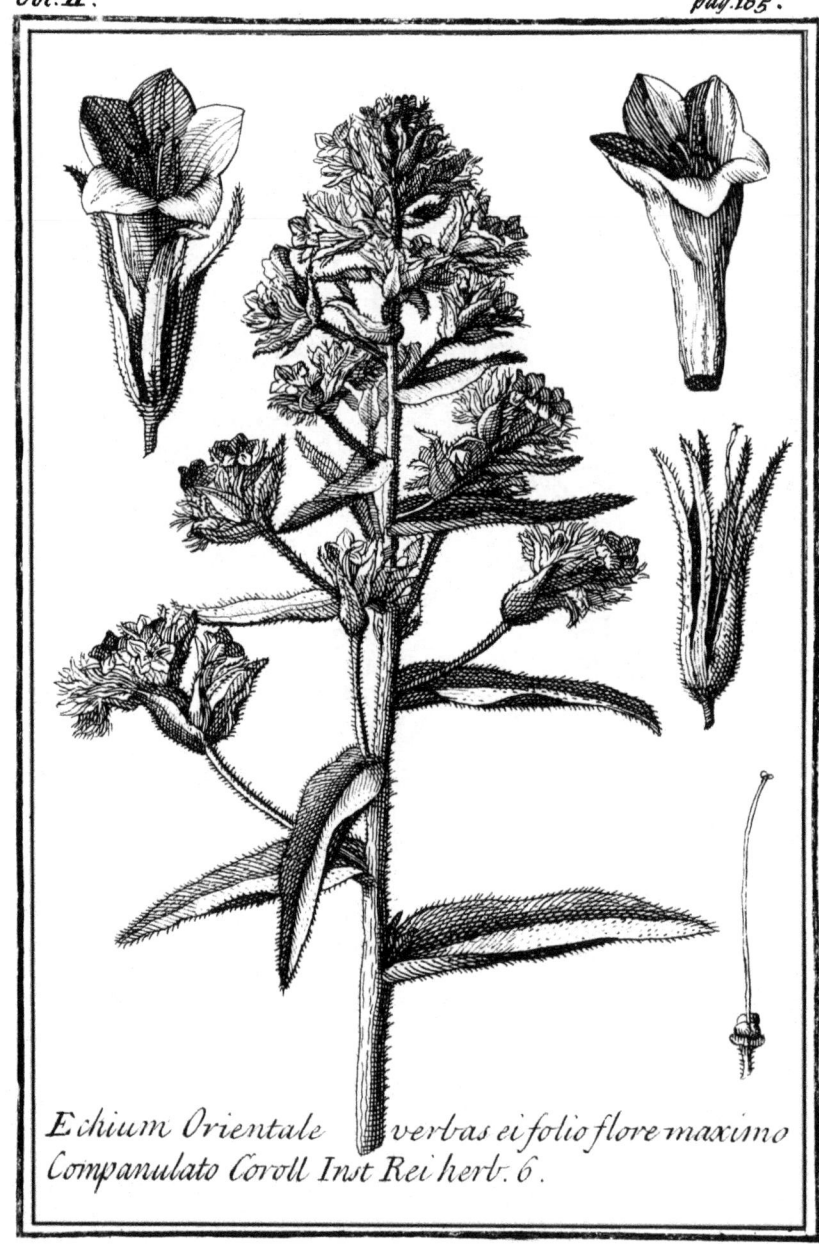

Echium Orientale verbas ei folio flore maximo Companulato Coroll Inst Rei herb. 6.

was a great deal of it alſo in the Bottoms, and we encamp'd cloſe to it. This Down was cover'd in ſome parts with that fine Species of *Violet with great Flowers*, yellow upon ſome Stocks, and deep Violet Colour upon others, and diverſify'd with yellow and violet upon ſome few, yellow ray'd with brown with the Standard Violet, and of a very agreeable Smell.

WE roſe about two of the Clock the 7th of *June*, and ſet out at three: we continued our Journey over bald Mountains among the Snow. The Cold was very ſharp, and the Fogs ſo thick, that we could not ſee one another at four Paces diſtance. We encamp'd about half an hour after nine in a Valley tolerably agreeable for Verdure, but very incommodious for Travellers. Not a Stick of Wood to be found, nor ſo much as a ſcrap of Cow-dung; and as we were pretty ſharp ſet, 'twas a diſmal Mortification to us to be unable to dreſs ſome Lambs that we had laid in, only for want of a Bruſh or two. The Bàſſa's Family liv'd that day upon nothing but Comfits. We diſcover'd nothing new. All the Down was cover'd with the ſame Violets: thus we ſpent the Day very mournfully; neither did the *Turks* reliſh this Faſt any more than us. On the 8th of *June* by Break of day we began to perceive that we were really in the *Levant*. From *Trebiſond* hither the Country look'd like the *Alpes* and *Pyrenees*; but now the Face of the Earth ſeem'd of a ſudden alter'd, as if a Curtain had been drawn, and a new Proſpect open'd to our view. We deſcended into little Valleys cover'd with Verdure, intermix'd with charming Streams, and full of ſo many fine Plants, ſo different from what we had been us'd to, that we knew not which to fall on firſt. About ten in the morning we arriv'd at *Grezi*, a Village which we were told is not above a day's Journey from the *Black Sea*; but the way is practicable only for People on foot. I was ſo ſtruck with a kind of *Echium* or Viper's-Bugloſs that I found in the Roads, that I cannot help giving a Deſcription of it here.

ITS Root is above a foot long, and two inches thick, accompanied with great whitiſh Fibres within, mucilaginous, ſoftiſh, cover'd with a brown Bark, and chapt. The Stalk, which is about three foot high, is as big as a Man's Thumb, pale-green, hard, ſolid, and full of Pulp, viſcous, and as it were ſlimy. The Under Leaves are fifteen or sixteen inches long, and four or five broad, pointed, whitish-green, soft, sweet, hairy, as it were sattiny a top, cottony beneath, heightened with a great Rib, which furnishes a Nervure pretty like that of the Leaves of the *Wolwort*: these Leaves diminish confiderably along the Stalk, where they are not above half a foot long, less cottony than the first, but much more pointed. From their Bosoms rife Branches about half a foot long, briftling with pretty ftiff Hairs like the top of the Stalk, accompanied with Leaves about an inch and half long. All these Branches are divided into little Slips, twin'd up like a Scorpion's Tail, laden with bigger Flowers than any hitherto obferv'd upon the Species of this kind. Each Flower is an inch and half high, towards the bottom 'tis a Pipe four or five lines diameter, and juft perceptibly crooked, which afterwards dilates it self in manner of a Bell, the Mouth whereof is divided into five equal parts, cut like a *Gothick* Arch. This Flower is pale-blue, approaching a little to Pearl-colour, but three of its Cuts are ftreak'd lengthways with two Stripes of deep Red upon a Ground of very bright Purple. From the inner Rims of the Pipe grow five white Stamina, crooked like a Hook, each laden with a yellow Summit. The Cup is almoft as long as the Flower, and flash'd in five parts almoft to the bottom, each of which parts is but about two lines broad, pointed, pale-green, roughen'd with very thick Hairs. The Piftile rifes from the bottom of this Cup, form'd by four Embryo's rounded and greenish from the middle, whereof grows a Thred almoft as long as the Flower, flightly hair'd, purple and forked. The Seeds, tho very backward, were pretty like thofe of a Viper. The Flower has no Smell: The Leaves have a graffy Tafte agreeable enough.

THE 9th of *June* we fet out at three in the Morning, and pafs'd thro Valleys very dry and very open. About nine we encamp'd beneath *Baibout* in the Plain, by the fide of a little River. *Baibout* is a fmall Town, very ftrong by its Situation upon a very fteep Rock. 'Twas reported that the Baffa would fojourn there five or fix Days, to hold a Seffions, and Prifoners were brought from various Parts; fo that we fpent the reft of the Day in running about to look for Plants: but we were deceiv'd, for we were forc'd to be gone a day afterwards, without having time to go up to the Town. Perhaps we might have found there

some

Onobrychis Orientalis fru   tescens, Spinosa,
Tragacanthæ facie Coroll. *Inst. Rei herb. 2 6*

fome Remains of Antiquity, or Infcriptions that might have inform'd us of its antient Name. By its Situation it feems to be fet down in our Maps by the Name of *Leontopolis* and *Juftinianopolis*, which was call'd *Byzane* or *Bazane*. We were as much furpriz'd as vex'd at hearing the Chamade, which gave us notice that we muft mount to be gone. Here is one of the fineft Plants that grows about *Baibout*, and which contributed not a little to comfort us for our hafty Departure.

'T I S a Bufh no more than a foot high, but ftretch'd in circumference to two or three feet, tufty, and extremely like the *Tragacantha*. Its Stalks towards the bottom are as thick as a Man's Thumb, white within, cover'd with a blackifh Bark, chapt, crooked higher up, divided into feveral Branches, bare, and divided into old Slips thorny and dry. The Summits of thefe Slips fupport young Sprigs crooked and branchy, ended in Pricks, pale-green, garnifh'd with Leaves rang'd upon a Stalk nine or ten lines long, whereon are ufually two or three pair of Leaves, oppofite to each other, 4 or 5 lines long, and lefs than one line broad, pointed at each end, a little folded gutter-wife. The Stalk ends in a Leaf of the fame nature. The top of the Prickles fuftains one or two Flowers, leguminous, purple, ray'd with a hairy Standard, rifing up about nine lines long, and three broad, hollow'd, and even indented. The Wings and the Under-Leaves are paler and fmaller. The Piftile comes to be a Fruit like that of our *Fenugreek*; but it is fleek, and we faw it not ripe. The Cup is reddifh, two lines long, flafh'd into five points. The Leaves have a graffy Tafte a little tartifh.

W E were oblig'd then to leave *Baibout* the 11th of *June*. We were told the Baffa had pardoned all the Prifoners. Many in our Caravan commended his Clemency ; others blamed him for not making fome Examples. The Rogues were made to pafs in review ; and if one may judge by their Looks, moft of them feem'd at leaft to deferve the Wheel. This day we gave a name to one of the fineft Plants in the whole *Levant* ; and becaufe M. *Gundelfcheimer* difcover'd it firft, we agreed that in Juftice it ought to bear his Name. By ill fortune we had nothing but Water to celebrate the Feaft; but this agreed the better with this Ceremony, for the Plant grows no where but in dry and ftony places. The Baffa's Mufick ftruck up juft at the inftant, which we took for a

good

good Omen : yet we were a long while before we could find a *Latin* Name equivalent to that gallant Man's. We concluded at laſt that the Plant ſhould be call'd *Gundelia.*

THE Stalk of the Plant is a foot high, five or ſix lines thick, ſleek, bright-green, reddiſh in ſome parts, hard, firm, branchy, accompanied with Leaves pretty like thoſe of the thorny *Acanthus,* ſlaſh'd almoſt to the Rib, and re-ſlaſh'd into ſeveral points, garniſh'd with very ſtrong Prickles. The biggeſt of theſe Prickles is half a foot, or eight inches broad, and about a foot long. The Rib is purple, the Nervure hairy, whitiſh, emboſs'd, cottony, the Ground of the Leaves bright-green, their Conſiſtence hard and firm ; they diminiſh to the end of the Branches, which ſometimes are cover'd with a little Down. All theſe parts ſuſtain Tops like thoſe of the *Fuller's Thiſtle,* two inches and a half long, and one and a half diameter, ſurrounded at their Baſis with a Row of Leaves of the ſame Figure and Tiſſure as the bottom, but only two inches long. Each top conſiſts of ſeveral Scales ſeven or eight lines long, hollow and prickly, among which are enchas'd the Embryos of the Fruit ; they are about five lines long, pale-green, pointed at bottom about four lines thick, ſet off with four Corners hollow'd at their Summities into five holes or beazles with notch'd rims, from each whereof riſes a Flower of one ſingle piece, half an inch long. It is a Pipe whitiſh or bright Purple, opening to a line and a half diameter, cleav'd into five points of a dingy Purple, which inſtead of widening like the broad end of a Funnel, rather come nearer and nearer to each other ; the inſide of the Flower is of a more agreeable Purple. From its ſides run off five Threds or Pillars, which ſupport a yellowiſh Sheath, ray'd with Purple, ſurmounted by a Thred yellow and duſty. Which ſhews that theſe Flowers are truly Fleurons that bear each upon a young Seed incloſ'd in the Embryos of the Fruit ; and theſe Embryos are divided into as many Boxes or Apartments as there are Fleurons. Moſt of theſe Embroy's prove abortive except the middlemoſt, which preſſing the others makes them periſh. All the Plant yields a very ſweet Milk, which clots into Grains of Maſtick like that of the *Carline* of *Columna.* The *Gundelia* varies, there are ſome Stocks which have hairy Heads, and Flowers of a deep red Colour.

W E

Gundelia Orientalis,   *Acanthi aculeati*
folio, capite glabro   *Coroll. Inst. Rei herb. 15*

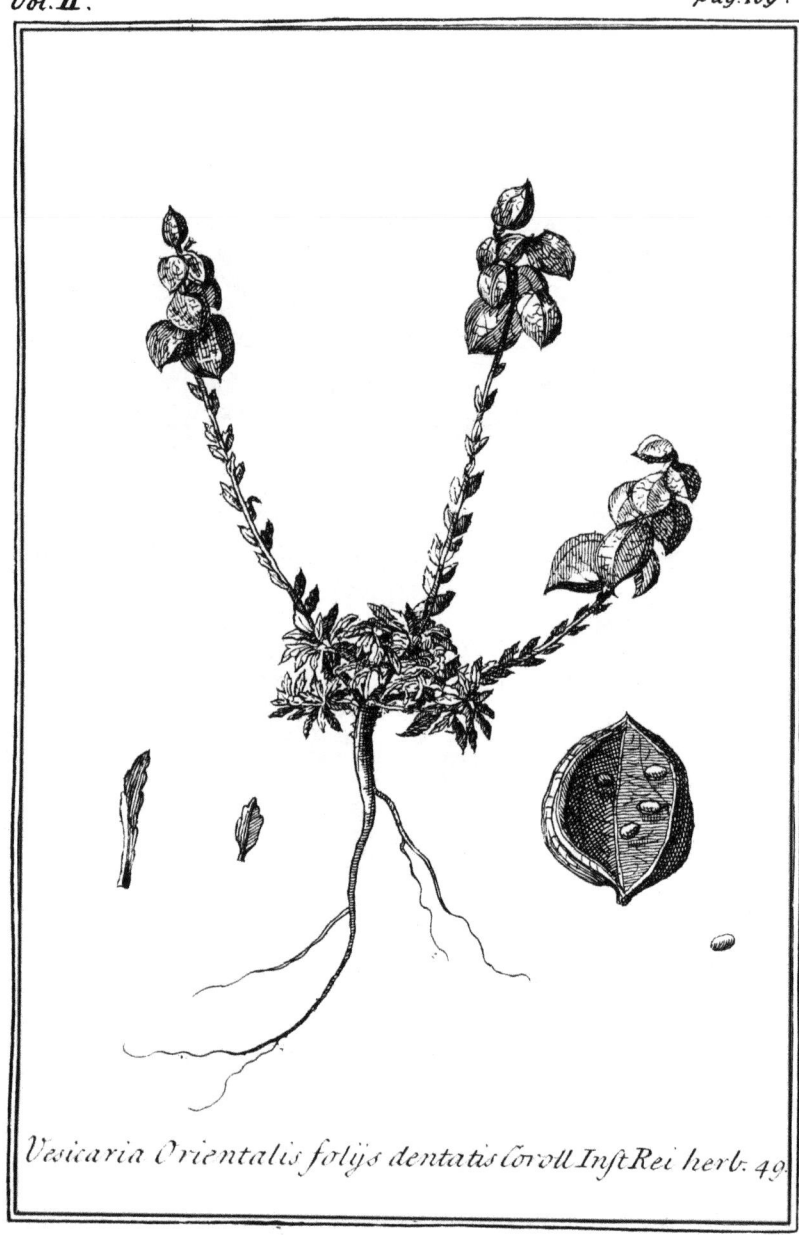

Vesicaria Orientalis folijs dentatis Coroll Inst Rei herb: 49.

W E set out this day about eight in the Morning, and travell'd thro nar-Lett. VI.
row Valleys uncultivated, bare of Trees, capable of infpiring nothing
but Melancholy. We encamp'd about Noon, and had no other Pleafure than that of determining another kind of Plant which we called *Ve-*
ficaria, becaufe of its Fruit. 'Tis a Bladder an inch long, and almoft
as broad, membranous, pale-green, travers'd lengthways by four Strings.
of a purplifh Colour, which by their Re-union form a little point at the
end of the Bladder, and by the way diftribute Veffels interlac'd like
Hurdles. This Fruit inclofes fome oval Seeds about a line and a half
long, each faftned by a String extremely fmall, which comes from the great
purple String. Moft of thefe Seeds were as yet either green or abortive This Fruit is nothing more than the Piftile of the Flower puff'd up
like a Bladder. The Flower confifts of four yellow Leaves placed like a
Nofegay, fuftain'd by a Stalk without Branches. The whole Plant is
but about four inches high, without reckoning the Root, which is two
inches long, reddifh, three or four lines thick at the Neck, divided
into fome Fibres a little hairy. It puts forth feveral Heads garnifh'd
with Leaves difpos'd in a Circle, often prefs'd downwards nine or ten
lines long, commonly one line broad, bright-green, neatly indented a-
bout the Rims almoft like thofe of *Buck-horn Plantane*. Thofe that are
along the Stalks are but about three or four lines long, and two broad,
and have very little Indenture. They diminifh to the top of the Stalk,
which is quite plain, and without Branches. If the Root of this Plant
were flefhy, it would be of the fame Genus as the *Leontopetalon*.

T H E 12th of *June* we fet out at three, and arriv'd at *Conac* by fix
in the Morning: What a pleafure was it to Men who languifh'd for nothing but Plants, to have a whole day before them to fearch after them?
We travell'd but three Miles in the aforefaid March of three hours, and
kept all along in the fame Valley, thro which winds a River that you are
oblig'd to crofs feven or eight times. The next day we fatigued ourfelves no more than the former, for the Caravan travell'd only from half
an hour after two till feven; and kept upon a very high Mountain, whereon are many of that kind of Pines which grow at *Tarare* near *Ly-*
ons. There is alfo upon this we are fpeaking of, a beautiful Species of
Cedar that fmells as ill as our *Sabin-tree*, and whofe Leaves perfectly

refemble

refemble thefe latter; but then 'tis a great Tree, and as big and high as our largeft Cypreffes. They made us be moving this day, I know not out of what whim, at eleven at night; and we arriv'd the 14th of *June* about feven in the morning, at a Village call'd *Iekmanfour*. The Moon fhone fo bright, that it invited the *Turks*, who had done nothing but fnore the live-long day, to profecute their Journey : But how could we fimple by Moon-light? We however omitted not to fill our Bags, our Merchants laughing all the while to fee us three groping about in a Country dry and burnt up in appearance, but notwithftanding enrich'd with very fine Plants. When it was Morning, we review'd our Harveft, and found ourfelves rich enough. Can any thing be more charming than an *Aftragalus*, two foot high, laden with Flowers quite from the bottom to the top of the Stalks?

THEIR Flowers are as thick as a Man's little finger, gutter'd, firm, folid, pale-green, cover'd with a white Down, garnifh'd with Leaves faftned to a Stalk a Span long, pale-green alfo and hairy, accompanied with two Wings at its Bafis, one inch long, and two or three lines broad, ending in a point. The Leaves are moft of them rang'd in pairs along this Stalk, which generally has 13 or 14 pair upon it. The biggeft, which are towards the Wings, are an inch long, and feven or eight lines broad, almoft oval, but a little narrower towards the top, deep-green, fleek, cover'd at top with white Hairs, and commonly folded gutterwife. They diminifh to the end of the Stalk, where they are but five or fix lines long. The Stock is branchy from the bottom, but from the Junctures of the Leave-ftalks it puts forth only Pedicules about two or three inches long, each laden with five or fix Flowers, difpers'd longways, and fuftain'd by a Tail two lines long, which rifes from the juncture of a Leaf pretty fmall, very thin, and extremely hairy. All thefe Flowers are yellow, fifteen lines long, with a thick Standard, which is hollow'd, almoft oval, feven or eight lines broad. The Wings and the Underleaf are much fmaller. The Cup is eight lines long, pale-green, membranous, about five lines broad, ftrew'd with white Hairs, and cut in five very fmall points. The Piftile is a Pyramidal Button, two lines thick, white and hairy, ending in a Thred of a dingy white, wrapt in a membranous Sheath, white, fring'd into Stamina with purple Summits. The

Piftile

Piftile comes to be a Fruit an inch long, eight or nine lines thick, terminating in a point four or five lines long. This Fruit is rounded behind, flat,
and ridgy on the other fide, cottony, divided into two Apartments, the
Partitions whereof are flefhy, three lines thick while the Fruit is yet
green. In each Apartment you find a Row of five or fix Seeds fhap'd
like little Kidneys, each faftned by a String. Thefe Seeds, when they
are ripe, are brown, as is alfo the Fruit. The whole Plant has an ill
Smell. It has rais'd Seed in the Royal Garden, where it thrives well, notwithftanding the Diftance and Difference of the Climates.

WE this day, for the firft time, difcover'd a very beautiful Species
of *Clary*, whereof I had only feen the Abortions fome Years before in
the Garden of *Leyden*. M. *Hermans*, Profeffor of Botanicks in the Univerfity of that Place, a very skilful Man, and who had obferv'd fuch
fine Plants in the *Eaft-Indies*, has given the Figure of this we are fpeaking of. *Rauvolfius*, Phyfician of *Ausbourg*, feems to have mention'd it
in his *Voyage into the* Levant, under the name of a *fine Species of Clary,
with narrow Leaves, hairy and deeply flafh'd.*

THE Root of this Plant is fharp at bottom, a foot long, the neck
of the Root twice as thick as a Man's Thumb, white within, cover'd
with a Bark of an Orange-red, or Saffron-colour. The Nerve of this Root
is hard and white, the Fibres are pretty large, and extend on the fides.
It puts forth one or two Sprigs a foot and a half high, towards the bottom as big as a Man's little Finger, purple, cover'd with a thick white
Down, accompanied with Leaves of a delightful Beauty, eight or nine
inches long, flafh'd almoft quite to the Rib in parts two or three inches
long, and half an inch broad, full of large Knobs all fhagreen'd and whitifh green. The Rib and Nervure are as it were tranfparent; this Rib
is two inches broad in its beginning, purple in fome parts, laden with a
very white Down, like the bottom of the Leaves. Thofe that grow afterwards are as long, and embrace a part of the Stalk by two rounded
Wings, but they diminifh in length towards the middle of the Stalk, where
they are two inches broad. Afterwards the Stalks are full of Branches
rounded and tufty, accompanied with Leaves about an inch long, cut
as it were into a *Gothick* Arch, the point whereof is very fharp; thefe
Leaves are not bunchy, but only vein'd and hairy. The Flowers grow

*

in

in rings, and by ftages along the Branches difpos'd in a plain row: nay, fometimes there is but one or two Flowers at each Verticillum. The Flower is about an inch long, a line and a half thick at the bottom, white, o-pening into two Lips, the uppermoft whereof is crooked like a Sickle, two lines thick, ftrew'd with very fhort Hairs, colour'd with a little caft of Orange, almoft imperceptible, hollow'd and rounded; the under lip is much fhorter, divided into three parts, whereof the middlemoft, which is the biggeft, is Orange-yellow, the other two are white and rifing like Ears. The Stamina are of the fame Colour, and interlac'd like the Di-vifions of the *Os Hyoides.* The Piftile confifts of four Embryo's fur-mounted by a Hair violet-colour'd, and forked at its Point, which winding about in the Sickle, juts out three or four lines. The Cup is half an inch long, ray'd, pale-green, hairy, parted into two Lips, one of which has three points pretty fhort, and the other only two, but much longer. The top of the Stalks is a little gluey, and fmells ill. The Root of this Plant is bitter. The Leaves have a graffy Tafte, and fmell rammifh like the *common Clary.*

ERUDITION, my Lord, muft be confefs'd to be of great help in lengthening out a Letter. The Country we are now in, would allow ve-ry large Scope to a Man more learned than me. How many great Ar-mies muft have pafs'd this way? Perhaps *Lucullus, Pompey,* and *Mithri-dates* would ftill know the Remains of their Camps. In fhort, we are in the *Great Armenia* or *Turcomania.* The *Romans* and *Perfians* protect-ed the Kings of it at different times. The *Saracens* poffefs'd it in their turn. Some believe that *Selim* added it to his Conquefts, after his Re-turn from *Perfia,* where he had won that famous Battel againft the great *Sophi Ifmael. Sanfovin* agrees that in *Selim's* time, who dy'd in 1520, there was one King of the *Greater,* and another of the *Leffer Armenia,* call'd *Aladoli. Selim* caus'd King *Aladoli's* Head to be cut off and fent to *Venice,* as a Mark of the Victory he had gain'd in the *Levant.* It is very like the *Turks* feiz'd the *Greater Armenia* at the fame time, that they might be able to go to *Perfia* all thro their own Dominions, with-out trufting the neighbouring Princes. Be this as it will, *Armenia* fell under the Dominion of the *Turks;* for the *Turkifh Annals* cited by *Calvi-fius,* tell us that *Selim* Son of *Selim* conquer'd *Armenia* in 1522.

ON

O N the 14th of *June* we were made to fet out two Hours after Midnight ; and we march'd till feven thro fruitful Meadows, fowed with all
manner of Grain.   We encamp'd clofe to the Bridge of *Elija,* upon one
of the Arms of the *Euphrates,* fix Miles from the City of *Arzeron* or *Arzerum,* which others call *Erzeron,* tho *Arzerum* is the true Name of it,
as I fhall fhew hereafter.   *Elija* is only a pitiful Village, the Houfes are
built of Mud, and moft of them entirely ruinate, and fallen down ; but
the Bath near the Village is what recommends this Place.   The *Turks*
call it *the Bath of Arzerum.*   The Building is pretty neat, octogonal,
vaulted, and pierc'd at top.   The Bafon, which is of the fame Figure,
that is to fay, confifting of eight fides, throws out two Gufhes of Water, almoft as thick as a Man's Body : this Water is frefh, and very tolerable for Heat ; and i' faith the *Turks* never let it ftand idle : they come
quite from *Erzeron* to bathe in it, and half our Caravan did not let flip
fo rare an Opportunity.

Next day we arriv'd at *Erzeron* .   'Tis a pretty large Town, five days ' Erzeron.
Journey from the *Black Sea,* and ten from the Frontiers of *Perfia.* *Erzeron* is built in a lovely Plain, at the foot of a chain of Mountains that
hinder the *Euphrates* from falling into the *Black Sea,* and oblige it to wind
to the South.   The Hills that edge this Plain were ftill cover'd with
Snow in many places : Nay, we were told that it had fallen the firft of
June, and we were very much furpriz'd to find our Hands fo numb'd
that we could not write at Day-break ; this Numbnefs continued an hour
after Sun-rife, tho the Nights were pretty gentle, and the Heats even troublefome from ten in the Morning to four in the Afternoon.   The Plain
of *Erzeron* is fruitful in all kinds of Grain.   The Wheat was lefs forward
than at *Paris,* not yet two foot high, fo that their Harveft is not till
September.   No wonder *Lucullus* fhould think it ftrange that the Fields
were quite bare in the middle of Summer, when he was juft come from
Italy, where they get in their Crop by that time.   He was yet more
furpriz'd to fee Ice in the Autumnal Equinox ; to hear that the Waters
by their extreme Coldnefs kill'd the Horfes in his Army ; that there was
no paffing the Rivers without breaking the Ice, and that his Soldiers
were forc'd to encamp among the Snow, which kept inceffantly falling.
Alexander Severus was no better pleas'd with this Country. *Zonaras* ob-

ferves, that his Army, in returning thro *Armenia*, was fo maul'd by the exceffive Cold, that they were oblig'd to cut off the Hands and Feet of feveral of the Soldiers, who were found half frozen on the Roads.

BESIDES the fharpnefs of the Winters, what makes *Erzeron* very unpleafant, is, the fcarcity and dearnefs of Wood. Nothing but Pine-wood is known there, and that too they fetch two or three days Jour-ney from the Town ; all the reft of the Country is quite naked. You fee neither Tree nor Bufh ; and their common Fuel is Cow's Dung, which they make into Turfs ; but they are not comparable to thofe our Tanners ufe at *Paris*, much lefs to thofe prepar'd in *Provence* of the Husks of the O-live. I don't doubt better Fuel might be found, for the Country is not want-ing in Minerals ; but the People are us'd to their Cow-dung, and will not give themfelves the trouble to dig for it. 'Tis almoft inconceivable what a horrid Perfume this Dung makes in the Houfes, which can be compar'd to nothing but Fox-holes, efpecially the Country-houfes. Every thing they eat has a touch of this Vapour ; their Cream would be admirable but for this Pulvillio ; and one might eat very well among them, if they had Wood for the dreffing their Butcher's Meat, which is very good.

THE Fruits brought hither from *Georgia* are excellent. That Coun-try is warmer and lefs backward, and produces in abundance Pears, Plums, Cherries, Melons. The neighbouring Hills furnifh *Erzeron* with very fine Springs, which not only water their Fields, but the very Streets of the Town. 'Tis very well for Strangers that their Water is good, for their Wine is the moft abominable ftuff that ever was touch'd. 'Twould be fome Comfort for all their Ice and all their Snow, and one might make a fhift to bear with their Stinks, if their Wine were tole-rable ; but it is ftinking, mouldy, tart, and fmells rotten : *Vin de Brie* would be reckon'd Nectar here. Their Brandy is no better ; it is mufty and bitter, and more than all this, it cofts no fmall Pains and Money too, before even thefe filthy Beverages can be got. The *Turks* affect more Severity here than any where elfe, and take mighty delight in furprizing and baftinading thofe that carry on fuch Trades : in my mind they are not much to blame, for 'tis very good fervice to the Publick, to hinder the Sale of fuch unwholefome Drugs.

THE

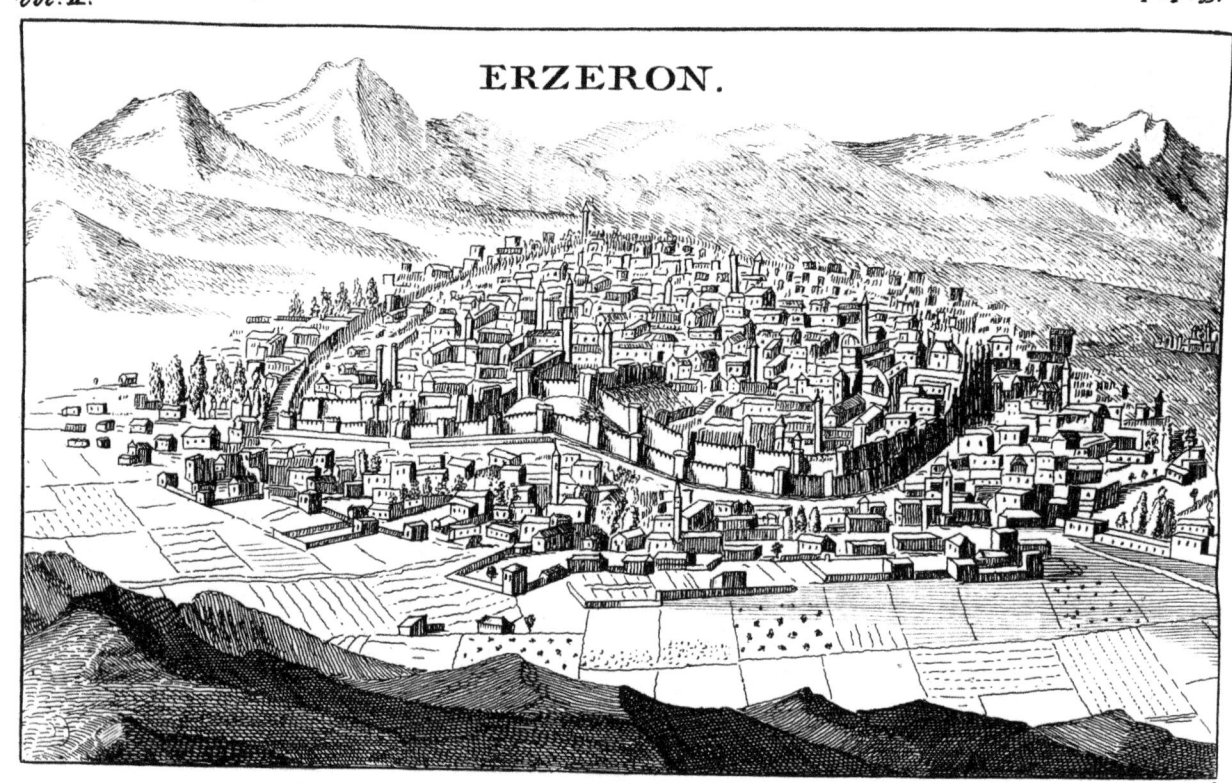

ERZERON.

A Profpect of ERZERON the Capital of Armenia.

106.

THE Town of *Erzeron* is better than that of *Trebifond :* the Inclofure of this firſt is of double Walls, defended by ſquare or pentagonal
Towers; but the Ditches are neither deep nor well kept up. The Beglerbey or Baſſa of the Province lives in an old Seraglio very ill built.
The Janizary-Aga dwells in a kind of Fort, in the higheſt part of the
Town. When the Baſſa or the moſt conſiderable Perſons of the Country go into this Fort, 'tis to leave their Heads behind them. The Janizary ſends them a Summons to attend there, by order of the Grand Signior: the Capigi arriv'd from Court ſhews them his Orders, and then
executes them without further Ceremony. 'Tis thought there are eighteen thouſand *Turks* in *Erzeron,* ſix thouſand *Armenians,* and four hundred *Greeks.* They reckon ſixty thouſand *Armenians* in the Province,
and ten thouſand *Greeks.* The *Turks* who are in *Erzeron* are almoſt all of
them Janizaries: they reckon about twelve thouſand there, and above fifty
thouſand in the reſt of the Province. They are moſtly Trades-people,
and are ſo far from receiving Pay, that the Majority of them give Money to the Aga, which purchaſes them the Privilege of being good for
nothing, and of committing all kind of Inſolences. The beſt ſort of People are forc'd to liſt themſelves in this Body; becauſe, beſides that
elſe they would not be welcome to the Governour, who is almoſt abſolute in the Town, they would be daily expos'd to the Violences of their
Neighbours, and not be able to obtain any Juſtice from the Officers.
The Grand Signior gives the true Janizaries of the Country but from five
to twenty Aſpers a day: the Aga pockets good part of this Money.

THE *Armenians* have a Biſhop and two Churches in *Erzeron.* They
have ſome Monaſteries in the Country, as the *Great Convent* and the
Red Convent. They all acknowledge the Patriarch of *Erivan.* As to
the *Greeks,* they have their Biſhop too in the Town, but they have only
one Church, and that a wretched poor one. They are moſtly Tinkers,
and inhabit the Suburb, where they work at making Utenſils of the Copper that is brought from the neighbouring Mountains. Theſe poor People make a dreadful Clattering night and day, for they are conſtantly at
their Forge; and the *Turks* are too fond of their Tranquillity, to ſuffer
the Anvil to be beat within the Town. Beſides theſe Utenſils, which
are carried into *Turky,* *Perſia,* and *Mogul* itſelf, they drive a great Trade

of

of Furs, and especially of those of *Jardava* or *Zerdava*, which are the Skins of a kind of Marten, pretty common in this Country. The deeper-colour'd the Skin is, the more it is valued: they make the most precious Furs only of the Tails, because they are blackish; and this is what makes them so dear, for a great many Tails go to the Lining of one Vest. They also bring to *Erzeron* abundance of Gall-Nuts, five or six days Journey from the Town, and they preserve the Oaks with great care, by the Bassa's order; the Wood besides would be too dear, if 'twere carry'd thither for burning.

THIS Town is the Thorow-fare and Resting-place for all the Merchandizes of the *Indies*, especially when the *Arabs* are upon the watch round *Aleppo* and *Bagdad*. These Merchandizes, the chief whereof are the Silk of *Persia*, Cotton, Drugs, painted Cloths, only pass through this Country. Very few of them are sold here by retail; and they would let a sick Man die for want of a Dram of Rhubarb, tho there were ever so many intire Bales of it. They sell nothing but the *Caviar*, which is a most odious Dish. 'Tis a common Proverb here, that if a Breakfast were to be presented to the Devil, he should be treated with Coffee without Sugar, Caviar, and Tobacco; I should add a Glass or two of *Erzeron*-Wine to the Bill of Fare. Caviar is only the Spawn of Sturgeon salted, which is prepared about the *Caspian* Sea. This Meat burns the Mouth with its high Seasoning, and poisons the Nose with its nasty Smell. The other Merchandizes before mention'd are carry'd to *Trebisond*, where they are shipt for *Constantinople*. We were surprized to see arrive at *Erzeron* so great a quantity of *Madder*, which they call *Boia*: it comes from *Persia*, and is used in the dying of Cloth and Leather. Rhubarb is brought hither from *Usbeq* in *Tartary*. The Worm-seed comes from *Mogul*. There are some Caravan-Masters, that from Father to Son meddle with nothing but carrying of Drugs, and that would think they degenerated from their Ancestors, if they troubled their heads about other Goods.

THE Government of *Erzeron* yields three hundred Purses yearly to the Bassa, whom we shall henceforth call the Beglerbey or Viceroy of the Province, to distinguish him from the other Bassas of the Country who are subject to him. Each Purse is 500 Crowns, as in all the other parts of *Turkey*; so that these 300 Purses amount to 150000 Crowns. They

arise,

arife, *Firft*, from the Merchandizes that come into the Province, or are Lett. VI.
carry'd out of it ; moft pay Three *per Cent.* and fometimes twice as
much. Great Duties are exacted upon the Species of Gold and Silver.
The *Perfian* Silk *Chorbafi*, which is the fineft, and the *Ardachi*, which is
the coarfeft, pay 80 Crowns every Camel-Load, which is from 800 to
1000 weight. *Secondly*, The Beglerbey difpofes of all Offices in the Cities
of the Province; thefe Offices are farm'd out according to the Cuftom of
the Country, and go to the higheft Bidder, as every where elfe. *Thirdly*,
Excepting the *Turks*, all that go out of the Province for *Perfia*, are obliged to pay in *Erzeron* at leaft five Crowns, tho they have no Merchandizes; which is a kind of Capitation conftantly impofed upon them.
Thofe that carry with them Gold and Silver only for the Expence of their
Journey, pay Five *per Cent.* for the Sum they export.

OUR Beglerbey at his arrival abolifh'd moft of thefe Duties, as thinking them tyrannical; perhaps his Succeffor has reftored or increafed them
fince his departure. Befides thefe Taxes, before the arrival of *Cuperli*,
they exacted the common Capitation of all Strangers, of what Nation
foever, when they enter'd *Erzeron*; and this Capitation was regulated
according to the Eftimation the *Turks* made of each Perfon. This Man,
quo' they, muft pay ten Crowns for his good Mien : this other having but
few things with him, fhall pay but five. Thus they fleeced poor Strangers with impunity, and the Miffionaries were worfe ufed than any of the
reft: that they might not be bit, the firft thing they did, was to uncover
the Heads of Paffengers, to fee whether they were fhaved or no ; fo that
thefe Apoftolical Men, bound for far Countries, were often obliged to
let their Caravan go without them, in hopes of getting fome Abatement,
or to ftay for fome great *Frank* or *Armenian* Merchant, that fhould be fo
charitable as to pay the Mony for them. There's no getting Juftice on
the Frontiers of fo great an Empire, when the Governours encourage
Extortion ; and the reafon why they encourage it, is becaufe they get
by it. When one fets out from *Conftantinople* for *Perfia*, the beft Precaution he can take, is, not only to obtain a Commandment from the *Porte*,
but alfo Letters of Recommendation from our Embaffador to the Beg-
lerbeys of the Frontiers through which he is to pafs. The *Italian* Religious are too cautious, to fail putting themfelves into our Embaffador's

Pro-

Protection. The King of *France* is much better known and efteem'd by the *Muffulmans*, than the Holy Father, whom they call barely the Mufti of *Rome*.

THE Miffionaries are very great Gainers by the death of *Fafullah-Effendi*, Mufti of *Conftantinople*, who was dragg'd through the ftreets of *Adrianople* in the laft Reign. 'Twas faid he had a fhare in all the Extortions that were made in the Province of *Erzeron*, of which he was Native, and where he had immenfe Poffeffions. That infatiable Man, who was abfolute Mafter of the Emperor *Muftapha*, was a declared Enemy of all the Religious, and efpecially of the Jefuits. They did not fail to inquire whether we were not Papas, that is, Priefts; but they did this only for form-fake: for befides that the Beglerbey honour'd us with his Protection, it is very certain we were not fhaved.

THE Province of *Erzeron* yields in Mony above 600 Purfes to the Grand Signior. Befides the 300 Purfes of the *Carach*, exacted from the *Armenians* and *Greeks*, he has alfo Six *per Cent.* Cuftom out of the Merchandizes. So that in the whole, thefe Merchandizes pay Nine *per Cent.* to wit, fix to the Grand Signior, and three to the Beglerbey. The Grand Signior alfo enjoys the Duty of *Beldargi* or Land-Tax, paid out of the Poffeffions of the Spahies.

THE Town of *Erzeron* is not upon the *Euphrates*, as the Geographers place it; but ftands rather in a Peninfula, form'd by the Sources of that famous River. The firft of thefe Sources runs a day's Journey diftant from the City, and the other a day and a half or two days Journey. The Sources of the *Euphrates* are Eaftward in Mountains not fo high indeed as the *Alps*, but cover'd with Snow almoft the whole Year round. Thus the Plain of *Erzeron* is inclofed between two beautiful Streams, that form the *Euphrates*. The firft flows from Eaft to South, and paffing along behind the Mountains at whofe foot the Town is fituated, runs Southward to a little Borough call'd *Mommacotum*. The other Stream, after having for fome time verged to the North, a little like that *des Gobelins*, goes through the Bridge of *Elijah*, and thence flowing towards the Weft along the Road of *Tocat*, is obliged by the Difpofition of the Ground to turn towards the South at *Mommacotum*, where it joins the other Branch, which is much more confiderable. Thefe two Branches are call'd *Frat*,

the

the Name of the River which they form. After their Junction, which is three days Journey from *Erzeron*, the *Frat* begins to be capable of carrying little Saicks, but its Channel is full of Rocks, and it is impoſſible to ſettle a Paſſage by water from *Erzeron* to *Aleppo*, without making this River navigable. The *Turks* leave the World as they find it, and the Merchants make the beſt ſhift they are able. Yet it would be a much ſhorter and ſafer way to go by water, for the Caravans are 35 days in travelling from *Erzeron* to *Aleppo*, and the Road is very dangerous by reaſon of the Thieves, who rob the Merchants at the very Gates of Towns.

THE Night-Robbers are ſometimes more prejudicial than the Day-ones. If good Watch is not kept in the Tents, they come privately and ſoftly, while the Folks are aſleep, and pull out Bales of Goods with hooks, without being perceiv'd by any body: if the Bales are faſten'd or laced together with Cords, they are ſeldom without a good Razor to cut them. Sometimes they empty them at a few paces diſtance from the Tents, but if they find them fill'd with Musk, they carry them clear off, and leave nothing but the Shell of the Bale. When the Caravans ſet out before Day-break, which they do generally, the Rogues mingle with the Drivers, and turn out of the way a few Mules laden with Goods, which they eaſily carry off in the dark. They ſeldom chuſe the worſt; for they know the Bales of Silks every whit as well as the Owners. Caravans ſet out every week from *Erzeron* for *Gangel, Teflis, Tauris, Trebiſond, Tocat*, and *Aleppo*. The *Curdes*, or People of *Curdiſtan*, who are ſaid to be deſcended from the antient *Chaldeans*, keep the field about *Erzeron*, till ſuch time as the great Snows oblige them to retire, and are conſtantly upon the catch for an opportunity to plunder theſe poor Caravaneers. Theſe are ſome of thoſe wandering *Jaſides*, that in reality have no Religion at all, but by Tradition believe in *Jaſid* or *Jeſus*; and are in ſuch fear of the Devil, that they pay him reſpect leſt he ſhould do them miſchief. Theſe Wretches ſtretch every year quite from *Mouſoul* or *New Niniveh* to the Sources of the *Euphrates*. They own no Maſter, and the *Turks* never puniſh them, even when they are taken up for Murder or Robbery; they only make them redeem their Lives with a Sum of Mony, and the whole matter is made up at the Coſt of the Perſons robb'd. Nay, it often happens that a Caravan ſhall enter into a Treaty with the

*

Thieves

Thieves who attack them, efpecially when they are out-number'd, or the Rogues put on murdering Faces; and then they come off fafe for a Sum of Money, and this is the beft thing they can do. Every Man muft live by his Trade : and it is much better in my mind to fhed the Blood of one's Purfe, than of one's Veins. Sometimes it ftands them not in above two or three Crowns a head. Befides, the Thieves too love ready Money better than any thing elfe; for not knowing readily where to find Chaps for their Goods, they often are no better than Incumbrances to them. At prefent all the Caravans of the *Levant* pafs by *Erzeron,* even thofe bound for the *Eaft-Indies*; becaufe the Roads of *Aleppo* and *Bagdad,* tho fhorter, are poffefs'd by the *Arabs,* who are revolted from the *Turks,* and have made themfelves Mafters of the Country.

THE 19th of *June* we fet out about Noon to vifit the Mountains to the Eaft of the Town. The Snow was fcarce melted upon them; and at fix we encamp'd fifteen Miles off, in fo backward a Country, that the Plants did but juft begin to peep out, and the Hills were only cover'd with a flight Turf: it is hard to account for the Lazinefs of this Climate. We lay under our Tents in a Valley in the middle of a Hamlet, where the Cottages ftood further diftant from one another than the Baftides of *Marfeilles.* The Water in which we had put our Plants to preferve them, in order to defcribe them next day, was frozen in the night two lines thick, tho 'twas under fhelter in a wooden Bowl. The next day, the 20th, after having fimpled to no great purpofe, becaufe of the Cold, which fuffer'd not the Earth to bring forth, we refolv'd to draw back towards *Erzeron* a different way from that we came. We therefore went to fee an antient Monaftery of *Armenians,* which is but one day's Journey from that Town, and which bears the Name of *St. Gregory.* The whole Country is bare, not the leaft Bramble to be feen any where about. This Monaftery is pretty rich, but I would as foon live at the foot of Mount *Caucafus,* for 'tis hardly poffible it fhould be colder. I believe that befides the foffile Salt, which is not fcarce hereabouts, the Earth is full of Sal Ammoniack, which keeps the Snows for ten Months upon Hills a little refembling Mount *Valerian.* It has been found by divers Experiments, that Sal Ammoniack makes the Liquors it is diffolv'd in extremely cold, and that rather by its fixed faline part, than by

✦

its

its volatile part, as appears by the Solution of the Caput Mortuum from Lett. VI.
which the Spirit and oily aromatick volatile Salt has been drawn; for
you feel a very confiderable Cold in the middle of Summer, if you lay
your Hand upon a Glafs Retort, whereon a Solution of that Caput Mortuum has been made.

THAT Night we lay at another Monaftery of *Armenians,* call'd
the *Red Monaftery,* becaufe the Dome, which is fhap'd like a dark Lanthorn, is befmear'd with Red: I cannot think of a truer Comparifon for it
than a dark Lanthorn; for this Dome ends in a Point or purfled Cone,
like an Umbrello half open. This Convent is but three Hours Journey
from *Erzeron*; and the Bifhop, who is reckoned the moft learned of all the
Armenians, makes his refidence in it: this Character of him muft not be
reckoned any thing prodigious, for Learning is no very frequent Commodity in *Armenia*; but as we were inform'd that he was much efteem'd a-
mong the *Curdes,* who according to their Cuftom were encamp'd about
the Sources of the *Euphrates,* we omitted nothing that might engage him
to go thither with us. 'Tis impoffible to be too cautious before one ven-
tures into thofe parts, for a *Curd* is a very obftreperous fort of an Animal; they are as unmanageable to the *Turks* as to any body elfe, and
will ftrip them to the Skin without any Ceremony, when they can get
an Opportunity. In fhort, thefe Highwaymen obey neither Beglerbey
nor Baffa; and you muft have recourfe to their Friends, when you would
obtain the honour of feeing them, or rather the Country where they make
their Abode. When they have eat up the Pafturage of one Country, they
remove into another. Inftead of applying their Heads to Aftronomy like
the *Chaldeans,* from whom they are ufually deriv'd, they ftudy nothing
but how they may rifle Paffengers, and follow the Caravans by the
Scent; while their Wives are employ'd in making Butter and Cheefe, bringing up their Children, and tending their Flocks.

WE fet out the 22d of *June* at three in the Morning from the Red
Monaftery. Our Caravan was not very numerous; we muft truft wholly
and folely to the Bifhop, or not think of feeing the Sources of the *Euphrates*: but after all, what did we venture? the *Curdes* do not eat Men,
they only ftrip them, and we had wifely provided againft that, by putting
on our worft Cloaths. Hunger and Cold therefore were all we had to

apprehend. As to the Bishop, he was an honest sort of a Man, and would never have expos'd us to shew our Nudities. We begg'd him to put into his own Box a few Sequins, that we had taken to bear our Expences. Thus secure of our Purse, he made provision of whatever we should have occasion for, and really seem'd to act with Sincerity, knowing full well that we were under the Beglerbey's Protection, and that we were publickly look'd upon in the Town as his Physicians. We had given Prescriptions gratis to all that belong'd to the Monastery; so that after all these Precautions, we boldly gave ourselves up to his Conduct. He put himself at the head of the Company, perfectly well mounted, as were also three of his Servants; and he order'd very good Horses for us too, and our Attendants. After half an hour's riding, we took up a venerable old Man of his Acquaintance at a pretty Village situated on that Branch of the *Euphrates*, which goes to *Elija*. They treated us with some Trouts which they caught on the instant; and nothing can be more delicious than these Fish when they are eat immediately upon being taken out of the Stream, and boil'd in Water into which you have thrown a handful of Salt. This old Man paid us abundance of Civilities, and after having made us promise to cure a Friend of his at our Return, (the old condition) he gave us to understand that he was a good Master of the Language of the *Curdes*, that he had some Friends in the Mountains to which we were going, and that we need fear nothing, being in company with the Bishop and him. We entred some fine Vallies, wherein the *Euphrates* serpentizes among wonderful Plants; and we were charm'd with finding here that beautiful Species of *Pimpernel with red Flowers*, which is one of the chief Ornaments of the Gardens of *Paris*, and which a long while ago was brought from *Canada* into *France*. What gave us most pleasure was, that the Plants were pretty forward, and we hop'd to find them in good condition in the Mountains; but as we went higher, we found nothing but Moss and Snow. The Forests are banish'd from them to the end of the world; yet the Country is agreeable, and the Streams which fall on all hands, make a pleasing Prospect. There are I know not how many Springs on the top of these Mountains; some flow directly down, others gush into little Basons edg'd with Turf. We chose one of the prettiest Green-swerds to spread our

Cloth,

Cloth upon, in order to refresh our selves with some of the Monastery Wine, which was better than all the Wine in *Erzeron*. Here we wash'd away the Terror, which the dreadful Name of *Curdes* had notwithstanding all our Care struck upon our Spirits; and dipt out Cup-fulls of Water from the Sources of the *Euphrates*, whose excessive Coldness was temperated by the Heat of our Nectar.

THERE was but one thing disturb'd our innocent Delights, which was, that every now and then certain Deputies from the *Curdes* rode up to us with their Lance in Rest, to reconnoitre what sort of Folks we were. I know not whether Fear or Wine did not make us see two instead of one; for in proportion as Fear laid hold of us, we run to our Cordial for Assistance. If it is ever allowable to drink more than ordinary, 'tis upon such an occasion as this; for had we not done it, the Water of the *Euphrates* had effectually frozen up our Senses. At length, as we thought the Deputation visibly grew more numerous, the Bishop and the old Man went forward some few Paces, beckoning to us to stay where we were. We were very glad to be excus'd from paying our compliments to the Embassadors. After the first Ceremonies, which did not last long, they all together mov'd toward us, and began to argue very gravely about I know not what Business. As People in fear always imagine themselves to be the Subject of Discourse, and besides as the *Curdes* honour'd us from time to time with their Looks, we also affected abundance of Gravity; and not doubting but the Bishop would let them know we sought for Plants, we pick'd up such as lay near us, and seem'd to talk about them, tho really we were speaking of the blessed Condition we had brought ourselves into; still jabbering in paltry Latin, for fear our Interpreters, who were us'd to our Dialect, should understand any thing we said.

THE Conference between the Bishop and the *Curdes* seem'd to us abominably tedious. 'Twas a great way from thence to the Monastery to go in one's Shirt; and who knows but these People, who are us'd to making of Eunuchs, might have taken it into their heads to have metamorphos'd us in the same manner, that we might have sold to more advantage? We were a little heartned, when our *Armenian* Druggerman came and told us the *Curdes* had made the Bishop a Present of a Cheese.

At the same time the old Man came and took a Flaggon of Brandy, which he gave them in return. We caus'd our People to ask him what they were doing: he answer'd smiling, that the *Curdes* were sad Fellows, but that we need fear nothing; for that the antient Friendship which was between them, and the Veneration they had for the Bishop, would secure us from all Dangers. And indeed after they had drank up the Brandy, they went their ways, and the Bishop return'd to us with a very pleasant Countenance. We did not fail to return him thanks for all the care he had been pleas'd to take to defend us from the Insults of those devouring Wolves, and then continued to make our Observations upon the Plants. There are very fine ones about these Sources. Their Concourse makes that Branch of the *Euphrates*, which we had almost constantly kept by the side of from the Monastery, and which runs to *Elija*. You may catch Trouts in it with your Hand, and we liv'd nobly upon them that day; but they were grown so soft the next day, that we would not touch them. Thus far we were very well satisfy'd with our Journey. We ask'd the Bishop if 'twould not be possible to go see the other Branch of the *Euphrates*, which joins the former at *Mommacotum*. He told us laughing that he did not know the *Curdes* of those parts, and that we should see nothing but Springs like those we just now come from. We very humbly thank'd him; but he had no occasion to throw us into new Apprehensions.

THIS good Man, out of the abundance of his Civility, as we afterwards judg'd, would needs go and take his leave of the *Curdes*, and distribute the rest of our Brandy amongst them: we should have approv'd very much of this, had not we been to go along with him, and venture among their Pavilions. They are great Tents of a kind of deep-brown Cloth, very thick and very coarse, which serves for a Cover to these portable Houses; the Compass whereof, which is the Body of the House, is a long Square inclos'd by Cane-Lattices of the height of a Man, lin'd within with good Mats. When they remove, they fold up their Houses like a Skreen, and lade it with their Implements and their Children upon Oxen and Cows. These Children are almost naked in the coldest Season; they drink nothing but Water half frozen, or Milk boil'd in the Smoke of Cow's Dung, which they save very carefully; for without

that,

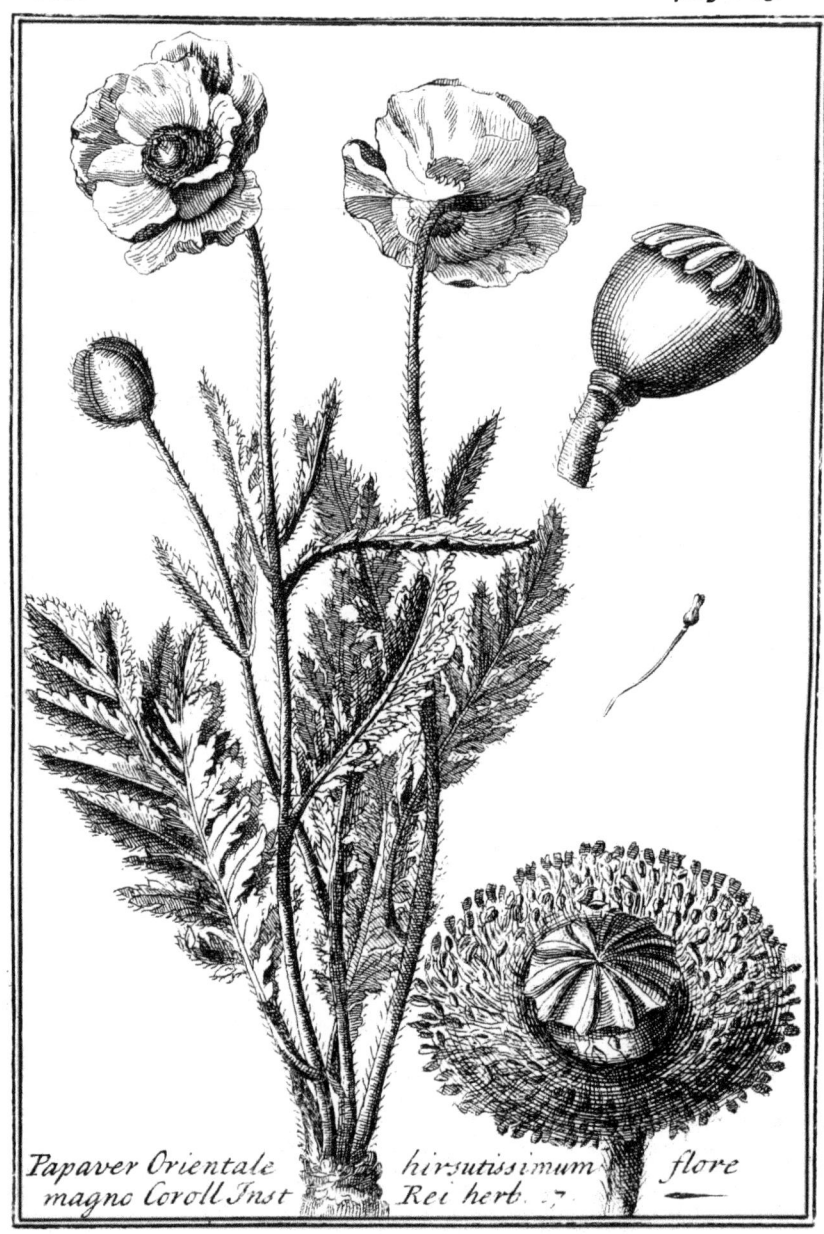

Papaver Orientale *hirsutissimum* *flore*
magno Coroll Inst *Rei herb.*

that, their Kitchen would be miserable cold.　Thus live the *Curdes*, driving their Herds and Flocks from Mountain to Mountain.　They stop at every good Pasturage; but about the beginning of *October* they are forc'd to decamp, and go into *Curdistan* or *Mesopotamia*.　The Men are well mounted, and take great care of their Horses; Lances are their only Arms.　The Women travel partly upon Horses, and partly upon Oxen. We saw a Troop of these *Proserpines*, who came out to look at the Bishop, and especially at us, who were taken for a sort of Bears that were led out to Airing.　Some had a Ring thro one of their Nostrils, and these they told us were betroth'd.　They seem'd strong and vigorous, but they are very ugly, and have a mighty fierce Air with them.　They have little Eyes, very wide Mouths, Hair as black as Jet, and a mealy ruddy Complexion.

Y E T even this is a Country that furnishes Matter for Learning. Who would think it, my Lord, among *Proserpines* and *Curdes*?　The Mountain wherein are the Sources of the *Euphrates*, must be one of the Northern Divisions of Mount *Taurus*, according to *Strabo*; and this Mount *Taurus*, with its Branches and its Oaks, possesses almost all *Asia Minor*. *Dionysius* the Geographer calls the Mountain that gives birth to the *Euphrates*, the *Armenian Mountain*: the Antients call'd it *Paryardes*.　*Strabo* expresseth himself more clearly in another part, when he positively says, that the *Euphrates* and the *Araxes* issue both from Mount *Abos*, which is a Parcel of Mount *Taurus*.　*Pliny* tells us that the *Euphrates* comes out of a Province call'd *Caranitide* in the *Greater Armenia*, which *Domitius Corbulo*, who had been upon the spot, calls Mount *Aba*; and which *Nutianus*, who also had seen the Country, names *Capotes*.　*Eustathius* upon *Dionysius Periegetes* calls it *Achos*.

M I T H R I D A T E S pass'd by the Sources of the *Euphrates* when he fled into *Colchis*, after being beaten by *Pompey*.　It is very probable that the Action happen'd in the Plain of *Erzeron*; for the two Branches of the *Euphrates*, recorded in History, may be call'd the Sources by Historians.　*Procopius* knew not these Sources; he imagines they come from the same Mountain as those of the *Tigris*.　There is, says he, a Mountain in *Armenia* five Miles and a half from *Theodosiopolis*, whence issue

sue two great Rivers; that which goes to the right, is call'd the *Euphrates,* and the other the *Tigris.* *Strabo* justly said that the Sources of these Rivers are two hundred and fifty Miles, or two thousand five hundred Stadia, distant from each other. *Pompey,* as we are inform'd by *Florus,* was the first that built a Bridge of Boats over the *Euphrates,* which he did in his Pursuit of *Mithridates.* 'Twas in all likelihood near the Elbow which this River makes, after its two Branches are join'd at *Mommacotum.* Some Years before, *Lucullus* had sacrific'd a Bull to this famous River, to obtain a favourable Passage.

'TIS generally believ'd that *Erzeron* is the antient City of *Theodosiopolis,* tho this is not over-certain; unless you suppose, as one indeed may, that the Inhabitants of *Artze* retir'd to *Theodosiopolis* after the Demolition of their Houses. *Cedrenus* relates, that in the Reign of the Emperor *Constantine Monomacus,* who dy'd towards the middle of the eleventh Century, *Artze* was a great Borough full of Riches, inhabited not only by the Merchants of the Country, but also by several other Merchants or Factors, *Syrians, Armenians,* and others of different Nations, who confiding much in their great Number and Strength, would not retire with their Effects to *Theodosiopolis,* during the Wars between the Emperor and the Mahometans. *Theodosiopolis* was a great and powerful City, in those times accounted impregnable, and situated close to *Artze.* The Infidels did not fail to besiege this Borough; the Inhabitants made a vigorous Defence six Days, intrenching themselves upon the tops of their Houses, from whence they incessantly flung Stones and Arrows. *Abraham,* the General of the Besiegers, finding such an obstinate Resistance, and apprehending that the Place might be reliev'd, caus'd it to be set on fire on all sides; thus sacrificing this wealthy Booty to his Reputation. *Cedrenus* tells us, that one hundred and forty thousand Souls perish'd in this Siege by Fire or Sword. The Husbands, says he, leap'd into the Flames with their Wives and Children. *Abraham* found in it abundance of Gold and Instruments of Iron, which the Fire could not consume. He also took a great many Horses, and other Beasts of Burden. *Zonaras,* with very little difference, gives the like account of the Destruction of *Artze,* but he does not mention *Theodosiopolis.* This Author only informs us that *Artze* had no Walls, and that its Inhabitants had fortify'd the A-

venues

venues of it with Wood; and I believe they us'd all they could find Lett. VI.
about the Country in that service, for the Species of it is now lost. As
the Town was reduc'd to ashes, and that this Passage is absolutely necessary for Trade, it is very probable the Remnant of those poor Inhabitants, and the foreign Merchants who afterwards settled here, that they
might not be in danger of the like Miseries, retir'd to *Theodosiopolis*, which,
according to *Cedrenus*, was close to it.

THE *Turks*, who perhaps thought *Theodosiopolis* too long and trou-
blesome a Name, gave it that of *Artze-rum*, that is to say, *Artze of the
Greeks*, or *of the Christians*; for *Rum* or *Rumili* in the *Turkish* Language   Ῥώμασσι-
signifies *Romania*, or *the Land of the Greeks*. They divide *Romelia* or *Rumili* into that of *Europe* and that of *Asia*; from *Artze-rum* comes *Arzerum*
and *Erzeron*, according to the Pronunciation of the Generality of the
Franks. We must take care not to confound this City of *Theodosiopolis*
with another of the same Name, which was upon the River *Abhorras* in
Mesopotamia, and which the Emperor *Anastasius* had fortify'd with good
Walls, as we are told by *Procopius*. The same Author makes mention
of the *Theodosiopolis* we have now been speaking of. 'Tis believ'd that
Orthogul, Father of the famous *Othoman*, the first Emperor of the *Turks*,
was the Taker of *Erzeron*; but this is not certain, for *Armenia* continued to have its Kings under *Selim* the first. The Similitude of Names
has made many imagine that *Erzeron* was the City of *Aziris*, which
Ptolemy places in *Armenia the Less*.

GIVE me leave, my Lord, to go from Erudition into Natural History. We observ'd in the Fields about this City a very fine Species of
Poppy, which the *Turks* and *Armenians* call *Aphion*, as they do the *common
Opium*: yet they do not extract Opium from the Kind we now speak of;
but by way of delicacy they eat the Heads of it when they are green,
tho very acrid, and of a hot Taste.

THE Root of this Plant is as thick as a Man's little Finger, and a
foot long, white within, brown without, fibrous, full of a Milk which
is of a dingy white, very bitter and very acrid. Usually the Stalks are
a foot and a half, or two foot high, three or four lines thick, strait, firm,
pale-green, bestrew'd with whitish Hairs, stiff, three lines long, unless
towards the top; where they are cover'd with short Hair. The Leaves

are

are a foot high, and are flaſh'd almoſt like thoſe of the *wild Poppy*, in ſeveral parts almoſt to the Rib. Theſe pieces are about two inches and a half long, and nine or ten lines broad, deep-green, and as it were ſhining upon certain Stocks, flaſh'd about the Rims with great Notches, pointed, and ending in a white Hair, like thoſe that cover the Leaves; and all theſe Hairs are as ſtiff and as long as thoſe of the Stalks. Each Stalk commonly ſupports but one Flower, the Button whereof, which is eighteen or twenty lines long, is cover'd with a Cup conſiſting of two or three membranous Leaves, hollow, whitiſh towards the edge, briſtling with Hairs. They fall when the Flower blows, and then you perceive that it conſiſts of from four to ſix Leaves, two inches and a half long, and three and a half broad, rounded like thoſe of other Poppies, and of the Colour of the wild Poppy, more or leſs deep, with a great Spot, which is alſo more or leſs obſcure. The inner Leaves are a little narrower than the outer, and ſtick hard againſt the Pedicule; nay, oftentimes they fall not till two days after the Stalk is cut. The middle of the Flower is fill'd by a Piſtile an inch long, oblong, ſpherical upon ſome Stocks, pale-green, ſleek, rounded toward the top like a Cap, purple, flaſh'd in a point near the edges, and ſet off with about a dozen Bands, deep violet-colour, duſty; which going out from the ſame Center, diſtribute themſelves in Radiuſſes, and terminate in one of the Points that are at the edges. This Piſtile is ſurmounted by a great tuft of Stamina in divers Rows, ſhining-grey, each laden with a Summit, deep-violet, duſty, a line and a half long, and half a line broad. The Plant yields a limpid Juice, but the Piſtile is full of a Milk of a dingy white, very bitter and very acrid like the Root. This Piſtile comes to be a Fruit or Cod. This fine Species of Poppy is mightily pleas'd with the King's Garden, nay, and with *Holland* too, where we have communicated it to our Friends. M. *Commelin*, a very able Profeſſor of Botanicks at *Amſterdam*, has publiſh'd the Figure of it.

THE 24th of *June* we return'd to *Erzeron*, where we were inform'd by M. *Preſcot*, who has been ten or twelve Years Conſul for the *Engliſh* Nation, that there were two Caravans ready to ſet out, one in three days for *Tocat*, and the other in ten or twelve for *Teflis*. We reſolv'd to go to *Teflis*, not only to have a Sight of *Georgia*, which is the fineſt

Country

Country in the World; but alfo to gather in our Return the Seeds of fo many fine Plants which we had obferved about *Erzeron*. Over and a-bove this, we were told, that there were a great many Thieves on the Road of *Tocat*, who would retire, according to their Cuftom, about the end of the Summer, becaufe then the Fields were burnt up by the great Heats, and yielded no more Forage. It is certain the Months of *June*, *July*, and *Auguft* are the moft favourable Seafon for Thieves: they every where find fufficient to keep their Horfes nobly, and this is what they have moft at heart; for thefe Gentlemen don't go a foot like Beggars. On the fide of *Tocat*, and in the *Turkiſh Georgia*, they reap at the end of *July*, whereas about *Erzeron* they don't cut the Corn till *September*. Of all the Caravans, this of *Teflis* is efteem'd leaft expos'd to danger.

WE did not lofe our time while this was getting together. When we were not upon the hunt, we went to have a little Converfation at the *Engliſh* Conful's, where there is always good Company. 'Tis the Rendezvous not only of the richeft *Armenian* Merchants, but of all man-ner of Strangers whatfoever. M. *Prefcot* is the moft of a Gentleman of any Man in the World, extremely good-natur'd, and prevented our Wifhes in every thing that might be a Gratification to us: I am even afraid the Natives abufe his Goodnefs, for they befet him continually. Tho he is not of the *Roman* Communion, yet he performs all manner of good Offices to the Miffionaries; he often gives them Lodging in his Houfe, and af-fifts them in their Entrance and Departure from the Country with a-bundance of Charity. We were told that three or four days Journey from the Town there were good Mines of Copper, whence they drew moft of that which is wrought in the *Greek* Suburb, and difpers'd all over *Turkey* and *Perfia*. They alfo affur'd us that there were Mines of Silver about *Erzeron*, as well as upon the common Road from that City to *Trebifond*. We could not fee thefe laft Mines, becaufe the Beglerbey took the better Road, which is a great way from it. As to thofe that are about *Erzeron*, we could find no body that durft be our Guide to them; the Beglerbey himfelf would not advife us to go near them, becaufe of the Jealoufy of the Natives, who imagine that Strangers go thither only to run away with their Treafure. We were told that there was fome Lapis Lazuli to be found among thofe of Copper, but in fmall quantities, and that it was too much mix'd with Marble. That which is found towards *Toulon* in *Provence*, in the Mountain of *Carqueirano*, has the same Fault; but certainly it is not the *Armenian* Stone, as many have fancy'd. The *Armenian* Stone, as appears by the Defcription of *Boot*, is of a sky-blue, very fmooth, but apt to crumble. Thofe about *Erzeron* and *Toulon* are very hard, harder even than Lapis Lazuli ; for properly fpeaking 'tis nothing but a fort of Marble naturally kneaded with Lazuli. Perhaps the fineft Lazuli is only a Species of Verdegreafe, or natural Ruft. Perhaps alfo 'tis Gold difguis'd by fome corrofive Liquor, as Verdegreafe is nothing but Copper difguis'd by Wine and the Skin of Grapes. Befides that Lazuli is found in Gold Mines, there feem to be in this Stone fome Threds of Gold as it were ftill uncorrupted.

WE one day enquir'd of Mr. *Prefcot*, in what Parts died Mr. *Vernon* a learned *Englifh* Mathematician, that had made very fine Aftronomical Obfervations in the *Levant*, and who is honourably mentioned by *Wheeler* and *Spon* : the Conful inform'd us he had often told him he would come to fome ill end with all his Knowledge, if he did not learn to keep his Temper. Mr. *Vernon* was a Man of admirable Vivacity, but he was too cholerick. In fhort, Mr. *Prefcot* prov'd a true Prophet, and our Mathematician died at *Hifpahan* of the Wounds he receiv'd in the Head, in a Quarrel he had with a *Perfian* one day after dinner. Mr. *Vernon* accus'd the *Mahometan* of having robb'd him of a very good Knife, *Englifh*-make ; the *Perfian* only laugh'd at him, whether he had taken the Knife or no ; the *Englifhman* was provok'd more at this than t'other. The Difpute grew warm ; from Words they came to Blows, and the *Perfian* wounded Mr. *Vernon* fo dangeroufly in the Head, that they were forc'd to tie him upon his Horfe, and carry him to *Hifpahan*, where he died fome days afterwards wanting Affiftance, for the *Englifh* were not then fettled in that City. At prefent they are very powerful there, and live like fo many Lords. Their Magnificence fometimes exceeds even to Profufion, efpecially when the Court pays them a Vifit.

WHILE our People were bufied in packing up our Bales, we often fimpled with a great deal of Pleafure, efpecially in the Valley of the *Forty Mills*, which is no more than a Walk from the City at the Entrance of two very fteep Mountains, from which run feveral fine Springs,

that

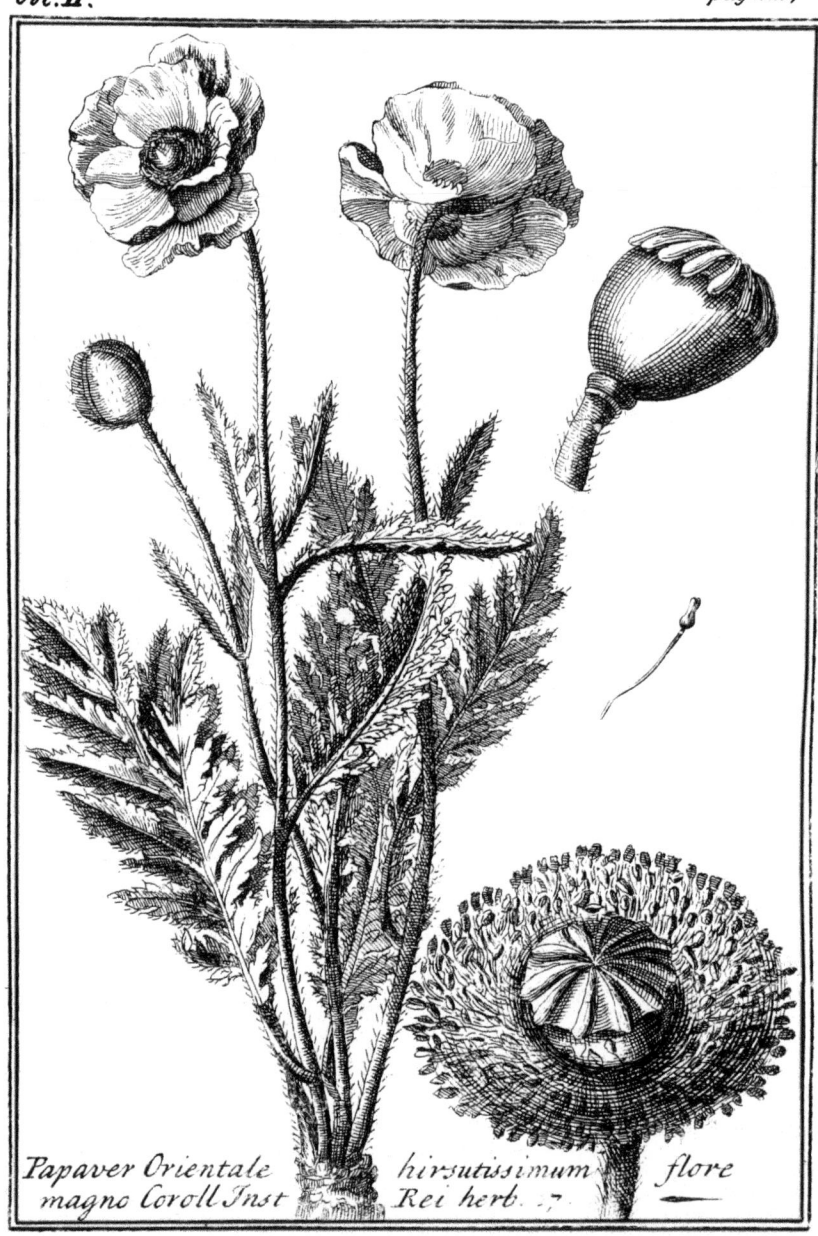

Papaver Orientale hirsutissimum flore magno Coroll Inst Rei herb.

that form a confiderable Stream, which not only turns a great many Lett. VI.
Mills, but alfo waters one part of the Country quite to the City. In
one of thefe Mills we had the Satisfaction to proceed to the Nomination
of one of the moft beautiful Genus's of Plants that is in all the *Levant*;
and accordingly we gave it the Name of a Gentleman very valuable
both for Learning and Virtue: I mean M. *Morin* of the Royal Academy
of Sciences, Doctor in Phyfick of the Faculty of *Paris*, who by fingular
good Fortune has rais'd this Plant from the Seed in his Garden of the
Abbey of *St. Victor*; I fay, by fingular good Fortune, for it would not
come up in the King's Garden, nor in fome others where I had caus'd it
to be fown. It feems to have been proud of bearing the Name of M.
Morin, who always lov'd and cultivated Botany with great Application.

THE Root of the *Morina* is thicker than a Man's Thumb, a foot
long, divided into great Fibres, brown, chap'd, but a little hairy. Its
Stalk, which is two foot and a half high, is firm, ftrait, fleek, purple at
firft, two or three lines thick, alfo reddifh, but hairy at the top,
ufually accompanied at each Joint with three Leaves pretty like thofe
of the *Carolina*, bright-green, fhining, four or five inches long, and about
one inch wide, flafh'd, wavy, garnifh'd with yellow Prickles, firm,
hard, four or five lines long. The Leaves diminifh a little towards the
top, and are fomewhat hairy beneath. From their Bofoms grow Flowers
by Stages, and in double Rows, an inch and a half long. Each
Flower is a crooked Pipe very flender towards the Bottom, where it
is white, and flightly haired; but it opens upwards, and parts into two
Lips. The upper is turn'd up, and about five inches long, and four
broad, rounded and deeply hollow'd inwards. The under is a little
longer, and flafh'd into three parts, rounded alfo. The opening of the
Pipe which is between thefe two Lips, is quite uncover'd. Two crooked Stamina that jut out almoft three lines, whitifh, and laden with yellowifh Apices, are faftned againft the upper Lip. The Thred of the
Piftile, which is a thought longer, ends in a greenifh Button. The Cup
is a Pipe three lines long, deeply cleav'd into two Tongues, rounded,
lightly channell'd. 'Tis from the bottom of this laft Pipe that the Flower rifes. There are often two forts upon the fame Stock, one quite
white, the others of a Rofe-colour with a touch of Purple, and whitifh

edges.

edges. All thefe Flowers have the fame Smell as thofe of the *Honey-Suckle*, and bear upon an Embryo of Seed. The Leaves of this Plant have at firft a faintifh graffy tafte, but afterwards one finds it fomewhat tartifh.

W E then went to kifs the Beglerbey's Veft, and to defire a continuation of his Protection. He had the goodnefs to return us thanks for the care we had taken of his Health, and of all his Family. He gave us un-ask'd the Letters of Recommendation which we wanted to the Baffa of *Cars*, and order'd us befides a very honourable Patent, wherein he prais'd our Capacity in matter of Phyfick, and gave good Teftimonies with relation to our Behaviour.

Journey into Georgia. W E fet out from *Erzeron* the 6th of *July* to go to *Teflis*, and came to *Elzelmic*, a Village to the North-Eaft, three Hours Journey from the Town. Our Caravan confifting of Merchants, whereof fome went to *Cars* and to *Teflis*, and others to *Erivan*, and fome few to *Gangel*, were in number but about two hundred Men, arm'd with Lances and Sabres; and fome had Fufees and Piftols. The Country of *Erzeron*, for half of the way to *Elzelmic*, is very dry; its Hills are quite bare. You afterwards enter into a Plain, fhut in to the right and left by Eminences, whereon was ftill a good deal of Snow. There fell fome about *Erzeron* in the night between the 2d and 3d of *July*.

T H E 7th of *July* we fet out at half an hour after three in the Morning, and encamp'd about ten near a Village call'd *Badijouan*, after having pafs'd by another, whofe Name I have forgot. There is not a Tree to be feen in all this part of the Country, which otherwife is flat, well cultivated, and water'd as abundantly as the Fields of *Erzeron*. Were it not for this, half of the Corn would be burnt up: yet this feems very ftrange, for from thefe very Fields which they are forc'd to water by Art, you fee the Snow upon the neighbouring Hills. On the contrary, in the Iflands of the *Archipelago*, where the Heats are ready to calcinate the Earth, and where it never rains but in Winter, the Corn is the fineft in the world. This plainly fhews that all Soils have not the fame nutritious Juice: That of the *Archipelago* is like a Camel, one drinking ferves it a long while. Perhaps Water is more neceffary to that of *Armenia*, to diffolve the foffile Salt wherewith it is impregnated, which would

*

deftroy

destroy the Contexture of the Roots, if the little Clods were not well Lett. VI.
moisten'd with a proportionable quantity of Liquid, and accordingly
they turn it deep up. Tho the Ground is not hard, they yoke three
or four pair of Oxen or Buffaloes to one Plough; which they certainly do
to mix the Earth more thorowly with the fossile Salt, which would
lie in too great quantities upon the Surface, and burn up the Plants.
On the contrary, in *la Camargue* of *Arles,* which is the fruitful Island form'd
by the *Rhone* below the Town, they only give the Earth a slight flourish, to avoid mixing it with the Sea-Salt that is beneath. With this
Precaution, *la Camargue,* where there is but half a foot of good Soil, is
the most fruitful part of *Provence*; and the *Spaniards* nam'd it *Comarca,*
by way of excellence, when the Earls of *Barcelona* were Masters of
it. *Comarca* in their Language signifies a fruitful Field. Thus the
word *Camargue* does not come from the *Camp of* Marius, as is
pretended, for that *Roman* General never did encamp in it. The great
Ditch that he cut to fortify his Camp, and to bring his Ammunition
from the *Mediterranean,* was, according to *Plutarch,* betweeen the *Rhone*
and *Marseilles.* The Footsteps of that Work are still to be seen on the
side of *Fos,* a Village near *Martigues,* which still retains the Name of
Marius's Ditch; and not that of the *Phocians,* a People of *Asia* above *Smyr-*
na, that settled at *Marseilles* during the Wars between the *Greeks* and the
Persians. A thousand Pardons, my Lord, for this Digression: we are so
us'd to go out of the way when we are simpling, that 'tis no wonder I
should sometimes wander in the Letters you permit me to write to you.

I RETURN to our Caravan. It set out the eighth of *July* about
nine in the Morning, and travell'd till one in the Afternoon over large
Champains, very negligently cultivated, but as we were inform'd, in
themselves excellent. We observ'd very fine Plants in them, as we also
did the day before; but that's all, for there's neither Town nor Village
near, and not the least Bush to be seen. Our Tents were pitch'd near a
Stream that turns a Mill, I know not for what use; for we met not one
Soul the whole day.

OUR Course the ninth of *July* was much more agreeable. The
they made us be moving at three in the Morning, we put in about ten,
after having pass'd over some low Mountains, whereon we saw Pine

of

of the fame Species as thofe of our Mount *Tarare*. The fhifting of the Scene affords no fmall Delight in travelling: nothing can be more tedious than marching along vaſt Plains, where all that is to be ſeen is Earth and Sky; and were it not for the Plants, I ſhould rather chuſe to be upon the Sea, I mean in calm Weather; for I muſt freely own, in a Storm one would give all one has in the world to be ſet down in the moſt diſagreeable Plain in the whole Univerſe. We encamp'd this day at *Coroloucaleſi*, a Village which in our Tongue might be call'd *the Tower of Corolou*. Our Harveſt was tolerably good; and as I have no uſe here for my Learning, for I know nothing either of *Corolou* or its *Tower*, you will give me leave to ſend you the Deſcription of a Plant, which is ſtill one of the higheſt Delights of Monſieur the firſt Phyſician. It has throve very well, and brought forth Flower and Seed to Perfection in the King's Garden; and in all probability will flouriſh there many Years.

IT is an Umbellifer, to ſpeak like a Botaniſt, the Root whereof goes a foot and a half down; it is as thick at the Neck as a Man's Arm, and divided into ſome other Roots of the thickneſs of a Man's Thumb, not very hairy, cover'd with a brown Bark, full of Milk, acrid and very bitter. The lower Leaves, which are about three foot broad, and as many long, are ſo ſlenderly cut, that one cannot compare them better, than to thoſe of another Species of this Genus, which *Moriſon* calls *Cachrys ſemine fungoſo, levi, foliis ferulaceis.* The Compariſon indeed ſeems to halt a little, for there is no Species of *Ferula* with ſuch ſlender Leaves; and without following *Moriſon*'s Example, I had better have compar'd the Leaves of this I am ſpeaking of to thoſe of *Fennel*. The Stalks of our Plant riſe to four foot high, as thick as a Man's Thumb, firm, hard, ſtrait, ſolid, cover'd with a Flower like that of *freſh Plumbs*, ſleek, channell'd, knotty, garniſh'd at the Joints with two or three Leaves, much ſmaller than the others; and from the Boſoms of theſe, towards the top grow three or four Branches, which form a Plant pretty much rounded. The Extremities of theſe Branches are laden with Umbellas or Cluſters half a foot diameter, conſiſting of unequal Rows that ſuſtain other Cluſters ſmaller, and as it were ſpherical, terminated by yellow Flowers of five, ſix, or ſeven Leaves, a line and a half long, with a point turn'd inward, which makes them ſeem as if they were hollowed. The Stamina and the Apices

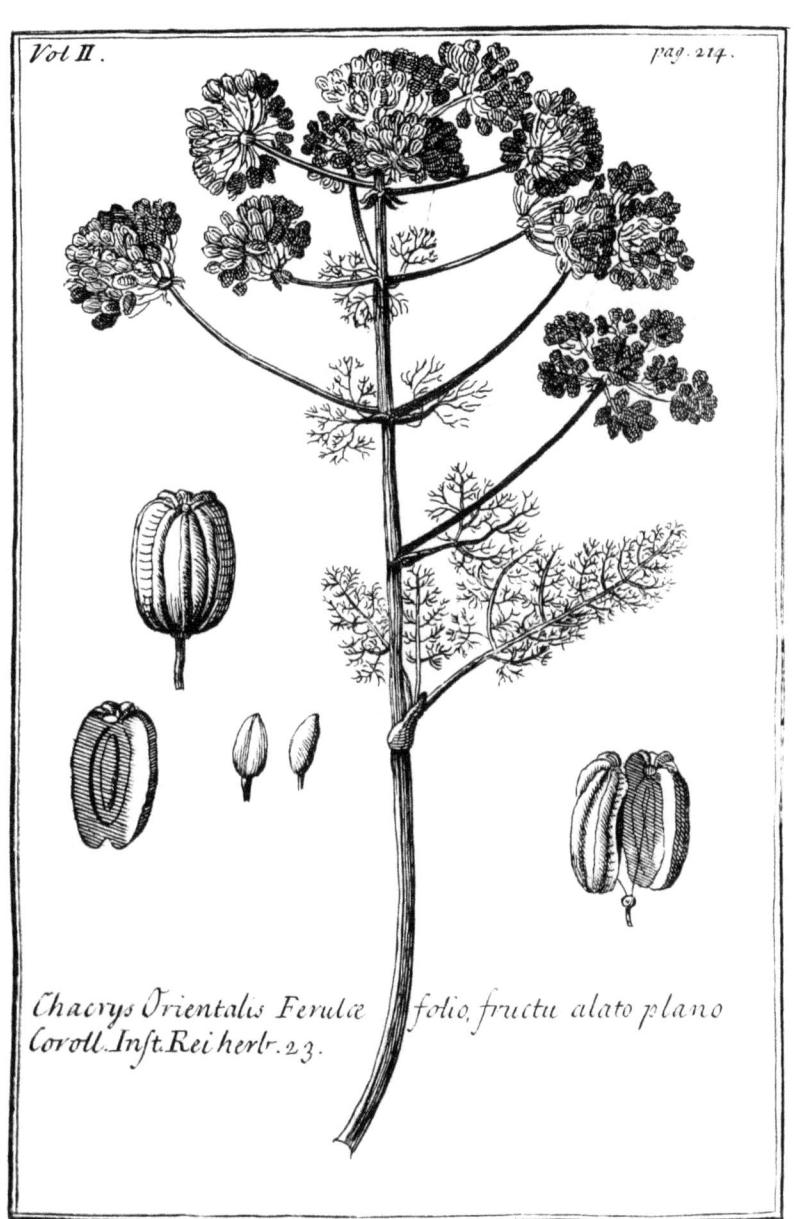

Chaerys Orientalis Ferulæ folio, fructu alato plano
Coroll. Inst. Rei herb. 23.

ces are of the fame Colour. The Cup, which at firſt is but two lines long, grows perceptibly as the Flowers paſs away, and afterwards becomes a Fruit about ten lines long and ſix broad, conſiſting of two parts, rounded at the back, garniſh'd lengthways with little Wings or Leaves, membranous, and white like the Fruit of the *Laterpitium*. We muſt neverthelefs refer our Plant to the Genus of *Cachrys*, becauſe the parts of its Fruit are ſpungy, three lines thick, and full of Seed thicker than a Barley-corn. The Leaves of this Plant are a little aromatick, but very acrid, and very bitter.

THE tenth of *July* we ſet out at three in the Morning, and travell'd till paſt twelve at Noon over agreeable Mountains well ſtock'd with Pines. Indeed we were not very attentive in examining the Nature of them, for we were from time to time alarm'd with the ſight of ſome Knots of Thieves arm'd with Lances and Sabres. However, they durſt not attack us, imagining we were the ſtronger, tho they happen'd to be very much deceiv'd, and might have had a good Pennyworth of us had they ventur'd. We had *Turks* enow indeed in our Caravan; but the *Armenians*, as we were inform'd by our Druggermans, began to talk about a Capitulation; and if the Thieves had not made off, they had infallibly ſent an Envoy to them to treat of a Ranſom. The next day we travell'd down hill into a good handſome Plain, where we encamp'd at *Chatac*, a ſorry Village upon a Stream that falls from ſome Hills where the Graſs was but juſt coming up. Scarce was there enough for Paſture in the very beſt Spots of Ground. The Ways here are edg'd with that fine Species of *Echium* with red Flowers, which *Cluſius*, the greateſt Obſerver of Plants of his Age, diſcover'd in *Hungary*. The Stalks grow three or four together, a foot and a half or two foot high, three lines thick, pale-green, ſpotted with deep red, brittle, roughen'd with white Hairs, garniſh'd with Leaves half a foot long, and but half an inch broad, of the ſame Colour and Contexture as thoſe of the *common Echium*, but much more briſtled of both ſides. They diminiſh to the top, and from their Boſoms almoſt from half way of the Stalk to the extremity grow ſlips an inch and a half long, crooked like a Scorpion's Tail, whereon reſt two rows of Flowers eight or nine lines high, turn'd in like a crooked Pipe, open and ſlaſh'd into five rounded parts, the undermoſt whereof are fhorter than the uppermoft. Thefe Flowers are of a Madder - co-lour, red but not bright. The Stamina, which jut out three ways, are a little more fhining, but their Apices are deep-colour'd. The Cup is a-bout half an inch, flafh'd into five parts, very narrow and very hairy. The Piftile is of four Embryos, which come to be as many Seeds, a line and a half long, brown, of the fhape of a Viper's Head.

T H E 12th of *July* we were jogging by four in the Morning, and travell'd till Noon in one of the fineft Plains imaginable. The Earth, tho black and fat, is not very productive, becaufe it freezes a-nights, and we often found Ice about the Springs before Sun-rife. As hot as it is in the Day-time, the Cold of the Nights puts the Plants terribly back; the Corn was not above a foot high, and the other Plants were not more forward than they are towards the end of *April* about *Paris.* The way of manuring thefe Lands is ftill more furprizing, for they will yoke you ten or a dozen pair of Oxen to one Plough. Each pair of Oxen has its Poftilion, and the Ploughman pufhes the Share along with his Foot befides; and this they do, to make deeper Furrows than ordinary. Experience has certainly taught them that it was neceffary to go very deep, either to mix the upper Soil which is too dry, with that beneath which is lefs fo, or to preferve the Seeds from hard Frofts; for were it not upon fome fuch Confiderations, they would not be at fo much Pains and Expence. We often enquir'd the reafon of our Guides, who barely told us 'twas their way in that Country. There are no Trees in thefe Fields but a few Pines, which they drag along the Roads to carry them into the Towns and Villages, by tying as many Oxen to them as are neceffary to pull them along: this did not furprize us. In *Armenia* you hardly meet with any thing elfe but Oxen and Buffaloes yok'd or with Loads on their Backs like Mules. Their Pines however, by the Confeffion of the Natives themfelves, begin to ftand very thin, and there are but few of them that will rife from Seed. I know not what they will do when they have cut down all the great Trees, for they can't build without them: I don't fpeak of building their better fort of Houfes, where they ufe Beams only to fupport the Co-verings; I mean their Cottages, which are their moft common Habitations, the four Walls whereof are made of Pines, rang'd pointways in right Angles one upon another up to the Covering, and faftned at the Corners

with

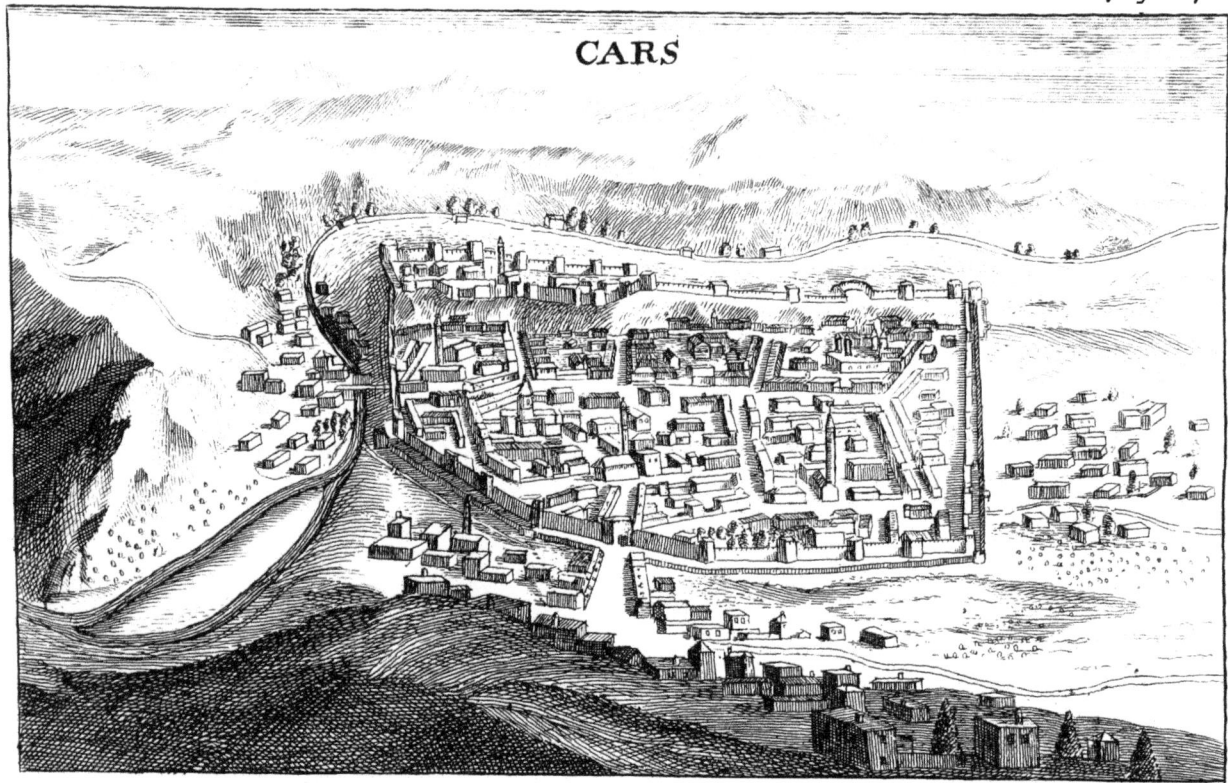

CARS

Prospect of Cars on the Frontiers of Persia.

with wooden Pins. We found no new Plant this day, and began to
be a little alarm'd at feeing among fome rare ones, which we had obferv'd
more than once, common *Mallows*, *Plantain*, *Pellitory of the Wall*, and
efpecially *Wallwort*, *Bank-Creffes*, and that Plant which is fold at *Paris* for
a Loofenefs, by the Name of *Thalitron*. We thought we were got into
Europe again ; yet we infenfibly arriv'd at *Cars*, after a March of feven
Hours.

C A R S is the laft Place in *Turky* upon the Frontiers of *Perfia*, which
the *Turks* know only by the Name of *Agem*. I was a little at a lofs
one day at the Beglerbey's, who ask'd me what Folks faid in *France* of the
Emperor of *Agem*. As Good-luck would have it, I remember'd to have
read in *Cornuti*, that the *Lilac* of *Perfia* was call'd *Agem Lilac*, and this
made me conceive that *Agem* muft fignify *Perfia*. But to return to *Cars*,
the Town is built upon a Bank, expos'd to the South-South-Eaft. The
Compafs is almoft fquare, and fomewhat bigger than half of *Erzeron*.
The Caftle of *Cars* is very fteep upon a Rock at the top of the Town.
It feems pretty well kept up, but 'tis defended only by old Towers. The
reft of the Place is like a kind of Theater, behind which is a deep Valley, fteep on every fide, and thro the middle of that runs the River. This
River does not go to *Erzeron*, as *Sanfon* believ'd ; on the contrary, it
comes from that great Plain, which is the way from *Erzeron* to *Cars*, and
falls from thofe Mountains where we firft faw Thieves. After having
winded about this Plain, it comes to *Cars*, where it forms an Ifland, running under a Stone Bridge, and follows the Valley that is behind the Caftle.
There it not only turns feveral Mills, but alfo waters the Fields and
Gardens. At laft it joins the River *Arpagi*, which flows not far from
thence ; and thefe two Rivers, join'd together by the Name of *Arpagi*,
ferve as a Frontier to the two Empires, before they fall into the *Araxes*,
which the *Turks* and *Perfians* call *Aras*. What may have deceiv'd *Sanfon*, is, that the *Araxes*, as will afterwards appear, has its Source in the
fame Mountain as the *Euphrates*. That Author places *Cars* at the Con-
flux of the two imaginary Branches of the *Euphrates*, which, according to
him, form a confiderable River that runs to *Erzeron*. Thefe Faults muft
be imputed to the bad Accounts that have been given him ; for *Sanfon*

was an excellent Man, and the firſt that drew good Maps in *France.*

C *A R S* is not only a dangerous Town upon account of Thieves, but the *Turkiſh* Officers alſo generally make great Exactions from Strangers. We deſir'd to ſee the Baſſa, upon occaſion of the Extortions we were threatned with. His Chiaia, to whom we were carried firſt againſt our Will, very fairly told us all our Patents ſignify'd not a Farthing, and that certainly we ſhould never be allow'd to go into the Country of *Agem.* And yet we had ſhewn him a Commandment from the Port, and a Paſſport from the Beglerbey of *Erzeron,* who is ſuperior to the Baſſa o *Cars.* Here follows the Analyſis the Chiaia was pleas'd to make of theſe Authorities. As to the Commandment of the Port, ſays he, 'tis the moſt venerable Patent in the world, (and he put it to his Forehead every moment) but the Town of *Cars* is not mentioned in it. I anſwer'd, it was impoſſible to put in a Sheet of Paper the Names of all the great Cities in their Empire. The Paſſport of the Beglerbey of *Erzeron* imports, ſays he, that you may come here, but it does not ſay you may go further. As I had got a Tranſlation of it made at *Erzeron,* I begg'd the Chiaia to read it over again, proteſting that the Beglerbey had made us believe that his Paſſport would remove all Difficulties that might impede our paſſing from *Cars* into *Gurgiſtan,* which belongs to the Emperor of *Agem,* and that this was what we really intended to do. After ſome Diſputes about this Paſſport, we told him we ſhould be very glad to kiſs the Baſſa's Veſt, and preſent him the Beglerbey's Letter. He anſwer'd, that he would take care to deliver the Letter; but he was ſure the Baſſa would never ſuffer us to go out of the Grand Signior's Territories: yet he would go and know his Pleaſure. Accordingly he left us very abruptly, to wait, as he ſaid, upon the Baſſa in his Apartment.

A F T E R having danc'd attendance a long while, we were told we ſhould run the riſque of lying in the Streets, if we did not make haſte into the Suburb where our Caravan-ſerai was. Tho the *Turks* and *Perſians* live together in as much Peace as can be wiſh'd, they nevertheleſs ſhut the Gates of their Town at Sun-ſet. Before we went, we deſired one of the Chiaia's Servants to tell him that we were forc'd to be gone, becauſe

becaufe it grew towards Night, but that we fhould be very glad to know Lett. VI
our Fate before we went, if poffible.   He fent us word that the Baffa his
Mafter having read and confider'd of the Beglerbey's Letter, could not allow us Paffage ; but that the next day he would call together the Mufti, the Janizary-Aga, the Cadi, and the chief Men of the Town, to read
it: that without this Precaution, the Baffa might forfeit his Head, if it
came to be known at *Conftantinople* that he had omitted to feize three
Franks, that perhaps might be the Great Duke of *Mufcovy*'s Spies.   All
thefe Ceremonies fretted us heartily : We apprehended they would be
tedious, and that what with one Scruple, and what with another, our
Caravan might go away without us, fo that we fupp'd very melancholy.
Two Emiffaries from the Chiaia had the Goodnefs next Morning to
rouze us at Day-break, to let us know in plain terms, that a Difcovery
had juft been made of our being Spies, that the Baffa was not indeed
inform'd of it as yet, fo that the thing might ftill be remedied, but that
we might affure ourfelves the Information came from a good hand.   As
we did not feem at all frightned at this, they added, that Spies were
condemn'd to the Flames in *Turky*, and that fome of the moft creditable
People in the Caravan were ready to declare, that upon pretence of
fearching for Plants we obferv'd the Situation and Walls of Towns, that
we took Draughts of them, that we enquir'd critically into the Strength
of the Garifons, that we would know what part the moft inconfiderable
Rivers came from : all which certainly was moft abominably criminal. This
was the Talk of him who feem'd the greateft Rogue of the two ; the o-
ther, who feem'd a little more moderate, faid, to be fure we never came
fo far to pick Straws.   We ftill infifted upon the good Teftimonies which
the Beglerbey of *Erzeron* gave of us in his Letter.   They replied, there
was no reading of that till the Cadi return'd from the Country, where
he was to ftay a day or two longer.   Upon this, we parted very
coldly.

BY good Fortune, as we were walking thro the Town, we met an
Aga of the Beglerbey of *Erzeron*, that was but juft arriv'd, and that knew
us immediately, having feen us vifiting the Sick in the Palace.   After
the firft Civilities, we told him the Trouble we were in.   Surpriz'd at our
Story, he went to the Baffa's Chiaia, and told him in our prefence that

there was no reafon for hindring our Paffage ; that the Beglerbey *Co-progli*, to whom we were recommended at *Conftantinople* by the Embaffador of the Emperor of *France*, honour'd us with his Protection ; that we had been permitted to accompany him from *Conftantinople* to *Erzeron*, that he had been fatisfied with our Advice and Prefcriptions ; and laftly, that Perfons fo well recommended by him, ought not to be receiv'd in that manner. He made a fign to us to retire, and gave us to underftand by his Servant, that we fhould have Satisfaction very fpeedily. We went to a Coffee-houfe to wait for the Decifion of this weighty Affair. A moment afterwards the fame Chiodars of the Chiaia, that had call'd us the Spies of the Great Duke of *Mufcovy*, and who were much rather Spies over us, for they kept us conftantly in view, came to inform us with a forced Joy, in hopes of getting fome fmall fpill of Money out of us, that all the Paffages of the Empire were open to us ; but that we had infallibly been ftopt, had it not been for the Beglerbey of *Erzeron's* Letter, or that at leaft they had made us pay a hearty Duty, as they do moft of thofe that go out of *Turky* into *Perfia*. They had fcarce finifh'd their Speech, when the Aga, our Deliverer, came out, and carried us to the Chiaia, who made us fmoke and drink Coffee. He told us we might go whenever we pleas'd ; that in confideration of the Beglerbey of *Erzeron*, he forgave us two Crowns which are due to him for all the Beafts of Burden that pafs that way ; and as he was told we were not Merchants but Phyficians, he made it his Bargain, that before we went we fhould cure an Aga of his Acquaintance that had a Fiftula *in ano*. As he faid this gravely, and we did not care to fall into his Nets again, we thank'd him for his Civilities, and told him we would take care of his Friend, and give him all the Affiftance we could during our Abode at *Cars* ; but added, that a Fiftula *in ano* could not be cur'd without cutting, and that we were fo unfortunate as not to have Inftruments to do it with.

WE retir'd to our Camp much better fatisfy'd than we were the day before. While we were at Table, one of the Servants of the Aga of *Erzeron* came and reprefented to us, that his Mafter had done us a very confiderable piece of Service ; that he did not exact any Gratuity from us, but that we knew the World better, than to go away without making

him

him some Present or other. We came off for thirty Pence for the Servant, and two Oques of Coffee which we sent his Master; heartily glad of escaping at so cheap a rate. And for fear of a second Greeting, we resolved to keep in the Fields, in quest of Plants, till the Departure of our Caravan: thus the *Turks* always fleece Travellers, especially upon the Frontiers; but we must say this in their behalf, they commonly take up with whatever you are pleas'd to give them.

'TIS a reasonable Conjecture whether *Cars* be not the antient City that *Ptolemy* sets down among those that are in the Mountains of *Little Armenia*. The Resemblance of the Names will support it, and there is no need of being perplex'd because that Author places it in *Little Armenia*. Besides that this might be a Fault of Inadvertency, the Divisions of *Armenia* have been so often alter'd, that there is great Confusion among the Authors that speak of this Country. One might suspect too that *Cars* is the Place which *Ptolemy* calls *Chorsa*, and which he says is in the *Greater Armenia*, only that he sets it down on the side of the *Euphrates*. This is what may have deceiv'd *Sanson*; but it is certain *Cars* is very far from that River, and I could rather forgive those that have propos'd it as a doubt whether *Cars* be not the City of *Nicopolis*, which *Pompey* built in the Place where he beat *Mithridates*, since that City is said to have been between the *Euphrates* and the *Araxes*. *Cedrenus* and *Curopalatus* call *Cars Carse*, *Leunclavius Carseum*. This last says, that in 1579, *Mustapha Bassa*, who commanded the Army of Sultan *Amurath* against the *Persians* and *Georgians*, fortify'd *Cars*, and provided it with necessary Ammunitions. It might be made one of the strongest Places in the *Levant*.

THE 12th and 13th of *July* the Caravan sojourned here to pay Customs. We departed next day at one in the Morning, because the richest of our Merchants, who had confess'd but part of the Money they were carrying into *Persia*, were willing to avoid any new Enquiries that the Officers might make. They mounted their Horses as soon as ever they were dispatch'd, and we travell'd over a great Plain, all the night-time, as dark as it was. About nine in the Morning we encamp'd near *Barguet*, a great Village, the Castle whereof, which is half ruinate, seems to have been a good Building in its time. We discover'd hardly any but common Plants, and especially abundance of yellow *Gallium* and *Gramen Sparteum*,

fparteum, pennatum, C. B. About noon we defcended into a pretty good
Valley, half a League from *Barguet.* Among fome fcarce Plants we obferv'd here a pretty fingular Species of *Betony,* whofe Seed has rais'd and
multiply'd in the King's Garden. It is chiefly diftinguifhable in the
length of its Leaves, which are half a foot long to one inch broad, and
Culture has not alter'd them. This Plant has been long known in *France,*
fince Monfieur the firft Phyfician found the Figure of it among the Plates,
which M. *de la Broffe* his great Uncle, and Intendant of the King's Garden, had caus'd to be grav'd. It is a pity thofe Plates did not appear
in time ; they are as big as thofe of the Garden of *Aifted,* and much
better grav'd. Monfieur the firft Phyfician, who has lately recover'd
them, gives us hopes of his making them publick.

I KNOW not by what Fatality it has happen'd that moft of the
great Works of Botany wrote in *France* in the laft Century, and which
would have done great Honour to the Kingdom, have never yet appear'd.
M. *Richer de Belleval,* Chancellor of the Univerfity of *Montpellier,*
had defcrib'd and caus'd to be engraven a vaft Number of fcarce
Plants that grow in the *Alpes* and *Pyrenees,* and that pafs daily for unknown Plants. It appears by the Plates which are in the hands of his
Heirs, that the *Bauhinuffes* never difcover'd any thing fo fine in thofe
times. The Work of F. *Barrillier* is buried at the Bottom of the Library
of the *Dominicans* in the Street of *St. Honoreus.* That indefatigable Man,
after having travell'd all over *Spain* and *Italy,* and laid out a great deal
of Money to get the fineft of his Difcoveries engrav'd, dy'd at *Paris,*
without having publifh'd any thing; and there is no likelihood of that
fine Collection's ever feeing the Light. The fame will happen to the
Labours of F. *Plumier* a Minim, unlefs you, my Lord, promote the E-
dition of it ; it may be faid, in praife of that Father, that he alone has
defcrib'd and drawn more *American* Plants than all that ever pretended
to treat of them befides put together. It is very eafy to make Books
of Plants, by publifhing the Figures of fuch as are cultivated in a Garden, and as are fent one in Seeds or Roots by a Correfpondent ; but F.
Plumier made four Voyages into *America,* and dy'd at *Cadiz,* juft as he
was going by your Orders to *Peru.* For my part, I flatter myfelf, my
Lord, that you will continue me the Honour of your Protection, and

caufe

Elephas Orientalis, flore magno
proboscide incurva Coroll Inst
Rei herb. 48

caufe to be grav'd the many beautiful Plants which I have obferv'd in my Voyages.

T H I S is one of thofe Digreffions that are allowable only in Letters; the epiftolary way of Writing will admit of every thing, and is wonderful convenient for Travellers, who cannot help ftraying a little out of the way fometimes in a long Journey. But I'll go back to the Caravan. The 15th of *July* we fet out at four in the Morning, and pafs'd over Plains pretty well cultivated, interfpers'd with fome agreeable Hills, whereon the Corn was much forwarder than about *Erzeron.* They fow a great deal of Flax, efpecially near the Villages, which are pretty frequent. About feven in the Morning we forded a little, tho not inconfiderable River, which, as we were inform'd, difcharg'd itfelf into the *Arpagi.* The great Caravan left us a League from this Place to go to *Gangel,* and we were in a pretty great Confternation to fee ourfelves reduc'd to fuch a fmall Company as three Merchants that were going to *Teflis.* A *Turkifh* Aga, encamp'd upon the Road, fent two Guards to learn who we were; but as they could not read, they only caft their Eyes upon our Paffports, and demanded for their pains fome Trouts which our Druggermans had caught. They made our Merchants pay ten Afpers per Load, and got each a piece of Soap to fhave himfelf with.

W E this day difcover'd in my mind the fineft Plant that the *Levant* produces. 'Tis a Species of *Elephas,* with great Flowers, the Trunk whereof turns in downwards.

I T S Root, which is about two or three inches long, is but a line and a half thick, hard, reddifh, hairy, and puts forth a Stalk nine or ten inches high, fquare, purple towards the Bottom, flightly haired, accompanied with Leaves oppofite crofs-ways, two and two, from an inch to fifteen lines long, and nine or ten lines broad, like thofe of the *Pedicularj*; yellow, hairy about the Edges, dented like a Battlement, vein'd. From their Junctures rifes a Flower on each fide, made like a Pipe behind, greenifh, but a line and a half or two lines long. This Pipe afterwards opens into two Lips, the uppermoft whereof is firft dilated into two kinds of Ears pretty much rounded, between which grows a Trunk or crooked Pipe nine lines long, one line thick, ending in an oval Lip, a line and a half diameter, curl'd, edg'd with little Hairs, and be-

yond

yond this juts out the Thred of the Piſtile. The Under-lip is an inch long, and an inch broad, and ſlaſh'd into three parts, the two ſide ones being ſhap'd like two great Ears. The under part is reſlaſh'd into three pieces. The ſide ones are rounded alſo, but the middlemoſt is only a little Beak very ſharp pointed. This whole Flower is of a Saffron-yellow, except the Bottom of the Upper-lip which is whitiſh. The Stamina are very ſhort and conceal'd under the Wings of the Upper-lip. Their Summits are two lines long, and a line broad, flatten'd, pale-yellow. The Upper-lip repreſents the Trunk of an Elephant when he is bending it to bring ſomething to his Mouth, whereas in the other known Species of this Genus this Lip turns up. The Cup is of one ſingle piece, three lines long, ſlightly haired; the Upper-lip is ob-tuſe, hollow'd. The under is more deeply cleav'd into two pieces. Each Flower is faſtned to a Stalk half an inch long, and very ſlender. The Piſtile, which is a Button ſomewhat oval, is but a line long, and comes to be a Fruit half an inch long, almoſt ſquare, with rounded Corners, pale-green, membranous, about two lines and a half thick, divided length-ways into two Apartments which open ſideways, and incloſe Seeds a line and a half or two lines long, and one line thick, channell'd lengthways, and of the Form of a little Kidney.

THE 16th of *July* we were moving at four in the Morning, and about eight encamp'd in a large fine Meadow, where our Tents were pitch'd for the firſt time in the Dominions of the King of *Perſia*. We lay the night before but one Hour's Journey from the Frontiers, which is taken from the top of a Hill, at the Deſcent whereof begins the *Per-ſian Georgia*, or the Country which the *Perſians* call the *Gurgiſtan*, that is to ſay, *the Land of the Georgians*; for *Tan* is an antient *Celtick* Word, ſignifying a *Country*, and this Word continues in uſe all over the Eaſt, where they ſay *Curdiſtan*, *Indoſtan*, &c. meaning the *Land of the Curdes*, that of the *Indians*, &c. We could ſee a great many pretty conſiderable Villages; but all this fine Country yields not one ſingle Tree, and they are forc'd to burn Cow's Dung. Oxen are very common here, and they breed them as well for their Dung as for their Fleſh. They will yoke fourteen or fifteen pair to one Plough, to turn up the Ground. Each pair has its Man to drive it, mounted like a Poſtilion: all theſe Poſtilions, who yawl and roar like Sailors in a Storm, make together a moſt in-

tolerable Confort. We had been accuftom'd to this Noife ever fince we
left *Erzeron*. Sure 'twas not this Ground in *Georgia* that is fpoken of
by *Strabo* to have been only glanc'd over with a Wooden Plow, inftead
of an Iron one.

THIS *Georgia* is an excellent Country. The moment you are got
into the King of *Perfia*'s Dominions, People come and prefent you with
all manner of Provifions, Bread, Wine, Fowls, Hogs, Lambs, Sheep.
They efpecially accoft *Franks* with a fmiling Countenance, whereas in
Turky you meet with none but ferious Fellows that furvey you gravely
from head to foot. What furpriz'd us moft, was, that the *Georgians* de-
fpife Money, and will not fell their things: Neither indeed do they give
them; but they truck with you for Bracelets, Rings, Necklaces of Glafs,
little Knives, Pins or Needles. The Girls fancy themfelves finer than ordinary, when they have five or fix Necklaces round their Neck, and
hanging down to their Breafts; their Ears alfo are fet off with them: and
yet all this together makes a very queer Show. We therefore fpread our
Wares upon the Grafs; and as we had been inform'd of their Cuftoms,
we laid out ten Crowns at *Erzeron* in what we thought would pleafe
'em, namely, in *Venetian* Enamels, which are exactly like thofe of *Ne-*
vers. We got a hundred for one by thefe Merchandizes, but you muft
not load yourfelf too much with them, for you have vent for them no
way but by Truck, and they give you nothing but Neceffaries for them,
and that too for no more than two days Journey; as if the antient Manners of the *Georgians* had been preferv'd only within that particular Country. Thefe People, as *Strabo* fays, are larger and handfomer than the reft
of Mankind, but their Manners are very fimple. They ufe no kind of
Money, no Weight, no Meafure, fcarce can they count above a hundred:
All their Traffick is by Exchange. We therefore trufted our little Treafure to thefe honeft People; they took what they pleas'd, but it is very
certain they did not abufe the Confidence we repos'd in them. They
gave us a Hen as fat as a Turkey, for a Necklace that coft but fix Blancs,
(Farthings) and a great Meafure of Wine for Bracelets of eighteen Deniers.
The Hogs run about freely, whereas in *Turky* they hunt them as unclean
Animals: it is faid they are much better in *Georgia* than any where elfe;
but the reafon I believe is, becaufe moft Travellers, who have generally

coming Stomachs, think every thing excellent: indeed their Gammons feem'd to us a new kind of Food, for we had eat none fince our Departure from the *Archipelago*. The *Georgians* look upon the *Turks* to be ignorant, and ridiculous in their abftaining from Hog's Flefh : the *Turks*, on the contrary, call the *Perfians Schifmaticks*, and the *Georgians Infidels*, becaufe they eat it without any fcruple.

A S to the *Georgian* Women, they did not furprize us, becaufe we expected to find them perfect Beauties, according to the Defcription commonly given of them. The Women with whom we exchang'd our Enamels were not at all difagreeable; nay, they might be counted Beauties in comparifon to the *Curdes*, whom we had feen towards the Sources of the *Euphrates*. Our *Georgians* had however an Air of Health that was pleafing enough; but after all, they were neither fo handfome nor fo well fhap'd as is reported. Their Skin is often perfum'd with the Vapour of Cow-dung; neither are thofe that live in the Towns any thing extraordinary, more than the others: fo that I think I may venture to contradict the Defcriptions that moft Travellers have made of them. We brought the Capuchins of *Teflis* to be of our Opinion; theyknow the Country better than Strangers, and have not yet been able to perfuade thefe Women to lay afide the ufe of the nafty Paint with which they fpoil their Faces, to keep up the antient Cuftoms of the Country. We were told that they ftole the moft beautiful Girls about fix or feven Years old, to carry them to *Hifpahan* or into *Turky*; the Parents of the Children and their neareft Friends often have a hand in thefe Doings. To avoid this Inconveniency, they marry them at feven or eight Years old, or fhut them up in Nunneries; fo that the Art of Ogling we had learnt at *Paris* was of no manner of ufe to us, for in all probability they had lately carried away all the Girls that were pretty to other Places. Here is the Picture of a *Georgian* Woman that we thought agreeable enough. The Cuftom of taking away the handfome People out of this Country is very far from being new. *Zonaras* obferves, that by the King's Order they us'd to make Eunuchs of the likelieft Boys, and then fell them to the *Greeks*; but to appeafe Seditions, it often cofts the Fathers their Lives.

W H A T

Women of
TEFLIS.

WHAT is moſt edifying upon the Frontiers of *Georgia*, is, that nothing is exacted from Strangers. You may go in and out of the King of *Perſia*'s Dominions when you pleaſe, without asking leave of any body whatſoever. The Merchants of our Caravan, which was grown ſomewhat more numerous by the way, aſſur'd us that they not only treated the *Franks* reſpectfully, but look'd upon them even with Fear and Veneration when they wore Hats and Coats; whereas in *Turky*, they would infallibly ſtone a Man that ſhould make ſuch a ſtrange Figure. There are but very moderate Cuſtoms on the Merchandizes that enter *Perſia*. About this Frontier we paſs'd the River *Arpagi*, which comes from *Cars*, or to ſpeak more properly, which receives the River of *Cars*, as was ſaid before. The *Arpagi* runs into the *Araxes*; the *Araxes* joins the *Kur*, and the *Caſpian* Sea receives all theſe different Waters. The *Arpagi* is reckon'd to abound with Fiſh the moſt of any River in the Country; ſome will have it that it ſerves as a Frontier to the two Empires: but it is not our buſineſs to decide this Queſtion, and at worſt, the whole difference is but a quarter of a League.

WE mounted our Horſes the 17th of *July* at half an hour paſt three in the Morning, and encamp'd about ten in a great Plain, after having paſs'd ſome pretty high Mountains, where our Teeth chatter'd in our Heads. The whole Country is full of Graſs; but all manner of Trees have been long baniſh'd out of it. Among the Plants which we obſerv'd, we diſcover'd a Species of *Aconite* like that which is call'd *Wolf's-Bane*. The Stalks of this we are now ſpeaking of form a Pyramid of Flowers about a foot and a half high. Each Flower is white. The Head-piece, which is fifteen lines high, is rounded at the end, and three lines broad. The Croſiers are purple. On ſome Stocks grow Flowers that approach to a dingy white.

THE 18th of *July* we ſet out at half an hour paſt four, and travell'd till Noon. The Change of Country ſurpriz'd us ſo agreeably, that we thought we were come into a new World. All round you, lay high Woods mix'd with Coppices, among which grew Oaks, Beech-trees, Elms, Lindens, Maples, Aſh-trees, Yoke-Elms, with great and ſmall Leaves. There are alſo Haw-thorns and Elders, Hazles, Pear, Plum, and Apple-Trees; Strawberries and Raſberries are far from being ſcarce. Who

could

could have expected to fee fo many fine things? They were reaping their Corn at the bottom of the Valley where we encamp'd. We this day firft faw Vines in this Country; and tho their Wine cannot be call'd good, yet it may be reckon'd Nectar in comparifon of that we drank at *Erzeron*. The next day's Scene was no lefs agreeable, for from three in the Morning till ten we journey'd in a Valley, which, tho narrow and fteep, was charming for its Verdure and its different Points of View. The Houfes are at the bottom, or half way up the fides; nothing but Woods run along the top of the Profpect; all the reft is taken up with natural Vineyards and Orchards, in which the Nut, Apricot, Peach, Plumb, Pear, and Apple-trees grow of themfelves. If this Valley be not that which *Procopius* defcribes between the Country of the *Tzans* and the *Armenian Perfia*, we cannot doubt its being one of thofe Parts of *Georgia*, wherein, according to *Strabo*, abound all manner of Fruits, which the Soil produces without Culture. They take no farther care of their Vines, fays that Author, than juft cutting them once every five Years. After you are paft the Country of the *Tzans*, *Procopius* fays you enter in a profound fteep Valley, which is one of the Appurtenances of Mount *Caucafus*, well peopled, where you eat all the kinds of Fruits that can be wifh'd for in Autumn. It is full of Vines, and after three days Journey, ends at the *Armenian Perfia*. It is very certain we were not far from Mount *Caucafus*. The Mountains that ftretch from *Cars* to *Teflis*, and towards the *Cafpian Sea*, are properly the *Mofchick* Mountains of the Antients, which, according to *Strabo*, take up *Armenia*, quite to the *Iberians* and the *Albaneze*. Be it as it will, the beautiful Valley we are now fpeaking of, concludes in an agreeable Plain pretty well cultivated, thro which runs a confiderable River that comes down from the Mountains; and which, as we were inform'd, runs into the *Kur*, on the fide of *Teflis*. It may be offer'd as a Conjecture, whether this be not the River which *Strabo* calls *Aragos*. The whole Country is fruitful in fine Plants. Here is a Species of *Caffida*, which by its yellow Flowers, and its Leaves flafh'd like the *Germander*, is diftinguifh'd from all the Species of this Genus.

ITS Root, which is reddifh, hard, ligneous, fometimes rifing like a Tuberculum, and garnifh'd with hairy Fibres, puts forth Stalks, crooked

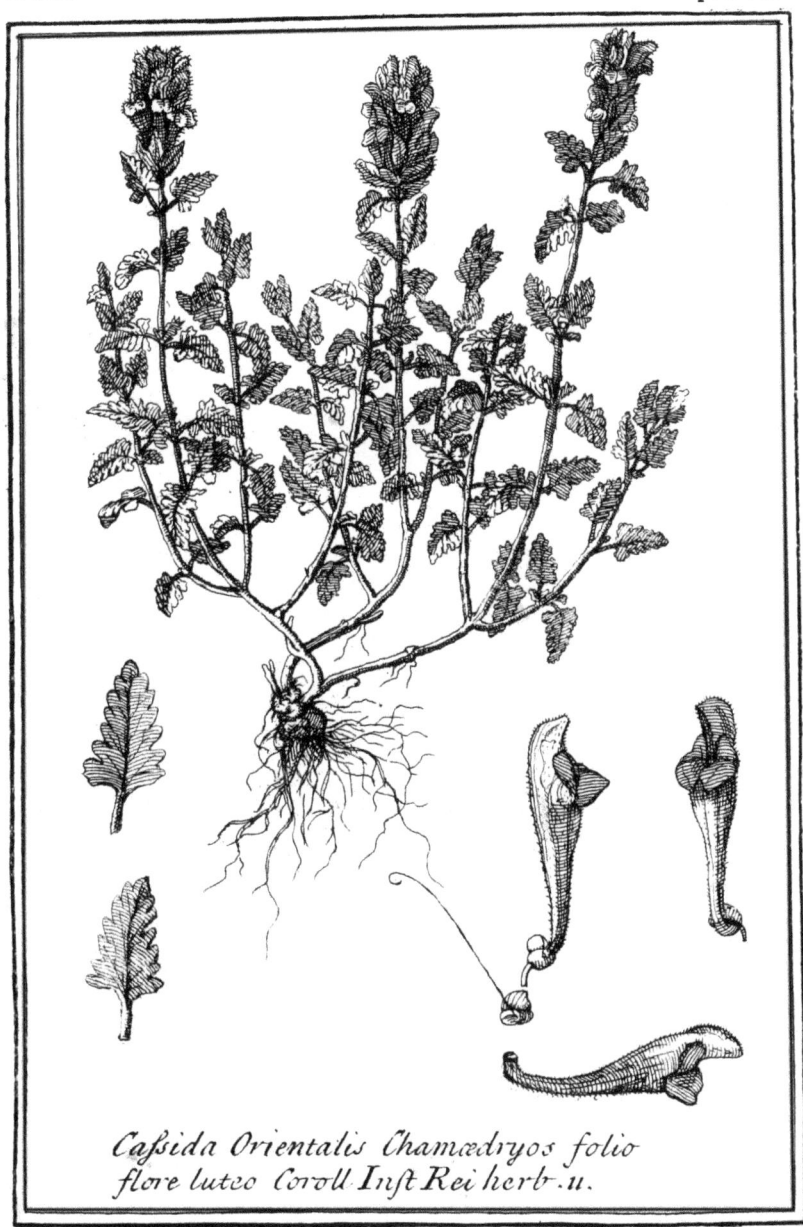

Cassida Orientalis Chamædryos folio
flore luteo Coroll Inst Rei herb. II.

to the Ground, then rifing ftrait again, which eafily multiply by Cluf- Lett. VI.
ters of Fibres, in the places where they lean down upon the Earth. The
Stalks are about eight Inches high, branchy from the very bottom, a
line thick, hard, tufty, accompanied with Leaves two and two, eight or
nine lines long, and four or five inches broad, deep-green, but white
within, flafh'd like thofe of the *Germander*, fuftain'd by a Tail three or
four lines long. They diminifh till they come towards the Summit, and
thefe Summits end in a Spike an inch and half long, garnifh'd with Leaves,
pale-green, feven or eight lines long, pointed, thick fet, not at all, or
very little indented. From the bofoms of thefe Leaves grow yellow Flowers about fifteen lines high, narrow'd at bottom into a Pipe, which is but
one line diameter, but opening higher, and cut into two Lips. The upper is a Head piece four lines high, garnifh'd with two little Wings
greenifh-yellow; the Under-lip is yellow alfo, three lines long, hollow'd, and approaches fomewhat to the form of a Heart. The Cup is
but two lines high, parted into two Lips, the higheft of which reprefents a
Scholar's Cap, at the bottom whereof is a Piftile or Pointal of four Embryo's
furmounted by a crooked Thred, ftretch'd out and parted in the Head-
piece of the Flower. The whole Plant is bitter. It loves a fat Soil
and a warm Climate. It is eafily rais'd in the King's Garden, and in the
Gardens of *Holland*, where I have communicated it to our Friends.

WE march'd all night the 20th of *July*, and arriv'd not at *Teflis* till
about Noon, after having refted for about an hour, three Miles diftance
from the City, upon a good agreeable Mountain. The Carriers generally fet out in the Night-time, to avoid the Couriers of the *Perfian*
Princes, who have a Privilege of taking any Horfes they find upon the
high Roads, except thofe of the *Franks*; for they think they fhould vi-
olate the Duties of Hofpitality, if they treated them as they do the
Natives. As there are no fettled Pofts, and thefe Couriers are fuppofed to be riding about Affairs of Confequence, no body grumbles at
their ufing the Horfes of private Perfons; fo that the difmounted Traveller is forc'd to walk afoot till he has caught his Horfe again. This
Fafhion is a little uncivil; but 'tis the Cuftom of the Country, and 'twould
be dangerous to refift.

AFTER

AFTER having pafs'd feveral flat Countries, you enter into fteep Paffes as you approach *Teflis*. This City is upon the Declivity of a Hill which is quite bare, in a pretty narrow Valley, five days Journey from the *Cafpian*, and fix from the *Black Sea*, tho the Caravans reckon it double the way. *Teflis* or *Tiflis* is at prefent the Capital of *Georgia*, known to the Antients by the Names of *Iberia* and *Albania*. *Pliny* and *Pomponius Mela* mention a People call'd *Georgi*. Perhaps *Georgia* retains that Name, or may be the *Greeks* call'd them *Georgi*, as much as to fay, *good Husbandmen*. The *Iberians*, as we are inform'd by *Dion Caffius*, inhabited the Lands on this and t'other fide the River *Kur*, and confequently were Neighbours of the *Armenians* that liv'd to the Weft, and of the *Albanefe* to the Eaft; for thefe latter poffefs'd the Lands that are beyond the *Kur*, quite to the *Cafpian Sea*. Thefe *Iberians*, a very warlike Nation, declar'd againft *Lucullus*, for *Mithridates* and *Tigranes* his Son-in-Law. *Plutarch* obferves, that they were never fubject either to the *Medes* or *Perfians*, nor even to the mighty *Alexander*; neverthelefs they were beaten by *Pompey*, who advanc'd within three days Journey of the *Cafpian Sea*, but he could not fee it, as defirous as he was of that fight, becaufe the whole Country was cover'd with Serpents, whofe Bite was mortal. *Artoces*, who then reign'd over the *Iberians*, endeavour'd to amufe *Pompey* with Pretences of feeking his Friendfhip; but *Pompey* entred his Dominions, and went ftrait to *Acropolis*, where the King kept his Court. *Artoces*, furpriz'd and frighted, fled beyond the *Kur*, and burnt the Bridge. The whole Country fubmitted to the *Romans*, who by that means became Mafters of one of the chief Paffages of Mount *Caucafus*. *Pompey* left Garifons there, and proceeded to a compleat Reduction of the Country along the *Kur*. Might not one imagine that *Teflis* is the antient City of *Acropolis* the Capital of *Iberia*, upon the River *Kur*? The Name and Situation of this Town are perfectly agreeable to this Notion.

POMPEY, without hearkening to any Propofals of Peace, purfu'd and defeated *Artoces*. This is probably the Battel mention'd by *Plutarch* in the Life of that illuftrious *Roman*, wherein, he fays, nine thoufand *Iberians* were kill'd on the fpot, and above ten thoufand taken Prifoners. This too is the fame *Artoces*, that to obtain Peace, fent *Pompey* his Bed, his Table, and the Saddle of his Horfe. Tho all this Furniture was of Gold,

Pompey

Pompey would not hear of any Accommodation, till he had got the King's Lett. VI. Son for an Hoſtage, order'd the Queſtors of the Army to put them into the publick Treaſury. *Appian* calls *Artocus* the King of *Iberia* ; *Eutropius, Arthaces,* and *Sextus Ruffus* calls him *Arſaces. Canidius Craſſus, Mark Anthony's* Lieutenant, made that General's Name conſiderable in Mount *Caucaſus,* to uſe *Plutarch's* Words. *Canidius* enter'd *Iberia* by the ſame Paſſage as *Pompey.* According to *Dion,* he ſubdu'd *Pharnabazes* King of *Iberia,* and *Zoberes* King of *Albania* : the ſame Hiſtorian relates that the Emperor *Claudius* reſtor'd *Iberia* to one of its Kings call'd *Mithridates.* That Name was common to ſeveral Kings of the *Pontus,* of the *Cimmerian Boſphorus,* and of *Iberia.* The *Mithridates* we now ſpeak of was dethron'd and ſlain by his Brother *Pharaſmanes*; but all theſe Revolutions concern us little. This which happen'd under *Conſtantine the Great,* is much more worthy our Attention.

GOD was pleas'd that the *Iberians,* whom at preſent we know by the Name of *Georgians,* ſhould be enlightned with the true Faith, thro the Miniſtry of a Chriſtian Slave. She converted them by her Miracles, and cur'd their King of a Suffuſion that fell upon his Eyes as he was hunting. *Socrates* adds, that theſe new Converts deſir'd Biſhops of *Conſtantine* to inſtruct them; and *Procopius* aſſures us they were the beſt Chriſtians of thoſe times. *Gyrgenes,* one of their Kings, being preſt by *Cavades* King of *Perſia* to conform to his Religion, implor'd the Aſſiſtance of the Emperor *Juſtin,* who ſucceeded *Anaſtaſius,* and this Affair kindled a War between the two Empires. Another of their Kings named *Zanabarzes* came to *Conſtantinople* in the time of *Juſtinian* to be baptized with his Wife, his Children, and ſeveral Noblemen of his Court. The Emperor gave him great Marks of Friendſhip and Eſteem.

THERE is a diſmal Alteration in this Point now. The Prince of *Georgia,* who in reality is no more than Governour of the Country, muſt be a *Mahometan*; for the King of *Perſia* will not give this Government to a Man of a Religion different from his own. The Name of the Prince of *Teflis* was *Heraclea,* while we were there ; he was of the *Greek* Rite, but they oblig'd him to be circumcis'd. They ſay this Wretch profeſs'd both Religions, for he went to the Moſque, and came to Maſs.

too at the Church of the *Capuchins,* where he would drink his Holiness's Health. 'Twas the most inconstant, irresolute Prince in the World; the People about him would turn his Opinion several times in an Affair as clear as the Day: here is an Instance of it which related to a Rascal, that in the Judgment of every Mortal deserv'd more than Death, were it possible to deprive a Man of any thing more precious than Life. A Nobleman came and represented to him the Enormity of the Crimes committed by this Villain; the Prince immediately commanded that the Hand should be cut off with which he had committed his Murders; but a Lady having besought his Clemency, and assur'd him that the poor Man's Children must starve, if he lost the Hand that got them Bread, the Order was presently revers'd. A Courtier after this told the Prince, that the Man really ought to have suffer'd Death for the Good of the People. Let him be executed then, says *Heraclea.* The Criminal's Wife throws herself at his Feet: Suspend the Execution, cries he. After the Wife was gone, one of the Prince's Favourites put him in mind that he would lose the Respect due to him, if he pardon'd such Crimes: Let him be punish'd, says he; and then the Executioner took him at his word, and cut off the Criminal's Hand: but the Prince, at the Sollicitation of another Favourite, who had receiv'd a Present from the Rogue's Relations, dispossess'd the Executioner of two Towns, which he own'd, for not having waited his last Pleasure. The Executioner in *Georgia* is very rich, and People of Quality exercise the Office: it is so far from being counted infamous, as in all other parts of the World, that here it reflects Glory upon a whole Family. They will boast what a number of Hang-men they have had among their Ancestors; and they build upon this Principle, that nothing is so noble as executing Justice, without which no Man could live safe. A Maxim worthy the *Georgians!*

GEORGIA is at present entirely at peace; but it has often been the Stage of the Wars between the *Turks* and *Persians. Mustapha* Bassa, who commanded the Army of Sultan *Amurath,* took *Teflis* in 1578. He wasted the whole Country with Fire and Sword, and carried away to *Constantinople* the Queen of *Georgia*'s two Sons, whereof one turn'd *Mahometan,* and the other dy'd a Christian. The *Persians* however came to

the

the Affiftance of the *Georgians*, and in a Battel left threefcore and ten
thoufand *Turks* dead upon the fpot. The War was rekindled in 1583,
but the *Turks* always came by the worft. M. *Chardin* gives a very long
and particular Account how *Georgia* fell into the hands of the *Perfians*;
and to him I fhall refer, for he feems to be an Author of great Exactnefs, only that he is a little too much prejudic'd in favour of the *Georgian* Women.

THE Prince of *Georgia* has above fix hundred *Tomans* in Lands, according to the way of reckoning in this Country; one *Toman* is worth twelve
Roman Crowns and a half, which make eighteen *Aflanis* or *Abouquels*,
which are a fort of Crowns coin'd in *Holland* for the *Levant*. The *Ea-
ftern* People call them *Aflanis*, from the Figure of a Lion, which they call
Aflan. This Coin is known in *Egypt* by the Name of *Abouquel*. The
Prince's Revenues confift in a Penfion of three hundred Tomans, which
the King allows him, and in the Cuftoms of *Teflis*, and the Entries of
Brandy and Melons; the whole amounts to near five hundred Tomans,
without reckoning what he exacts under pretence of treating fuch great
Men as pafs thro *Teflis*. The Country provides him with Sheep, Wax,
Butter, and Wine. As to the Sheep, he has one every Year for every
Fire-hearth, which amount to forty thoufand Sheep; for tho there are fix-
ty thoufand Houfes in *Georgia*, there are Sheep bred but in forty thoufand.
Of Wine they give the Prince forty thoufand Load; one Load weighs
forty Batmans, and the Batman is fix Oques.

THE *Sequins* of *Venice*, which are current all over the *Eaft*, are worth
at *Teflis* fix *Abagis* each, and three *Chaouris* or *Sains*. The Sequin is
worth feven Livres ten Sous *French* Money, fo that the Abagi is worth
about two and twenty Sous: four Chaouris make one Abagi. This Coin
feems to have retain'd the Name of thofe antient Inhabitants of *Iberia*
that were call'd *Abafgians*. 'Tis true, they write it *Abaffi*, tho it is pro-
nounced *Abagi*, that is to fay, Money coin'd in the Name of King *Abas*.
Thus the Chaouri comes to five Sous fix Deniers: An Ufalton is worth
half an Abagi or two Chaouris, that is to fay, eleven Sous. A Chaouri
or Sain is worth ten Afpers of Copper or Carbequis, forty of which make
an Abagi. Laftly, a Piafter is worth ten Chaouris and a half.

T H E *Georgians* and *Armenians* pay Capitation to the King of *Perſia* at the rate of ſix Abagis a head. This Capitation is farm'd out at three hundred Tomans. They preſent the King, by way of Homage, four Hawks every Year, ſeven Slaves every three Years, and four and twenty Loads of Wine : but much more than this is ſent him ; and beſides, moſt of the handſome Girls are ſet apart for his Seraglio. The *Georgians* are great Sots, and drink more Brandy than Wine ; the Women carry this Debauchery even further than the Men, and when I have ſaid thus much, I'll leave any body to judge whether they have cruel Hearts. This Exceſs is perhaps what has ſpoil'd the fine Breed of *Georgia,* for nothing more contributes to the Procreation of handſome Children than a regular Life, for which reaſon the *Turks* are generally handſome. Very few among them are lame or crooked, eſpecially in thoſe Countries that lie a little diſtant from the Sea-Coaſt, where the *Franks* have little to do ; for theſe Gentlemen are accus'd of being very incontinent where-ever they find an Opportunity.

T H E R E is great Debauchery in *Teflis* among the Chriſtians : 'tis true they have nothing but the Appellation of Chriſtians ; and indeed the *Jews* and *Mahometans* live not a whit more ſoberly. Wine is the Source of all theſe Diſorders ; it would be good Policy to forbid the Uſe of it to thoſe who are well in health, and to allow it only to ſick People. *Chardin* juſtly obſerv'd that there is hardly any Nation where they drink ſo much Wine as in *Georgia* ; rich and poor, all in general, ſwallow it without any Moderation : this ſtupifies their Senſes, and makes 'em the more patient under the Yoke of their Lords, who treat them tyrannically. They not only ſtand over them with Sticks, to force them to work, and take their Children from them, to ſell them to their Neighbours, when Money runs low with them ; but even pretend to have Power of Life and Death over their Subjects. The white Wine of *Georgia* is tolerably good ; that which they ſend to the Court of *Perſia* is a red Wine that has ſomewhat of the taſte of Cote-rotie, but it is ſtronger and more heady. The Vines in this Country grow round Trees, and creep up them as in *Piemont* and ſeveral Parts of *Catalonia.* The *Mahometans* drink Wine, or let it alone, according to the taſte of the King. If their Prince does

not

TEFLIS,
The Capital of Georgia.

not love it, they are forbid to touch it; but in this cafe 'tis with great regret, that they follow the Fashion of the Court.

T E F L I S is a pretty large Town, and very populous; the Houses are low, dark, and for the most part built of Mud and Bricks; and even these are superior to the Houses in the rest of the Province, where they are very far now-a-days from keeping up to the Description given us of them by *Strabo*: *Most part of* Iberia, says he, *is well inhabited*; *it contains large Towns, and Houses cover'd with Bricks*; *their Architecture is good, as is also that of the publick Edifices and Squares.* At present the Walls of *Teflis* are hardly higher than those of our Gardens, and the Streets are ill pav'd. The Citadel is in the highest part of the Town, upon a fine Situation, but the Inclosure is almost ruinated, and defended by very forry Towers. The whole Garison consists in a few wretched *Mahometan* Trades-people, who are paid for keeping Guard in it. There they lodge with their whole Families, and know nothing in the world of the Management of their Arms. The Place serves for a Refuge to People deep in Debt, or in fear of Prosecution for their Villanies. The Place for Exercise, which is before it, is handsome and spacious, and serves for a Market-place, where you may buy the best Wares that the Country produces. When you come from *Hispahan* to *Teflis*, you must pass thro the Citadel; so that the Prince of *Georgia*, who, according to the *Persian* Custom, must go without the City to receive the King's Orders or Presents, is forc'd to go thro that Citadel, where the Governour might easily seize him, if he had Orders for that purpose.

THE City stretches from South to North. The Citadel is in the middle. It might be made a considerable Place; for the side of the Mountain on which it is situated is very steep, and the River *Kur*, which runs along it, is not fordable. The Circuit of the Town takes up the side of this Mountain, and makes a kind of Square, the sides whereof descend to the very bottom of the Valley; but half the Walls are ruin'd, and scarcely so good as those of the *Bois de Vincennes*, whatever M. *Chardin* may say to the contrary. The Prince's Palace, which is below the Citadel, is very antient, and tolerably well laid out, considering what Country it is in. The Gardens, the Volaries, the Dog-Kennel, the Falconry, the Square and Bazar, which are before it, are worth feeing. They carried us into a new Hall, which was agreeable enough, tho built of nothing but Wood. It has Windows on every fide, which are glaz'd with great Squares of blue, yellow, grey, and other colour'd Glaffes. There is alfo fome *Venice*-Glafs among the reft ; but the Pieces are but fmall, and not comparable in Beauty to thofe of *Paris*. The Cieling confifts of Compartments of gilded Leather. The Womens A- partment, we were told, was much finer even than this ; the Key, by I know not what Accident, was out of the way, or elfe they feem'd very well inclin'd to fhew us that too. The Court was then out of Town. The Prince was faid to be a little out of Order, and this was one of the chief Reafons of our leaving *Teflis*, for fear he fhould take it in his head to detain us with him, to take care of his Health, which would be no- thing uncommon in the *Levant*.

FROM the Palace we went to fee the Baths, which are not far from it. They are very fine Springs, and about as hot as thofe of the Wa- ters of *Elija* near *Erzeron*. In the Baths of *Teflis* there are Waters both lukewarm and cold, befides the hot. Thefe Baths are well kept up, and are almoft the only Diverfion of the Citizens of the Town. Their Trade confifts moftly in Furs, which they fend into *Perfia* or to *Erzeron*, for *Conftantinople*. The Silk of the Country, and that too of *Schamaki* and *Gangel*, do not pafs thro *Teflis* ; which they forbear, to avoid the exceffive Impofts that would be laid upon them. The *Armenians* go and buy it upon the fpot, and have it carried to *Smyrna*, or fome other Port of the *Mediterranean*, to fell it to the *Franks*. They fend every Year, from the Country about *Teflis*, and other Parts of *Geargia*, about two thoufand Camel-load of the Root call'd *Boia* to *Erzeron* ; from thence it goes to the *Diurbequis*, where it is us'd in dying the Stuffs which they make there for *Poland*. *Georgia* alfo remits great quantities of the fame Root to *Indoftan*, where they make the fineft painted Stuffs. We fail'd not to take a Walk in the Bazar of *Teflis*, where you fee all manner of Fruits, and efpecially Plums and excellent Summer Bon-Chretien Pears. We alfo went to fee the Prince's Country-Houfe, which is in the Suburb as you come from *Turky*. This Houfe is diftinguifh'd by an Eftrapade that is before the Door ; the Gardens are much better planted, and more artfully laid out than thofe in *Turky*. In thefe Gardens it was that we faw

with

with admiration that fine Species of *Perſicary* or *Arſe-ſmart*, with Tobacco Lett. VI.
Leaves, whereof I have given a Figure and Deſcription in one of the Vo-
lumes of *the Hiſtory of the Royal Academy of Sciences.* M. *Commelin* has mention'd it in his *Treatiſe of rare Plants.* As the Seed was not then ripe; we deſir'd an *Italian* Capuchin, who had finiſh'd his Miſſion at *Teflis,* and was to
return by the way of *Smyrna,* to get us ſome of it in its Seaſon : this Father
has communicated it, as well as we, to the Curious in *Holland* and *England.*
We alſo found of it in the Gardens of the Monks of the *Three Churches.*

THE Grand Viſier's Houſe is the fineſt in the City. It was hardly
finiſh'd when we arriv'd at *Teflis.* The Apartments are upon a line, but
low, according to the Faſhion here, with Frizes of Flowers very ſadly
done, as are alſo the Hiſtory - Paintings, in which the Figures are ill
drawn, ill colour'd, and worſe group'd. The *Perſians,* tho *Mahometans,*
are pleas'd with Pictures, and they paint in Freſco at *Teflis* upon beaten Plaiſter, in a manner agreeable enough. Plaiſter is very common
here, and Wood too, tho their ordinary Fuel is Cow's Dung. 'Tis believ'd there are about twenty thouſand Souls in the City, to wit, fourteen thouſand *Armenians,* three thouſand *Mahometans,* two thouſand *Georgians,* and five hundred *Roman* Catholicks. Theſe laſt are converted *Armenians,* declar'd Enemies to the other *Armenians* ; the *Italian* Capuchins
could never reconcile them.

WE lodg'd with theſe good Fathers, who are very much belov'd in
Georgia, where they are Phyſicians both for Body and Soul. They do
not want Employment, for there are but three of them, two Fathers
and one Brother. The Congregation of the *Propaganda* gives them at
preſent but twenty five *Roman* Crowns a Man, which is about a hundred
French Livres ; but then they are allow'd to practiſe Phyſick, which 'tis
ſuppos'd they underſtand, tho in reality they have but very ſlight Notions of it. If the Patient dies, or is not cur'd, the Doctor has not a
Farthing ; if he recovers, which happens merely by chance, they ſend
Wine to the Convent, Cows, Slaves, Sheep, &c. Their Convent is pretty ; they entertain almoſt all the *Franks* that paſs thro *Teflis* ; and
their Hoſpital belongs to the F. Capuchins of *Romania.* The Superior
of the Houſe aſſumes the Title of *Prefect of the Miſſions of Georgia.*
The Theatins, who were in *Colchis* or *Mengrelia,* receiv'd from the ſame

Con-

Congregation a hundred Crowns a Man, and were become Lords of a Town. There is now but one of their Fathers refiding there; the reft are retir'd. The Patriarch or Metropolitan of the *Georgians* acknowledges the Patriarch of *Alexandria*, and both agree that the Pope is the firft Patriarch in the world. When that of the *Georgians* comes among the Capuchins, he drinks to the Pope's Health; but he will own him no otherwife. The King of *Perfia* names the Patriarch of *Georgia*, without exacting either Prefent or Money. He of the *Armenians*, on the contrary, who refides at *Erivan*, expends above twenty thoufand Crowns in Prefents to obtain his Nomination, and yearly provides all the Wax that is burnt in the King's Palace. This Patriarch is very much defpis'd at Court, as indeed the *Armenians* are too: they are look'd upon as a Pack of Slaves, that will never dare to endeavour at fhaking off the Yoke.

THE King of *Perfia* is forc'd to be at more charge in *Georgia*, than the Profits arifing to him from thence will pay. To make fure of the *Georgian* Nobility, who are Mafters of the Country, and might give themfelves up to the *Turks*, he bribes them with handfome Penfions. The *Turks* would receive them with open Arms; and the *Georgians*, who are a well-made People, and very fit for War, are not a little inclin'd to change their Lord. Before the Court of *Perfia* could be inform'd of their Revolt, they might not only join themfelves to the *Turks*, but alfo to the *Tartars* and *Curdes*. There are in *Georgia* a dozen confiderable Families that live in a good Underftanding, with relation to their common Interefts. They are divided into feveral Branches, fome have two hundred Fire-hearths, others from five hundred to a thoufand or two thoufand; nay, there are that poffefs even to feven or eight thoufand Fire-hearths. Thefe Fires are fo many Houfes, which make Villages, and each Fire pays a Tenth to its Lord. Each Fire fends a Man in time of War; but the Soldiers are not oblig'd to march more than ten Days, becaufe they can carry Provifions for no longer Term; and they retire when thofe grow fhort, fuppofing Care has not been taken to lay in Stores to furnifh them.

ANY Man may make Gun-powder at *Teflis* for his own Ufe: they bring the Sulphur from the *Gangel*, and the Nitre is found in the Mountains

near

near *Teflis*. Foffile Salt is very common in the Road of *Erivan*.   Olive-Oil Lett. VI·
is very dear here : the People eat and burn nothing but Linfeed Oil ; all
the Fields are cover'd with this Plant, but they cultivate it only for the
Seed, for they throw away the Stalk without beating it to fpin : What a
Lofs is here? it would make the fineft Stuffs in the world ; perhaps
indeed thofe Stuffs might prejudice their Trade of Cotton-Stuffs.   The
Kur carries Plenty thro all thefe Countries ; it runs thro the middle of
Georgia, and its Head is in Mount *Caucafus*.   *Strabo* was well acquainted
with its Courfe.   It was here that the Kings of *Iberia* and *Albania*, according to *Appian*, plac'd themfelves in Ambufh with threefcore and ten
thoufand Men, to ftop the Progrefs of *Pompey* ; but that General continued
a whole Winter upon the Banks of the River, and cut to pieces the *Albaneze* tnat durft pafs it in his prefence.   This River receives feveral others,
befide the *Araxes*, which is the biggeft of all : afterwards it difcharges it
felf in the *Cafpian Sea* by twelve Mouths, all navigable.   *Plutarch* doubts
whether the *Kur* mingles with the *Araxes* ; but without troubling our
felves here with the Opinions of the antient Geographers, *Olearius*, who
had been on the fpot, affures us it does, in his *Journey into* Mufcovy,
Tartary, *and* Perfia.

    T O finifh this Epiftle, my Lord, it remains only that I give you an
Account of what I have gather'd in this Country, touching the Religion
of the *Georgians*, if we may do them the honour to fay they have any
Religion at all.   Ignorance and Superftition are fo general among them,
that the *Armenians* know no more of the matter than the *Greeks*, and the
Greeks are as ignorant as the *Mahometans*.   Thofe whom they here call
Chriftians, place the whole of their Religion in fafting ftoutly, and above
all, in obferving the great *Lent* fo ftrictly, that the very Monks of *Trappe*
would go near to be ftarv'd in it.   Yet not only for the fake of Example,
but alfo to avoid Scandal, the poor *Italian* Capuchins faft without Neceffity as often and as feverely as the Natives.   The *Georgians* are fo very fuperftitious, that they would be chriftned anew, if they had broken
but one of their Fafts.   Befides the Gofpel of Chrift, they have their
little Gofpel, which is fpread among them in Manufcript, and contains nothing but Extravagancies : for inftance, how that *Jefus Chrift, when a
Child, learnt the Trade of a Dyer , and that being commanded by a Lord to go*
of

of an Errand, he ſtaid too long before he came back; whereupon this Lord grow-
ing impatient, went to his Maſter's to enquire for him. Jeſus Chriſt return-
ing ſoon after, was ſtricken by this Man; but the Stick with which he did it,
bloſſom'd immediately: this Miracle was the Cauſe of the Converſion of this
ſame Lord, &c.

WHEN a *Georgian* dies, if he does not leave a good deal of Money behind him, which they ſeldom do, the Heirs take two or three of their Vaſſal's Children, and ſell them to the *Mahometans*, to pay the *Greek* Biſhop, who has a hundred Crowns for one Maſs for the Dead. The *Catholicos* or *Armenian* Biſhop lays on the Breaſt of thoſe that die in his Communion a Letter, whereby he deſires St. *Peter* to open them the Gate of Paradiſe; and then they put him into the holy Linen. The *Mahometans* do the like for *Mahomet.* When a Man of Note is ſick, they conſult the *Georgian, Armenian,* and *Mahometan* Diviners: theſe Fellows commonly ſay that ſuch a Saint or ſuch a Prophet is angry; and that to appeaſe his Wrath, and cure the Patient, they muſt ſlay a Sheep, and make divers Croſſes with the Blood. After the Ceremony is over, they eat the Fleſh of it, whether the ſick Perſon recovers or no. The *Mahometans* have recourſe to the *Georgian* Saints, the *Georgians* to the *Armenian,* and ſometimes the *Armenians* to the *Mahometan* Prophets; but they all hang together to create Coſts for the Patient, and uſually chuſe their Saints, according to the Inclination or Devotion of the Kinſ-folks.

THE Women and Girls are deeper inſtructed in their Superſtitions than the Men. They breed up moſt of the *Georgian* Girls in the Mo-naſteries, where they learn to read and write. They are firſt receiv'd No-vices, and then profeſs themſelves; after which they may perform the Aurial Functions, as baptizing and applying the holy Oils. Their Religion is properly a Mixture of the *Greek* and of the *Armenian.* There are ſome *Mahometan* Women in *Teflis,* who are Catholicks in their Hearts, and theſe are better Catholicks than the *Georgian* Women, being well inſtructed. The Viſier's Daughter, at the time that we were there, the Wife of the Prince's Phyſician, and ſome others, as we were aſſur'd by the Capuchins, had been baptiz'd in ſecret. Thoſe Religious confeſs them, and give them the Communion, viſiting them under pretence of preſcribing them Remedies

for

for fome feigned Diftemper; and they fometimes come to their Church, where they keep ftanding, not daring to give any Token of their Faith.   In the laft Revolt of Prince *George*, who ftir'd up the whole Coun- try againft the King of *Perfia* about twenty Years ago, the Soldiers took up their Lodgings in the Houfes of the Citizens of *Teflis*, and even in the *Greek* and *Armenian* Churches; but paid great Refpect to the *Latin* Church, where the *Mahometans* themfelves begg'd entrance as a Favour.

T H E R E are five *Greek* Churches in *Teflis*, four in the City, and one in the Suburb; feven *Armenian* Churches, two Mofques in the Citadel, and a third which is deferted.   The Metropolis of the *Armenians* is cal- led *Sion*, it is beyond the *Kur* upon a fteep Rock ; the Building is ve- ry folid, all of hewn Stone, terminated by a Dome, which is an Honour to the Town.   The *Tibilcle* (as they call the Bifhop of *Teflis*) dwells clofe to it.   The Chriftian Churches have not only Bells, but even Croffes upon the tops of the Steeples.   This is wonderful in the *Levant*.   On the contrary, the *Muezins* or *Mahometan* Chanters dare not give out their times of Prayer in the Minarets of the Mofques of the Citadel, for fear of being fton'd by the People.   The Church of the Capuchins is fmall; but 'twill be pretty enough when compleated.

I am, My L o r d, &c.

LETTER VII.

To Monseigneur the Count de Pontchartrain, *Secretary of State, &c.*

MY LORD,

E can no longer defer giving an account of the Observations we have made in our Walks thro the Terrestrial Paradise. 'Tis owing to your Lordship, that we have the Happiness of visiting this Place; and we ought not to satisfy ourselves with acknowledging this in a common manner: but indeed, every Letter I have the honour to write, would bring you fresh Expressions of our Thankfulness, had you not been pleas'd so strictly to forbid it. We hope however that your Lordship will pardon us for once, and upon this extraordinary Occasion. I persuade myself, that all who shall read with attention, what I am about to write concerning this Place, will agree with me, that if it is possible at this time to assign the Place where *Adam* and *Eve* first appear'd upon the Stage of the World, it was undoubtedly this in which we now are, or that from whence we last came.

¹ Gen. 2. ver. 10-15. I F we follow the Letter of that Passage, ¹ wherein *Moses* describes the Situation of the Terrestrial Paradise, nothing seems more natural than the Opinion of M. *Huetius,* the antient Bishop of *Avranches,* one of the most learned Men of his time. *Moses* assures us, that a River went out of that delightful Place, and divided it self into four Channels, the *Euphrates,* the *Tygris, Pison,* and *Gihon.* But no such River can be found in any part of *Asia,* except this of *Arabia*; that is to say, the *Euphrates* and *Tygris* join'd together, and divided into four great Channels, which empty themselves

into

into the Bay of *Perſia*. *Huetius* therefore ſeems indeed to have fully ſatisfy'd the Letter of the Text, in fixing Paradiſe in this Place : but notwithſtanding this, his Notion cannot be maintain'd, it being ſo very ma-
nifeſt from the *Greek* and *Latin* [1] Geographers and Hiſtorians, that the
Euphrates and *Tygris* formerly ran in ſeparate Beds ; and likewiſe that there
was a Deſign to make a Canal of Communication between the two Rivers ; and that afterwards ſeveral Canals were actually made, by Command of the Kings of *Babylon*, of *Alexander the Great*, and even of *Trajan* and *Severus*, for the facility of Commerce, and to render the Country more fruitful. There is no reaſon therefore to doubt but theſe Branches of this River of *Arabia* were made by the Art of Man, and conſequently were not in the Terreſtrial Paradiſe.

THE Commentators upon *Geneſis*, even thoſe who are moſt confin'd
to the Letter, don't think it neceſſary, in order to aſſign the Place of Paradiſe, to find a River which divides itſelf into four Branches, becauſe of
the very great Alterations the Flood may have induc'd ; but think it e-
nough to ſhew the Heads of the Rivers mention'd by *Moſes*, namely,
the *Euphrates*, *Tygris*, *Piſon*, and *Gihon*. And thus it cannot be doubted
but that Paradiſe muſt have been in the way between *Erzeron* and *Teflis*,
if it be allow'd to take the *Phaſis* for *Piſon*, and *Araxes* for *Gihon*. And
then, not to remove Paradiſe too far from the Heads of theſe Rivers, it
muſt of neceſſity be plac'd in the beautiful Vales of *Georgia*, which furniſh *Erzeron* with all kinds of Fruits, and of which I gave an account in
my laſt Letter. And if we may ſuppoſe the Terreſtrial Paradiſe to have
been a Place of conſiderable Extent, and to have retain'd ſome of its
Beauties, notwithſtanding the Alterations made in the Earth at the Flood,
and ſince that time ; I don't know a finer Spot to which to aſſign this
wonderful Place, than the Country of the *Three-Churches*, about twenty
French Leagues diſtant from the Heads of *Euphrates* and *Araxes*, and near
as many from the *Phaſis*. The Extent of Paradiſe muſt at leaſt reach to
the Heads of theſe Rivers ; and ſo it will comprehend the antient *Media*,
and part of *Armenia* and *Iberia*. Or if this be thought too large a Compaſs, it may be confin'd only to part of *Iberia* and *Armenia*, that is, from
Erzeron to *Teflis* ; for it can't be doubted that the Plain of *Erzeron*, which
is at the Head of *Euphrates* and *Araxes*, muſt be taken in. As to *Paleſ-*

tine,

[1] Plin. Hiſt. Nat. l. 6. c. 26. Polyb. Hiſt. l. 5. Strab. Rer. Geogr. l. 16. Appian de Civ. Bel. l. 2. Arrian de Exped. Alex. l. 7. Ptolem. Geogr. l. 5. c. 17. Ammian. Marcell. l. 24. c. 21. Zoſim. l. 3. c. 24.

tine, where some would perfuade us Paradife lay, to me it feems trifling to attempt to make four Rivers of *Jordan,* which is itfelf but a Brook or Rivulet : and befides, this Country is very dry and rocky. Our learned Men may judge as they pleafe ; but as I have never feen a more beautiful Country than the Neighbourhood of *Three-Churches,* I am ftrongly perfuaded that *Adam* and *Eve* were created there.

WE fet out for this fine Place on the 26th of *July,* and encamp'd at four Hours Diftance from *Teflis,* to join a Caravan defign'd for *Three-Churches,* which affembled upon a large Plain at the end of the Vales of *Teflis.* This Plain is agreeably cover'd with Orchards and Gardens. The River *Kur* runs crofs it from North-North-Eaft to South-South-Eaft, which was likewife nearly our Courfe. The Merchants of the Caravan furnifh'd themfelves here with a fine fort of Reeds, which grew about our Camp, and are very fit for writing in their manner. It is a Species of Cane, which grows about the height of a Man; its Stalk is not above three or four lines in thicknefs, and folid from Knot to Knot, or rather fill'd with a whitifh Pith. The Leaves are about a foot and a half long, and eight or nine lines broad, and cover the Knots with a kind of hairy Sheath; but the reft is fmooth, of a bright green, and lying in Folds or Gutters, white at bottom. The Pannicle or Clufter of Flowers was not full blow , but was whitifh and filky, like the common Reeds. The People of the Country ufe thefe Reeds for Writing ; but the Strokes they make with them are broad and thick, having nothing of the Beauty of our Characters made with a Pen.

THE 27th of *July,* at eleven a Clock in the Night, we left this Place, and travell'd till fix in the Morning, thro moorifh Plains. In the Night we loft our River, and were upon the Approach of Day fo much furpriz'd, that we could not guefs which way it lay. However, it running into the *Cafpian Sea,* it muft of neceffity have turn'd gradually toward the Eaft, as muft likewife the *Araxes,* which joins itfelf to the *Kur*; but it muft be a great way from *Erivan,* feeing in all our Journey we heard no more mention made of the *Kur.* We refted this Day till eight a Clock, and then travell'd till about half an hour after twelve at Noon, to reach *Sinichopri,* a Village which has a handfome Stone Bridge, and a fort of Fort which is now abandon'd. About two a Clock we departed from

hence,

hence, to encamp on the Mountains which are well cover'd with Herbage ; Lett. VII where, with Surprize, we saw the most common Plants intermingled with ⌒⌒⌒° some few others that were very rare. Who would have expected to meet with *Nettles, Celandine* and *Melilot* in the way to Paradise? All which however we found there, as likewise common *Marjoram* and *Mallows*. The *white Dittany* is admirably fine, which grows at the entrance upon these Mountains, from whence there came a certain Freshness which gave us a great deal of Pleasure.

WE were not more happy in Plants the Day following, *viz.* 28 *July*, and I began to doubt whether we were going towards Paradise, or had turn'd our backs upon it, and were going from it ; for after having travell'd from Two a-Clock in the Morning till Seven, upon Mountains cover'd with Woods and Pasturage, we found nothing in the great Roads but *Millet*, *black* and *white Horehound, Burdock, Centaury the leffer, Plantain*, with *Nettles* and *Mallows*, as the Day before. As Fatigue and Trouble is not wont to increase the Appetite ; and being destitute of any other matter of Learning wherein to employ our selves ; and withal having reason to expect to meet with nothing in our suppos'd Paradise, but the Brambles and Thistles which God caus'd to spring up there after the Fall of the first Man ; we should have spent our Time here very ill, if we had not met with an admirable sort of *Ciboulette*, whose Flower smells like *Storax in Tears*. Its Leaves and Roots, which smell like *Spanish Chibouls*, gave us a good Stomach to the remainder of our Provisions.

THE Root of this Plant is almost round, tolerably sweet, and of a Scent between Garlick and Onion. The Suckers or Off-sets which grew by them, form a Head of an inch diameter. The Stalk grows to two feet and a half high, and two or three lines thick ; is solid, smooth, cover'd with a Flower or Powder like that on Plumbs fresh gather'd, and furnish'd with Leaves of a foot and half long, hollow, and three lines wide. At the end of this Stalk is a round Head of an inch and half diameter, whose Flowers, which stand on little Feet or Stalks of four lines in length, consist of six Leaves of two lines long, rais'd on the Back, shining, of a dark red Colour, but brighter toward the Edges. Thro the middle of the Leaves run so many purple Threds, about one line longer than the Leaves themselves, and adorn'd with Tops of the same colour. The Pestle or Pointal

is

is three-corner'd, greenifh, and grows as in other kinds of *Onions*, that is to fay, in three Apartments; but the Plant was not come to Perfection when we faw it, and therefore cannot be more accurately defcrib'd.

WE fet forward about midnight the 29th of *July*, and paffed feveral rugged Mountains; on which we perceiv'd, at break of day, Forefts filled with *Savines* as high as *Poplars*: They differ from thofe we defcribed in the Tenth Letter, in this, That their Leaves which are of the fame make as Cyprefs Leaves, are not faftened together, but come out of the fides, and are ranged three by three as in Stories. The Shells or Husks of thefe Leaves are one line and half in length, ending with a Prickle, of a bright green above, but white and yellowifh below. Thefe Trees were all laden with green Fruit, of about half an inch diameter.

WE encamp'd this morning from feven a-Clock till eleven: And afterwards we put forward till half an hour after one, when we reached *Dilijant*, a handfome Village. The Guards pofted upon the great Road, pretended, that paffing from *Georgia* into the Territories of *Cofac*, a fmall Country between *Georgia* and *Armenia*, we muft pay a Sequin *per* Head; but knowing the *Perfians* to be a poor filly fort of People, we began to be rough with them, and clapped our Hands to our Sabres. At length what with our making a mighty Noife, and talking a Language they did not underftand, no more than we did theirs, they let us pafs quietly. So true it is in all Countries, that they who make moft Noife, and are moft numerous, are always in the right. However, the moft confiderable People of the Place, who were drawn together by reafon of the Difturbance, having affured our Guides that all Horfemen who pafs this Way, are wont to pay an *Abagi per* Head, we voluntarily did fo too: Upon which the Guards made more Excufes, and return'd us more Thanks by far than we had deferv'd of them. They told us that this Tax was laid on Travellers for the Security of the Roads; and that this was the ufual Method in many Provinces of *Perfia*, where the Governors maintain Guards for the publick Safety; the King not permitting them to raife thefe Taxes, but on condition that they be accountable for all Robberies which fhould be committed. The Inhabitants of *Cofac* are counted a very bold and daring People; and pretend to be defcended of the *Cofacks*, who inhabit the Mountains on the North Side of the *Cafpian* Sea.

Sea. The People of *Dilijant*, who were got about us, asked why we were not drefs'd like *Franks*, and did not wear Hats : We anfwer'd, That we came from *Turky*, where Perfons in that Drefs were but ill treated ; which made 'em laugh. They offer'd us pretty good Wine ; and we continu'd our Courfe for an Hour beyond the Town, and encamp'd on the Top of a Mountain cover'd with *Chefnuts*, *Elms*, *Afh-Trees*, *Sarvice-Trees*, and *Yoke-Elms*, with great and fmall Leaves.

W E hop'd to have pafs'd this Night in a Lodging agreeable enough ; but, tho it was a very dark Night, our Guides made us leave this Place about eleven a Clock, to travel all the remaining Part of it over hideous Mountains. In Snowy Seafons few People venture to go this Way. I trufted entirely to my Horfe ; which was much better than to pretend to guide him. *Automata*, which naturally follow the Laws of Mechanifm, extricate themfelves out of Dangers, on all fuch Occafions, much better than the ableft Mechanick, who fhould go about to make ufe of the Rules he has learn'd in his Study, even tho he were a Member of the *Royal Academy of Sciences*. At length, about Five in the Morning, *July* 30. we found our felves upon a Plain near *Charakefis*, a poor Village, ftanding upon a fmall River. Here we became Mafters in our Turn, as in reafon we ought to be ; and oblig'd our Guides to ftop, that we might refrefh our felves with Sleep. But how fhort was our Repofe ! The Demon of Botany, who poffeffed us, foon raifed us : But we made no great Advantage here, and therefore were forry we had ftopp'd. The River *Zengui*, which comes from the Lake of *Erivan*, and paffes by this Town, winds about here ; but is not very large.

J U L Y 31, at Five in the Morning, we fet forward, and travell'd over Mountains very agreeable, but without Trees : And we began to fmell the Smoke of Cow-Dung, as we drew near to *Bifni*. This Scent incommoded us very much in a Convent of *Armenian* Monks, where we din'd. Their Court was full of a fine Kind of *Creffes*, which *Zanoni*, without any reafon, took for the firft fort of the *Thlafpi* of *Diofcorides*. Thefe good Monks received us very handfomely ; but we were not fo agreeably entertain'd by them, as by the *Greek* Monks. The *Armenians* are more grave : And befides, we could not fpeak one Word to them ; whereas we made a fhift to ftammer out a little of the vulgar *Greek* to the Caloyers.

loyers, whose Vivacity is very pleasant and diverting. The Convent at *Bisni* is the best built of any we saw in these Parts; it is strong, and built of good hewn Stone. The Ruins about it show there has been once a considerable Town, tho the Village be small at present. We should have taken it for *Artaxata*, but that it lies upon the River *Zengui*. One would guess the Monastery to be of seven or eight hundred Years standing. We went from thence about Noon; and travelled over another Mountain, to a Monastery of the *Armenians* at *Tagovat*, a smaller Village than *Bisni*, at the Entrance of the great Plain of *Three-Churches*, where we pretend to find Paradise.

WE set out the next Day at Three in the Morning, very impatient to see this famous Borough of the *Armenians*, visited with more Devotion than the *Romipetes* visited *Rome* in the Time of *Rabelais*. *Three-Churches* is but six Hours from *Tagovat*. The *Armenians* call this Borough *Itchmiadzin*, that is, *The Descent of the only Son*; because they believe that our Lord appeared to St. *Gregory* in this Place, as we were told; for we don't understand one Word, either of the vulgar or learned *Armenian* Tongue. Tho we were not much acquainted with the *Turkish* Language, yet being able to count ten, we easily understood that *Utch*, which signifies *Three*, being joined to *Klissé*, a Corruption of *Ecclesia*, signified *Three-Churches*, as the *Turks* call this Place. But it had been more properly call'd *Four-Churches*; for here are four, which seem to have been built a great while. The Caravans stop here to perform their Devotions, that is, to confess themselves, communicate, and receive the Patriarch's Benediction. This Convent consists of four Sides, built like Cloisters, in a very long Square. The Cells of the Religious, and the Chambers for Strangers, are all of the same Make, having each a little Dome in the Form of a Bonnet all along the four Cloisters: So that this may be look'd on as a large *Caravanseria*, in which the Monks have their Lodgings. The Patriarch's Apartment, which is to the right of the Entrance into the Court, is a Piece of Building higher, and better built than the rest. The Gardens are handsome, and well kept; and indeed, in general, the *Persians* are much better Gardeners than the *Turks*. In *Persia* the Trees are planted by Line; and their Parterres are well disposed and manag'd: The Compartments are well laid out; and

A View of Mount ARARAT from Three Churches.

and the Plants are very neatly difpos'd and fet out. Whereas in *Turky* thefe are all in the greateft Confufion. The Enclofure of the Patriarch's Garden, as likewife of moft of the Houfes of the Borough, is nothing elfe but Mud dried in the Sun, and cut into large thick Pieces; which are laid one upon another, and join'd together with a temper'd Earth inftead of Mortar. The Walls of Parks about *Madrid* are the fame. The *Spaniards* call thefe baked, or rather Sun-dry'd, Pieces of Earth *Tapias*.

THE Patriarchal Church is built in the Middle of the great Court, and confecrated to St. *Gregory the Enlightner*, who was the firft Patriarch, in the Reign of *Tiridates* King of *Armenia*, under *Conftantine* the Great. The *Armenians* believe that the Palace of this King ftood where the Convent now does; and that Jefus Chrift appeared to St. *Gregory* in the Place where the Church ftands. They keep here an Arm of this Saint, a Finger of St. *Peter*, two Fingers of St. *John Baptift*, and a Rib of St. *James*. The Building is very ftrong, and of fine hewn Stone; the Pillars and Arches are very thick: But the whole is dark and clofe, and not well illuminated. Within, at one End are three Chappels, whereof the middle one only is furnifhed with an Altar, the others ferving for a Veftry and a Treafury. Thefe are fill'd with rich Ornaments and fine Veffels for the Service of the Church. The *Armenians*, who don't pretend to much Magnificence but only in their Churches, have fpar'd no coft to enrich this with all the fineft Manufactures of *Europe*. The facred Veffels, the Lamps and Candlefticks, are of Silver and Gold, or Silver gilt. The Pavement of the Church and Chancel is cover'd with fine Carpets. About the Altar the Chancel is hung commonly with Damask, Velvet, and Brocade: Which, however, is the lefs to be wonder'd at, becaufe the *Armenian* Merchants, who trade to *Europe*, and are very rich, make great Prefents to this Church: But it may well be wonder'd that the *Perfians* fuffer fo much Riches to lie there. The *Turks*, quite contrary, don't fuffer the *Greeks* to have fo much as one Silver Candleftick in their Churches; and nothing is poorer than that belonging to the Patriarch at *Conftantinople*. The Monks of *Three-Churches* pride themfelves in fhewing the Riches they have received from *Rome*, and ridicule all Talk of a Reunion. Several Popes have fent them whole Chappels of Silver, without being able to do any thing: For the Patriarchs have hitherto only amufed the Miffionaries; it being no hard matter to deceive thofe who are down-right and honeft in all their Defigns. The reuniting of Religions is a Miracle, which the Lord will work when he fhall fee proper. 'Tis from Heaven we muft wait the Converfion of Schifmaticks, who are vaftly more numerous than thefe Roman *Armenians.* Thefe unfortunate Schifmaticks would, by their Intereft and Money, depofe that Patriarch who fhould attempt a Reunion. The Hatred they bear to the *Latins* feems irreconcileable. In fhort, be it through Envy or Intereft, the Schifmatical *Armenians* or *Greek* Priefts will bear great Sway in that Church; and the Patriarchs are oblig'd to give way, left the Populace fhould throw off their Authority.

THE Architect who form'd the Plan of this Patriarchal Church, was a good Mafter, according to a Tradition which prevails among the *Armenians,* who fay that Jefus Chrift himfelf drew the Plan in the Prefence of St. *Gregory,* and commanded him to fee the Church built according to it. Inftead of a Pencil, they fay the Lord made ufe of a Ray of Light, in the midft of which St. *Gregory* was at Prayers upon a great fquare Stone, of about three Feet diameter, which they ftill fhow in the Middle of the Church. If this Story be true, the Lord has made ufe of a very fingular Order of Building; for the Domes and Steeples are in the fhape of a Tunnel turn'd upfide down, with a Crofs on the Top.

THE two other Churches are without the Monaftery, but are now gone to ruin; and Divine Service has not been perform'd there a great while. That of St. *Caiana* is on the Right of the Convent, as we enter at the great Gate, but not at that of the Refectories. The other Church, which is on the Left, and confiderably farther from the Houfe, bears the Name of St. *Repfima.* The *Armenians* pretend that *Caiana* and *Repfima* were two *Roman* Virgins, who were martyr'd in the Places where the Churches are built. St. *Caiana* they will have to be defcended of I know not what Family of *Caius*: But they are more put to it to find the Genealogy of *Repfima,* which is not a *Roman* Name. However, their Chronicles fay that they were both *Roman* Princeffes, who came into the Eaft to fee St. *Gregory*: At which, *Tiridates,* King of *Armenia,* being offended, he caufed *Caiana* to be put into a Well full of Serpents, expecting that fhe would have been foon kill'd; whereas the Saint

was

Armenian Monks.

was not hurt, but the Serpents died, and fhe liv'd there in good Health Lett. VII. for the fpace of forty Years. But how will this agree with the Sequel of the Hiftory? For they add, That King *Tiridates* falling in love with her, and not being able to prevail with her, nor any of her Companions, who were very beautiful, and, according to the Chronicle, forty in Number, caufed them all to be martyr'd.

THE Country about *Three-Churches* is admirably fine; and I don't know of any which can give us a better Idea of Paradife. 'Tis full of Rivulets, which render it extremely fruitful: And I queftion whether there be any other Country in the World where one may gather fo many Commodities all at one time. Befides great Quantities of all Sorts of Grain, there are Fields of a prodigious Extent cover'd with Tobacco. It would be a pleafant Queftion in Botany, Whether this Plant grew in the Terreftrial Paradife, which is now fo acceptable to the Generality of the Inhabitants of this Place, that they can't be without the conftant Ufe of it? However, originally it came from *America*; but it grows altogether as well in *Afia*. The reft of the Country of *Three-Churches* abounds with Rice, Cotton, Flax, Melons, Paftiques, and fine Vineyards. There wants nothing but Olives: And I don't fee where the Dove which went out of the Ark could find an Olive-Branch, if the Ark be fuppofed to have refted upon Mount *Ararat*, or any of the Mountains in *Armenia*; for this Sort of Trees is not found hereabouts, where the Species muft be loft: And yet Olives are known to be a kind of Trees which never die. The *Ricinus* is much cultivated about this Monaftery, of which they make an Oil to burn, that of Linfeed being ufed in their Kitchins. 'Tis perhaps for this Reafon that the Pleurify is fo rare in *Armenia*, notwithftanding the Climate is very uncertain, and by confequence apt to produce that Diftemper. *Gefner* has obferv'd, that Linfeed-Oil, drank inftead of Oil of Sweet-Almonds, is an excellent Remedy for a Pleurify.

AS to the Melons, there are not better in all the *Levant* than thofe of *Three-Churches*, and the Country thereabouts. We loaded one of our Horfes for thirty *Sols*: And in that large Quantity we met with feveral which were far fuperior to thofe at *Paris*. But that which was moft extraordinary, was, that they fatten without ever doing any harm;

K k 2

on

on the contrary, the more we eat of them, the better we were. Thofe which are called *Paftiques*, or *Water-Melons*, even in the Heat of the Day are like Ice, tho they be laid on the Ground in the middle of a Field, where the Earth is hotteft. They are not cultivated in watry Places, as has been here believed; but they are call'd Water-Melons becaufe the Meat of them does not only melt in the Mouth, but fends out fo great a Quantity of Water, that one half of the Fruit is loft, efpecially when it is bit with the Mouth to peal it; which is the Way us'd by the People of the Country, who ordinarily eat them as Apples. Our *Butter-Pears* and *Moüille-bouche* are perfectly dry in comparifon of thefe Melons, which would be the moft delicious Fruit in the World, if they had as good a Smell and Tafte as the other Melons. The Meat of thefe Water-Melons becomes more firm as they grow riper, and indeed, to fpeak properly, does not melt at all; but this delicious Water, which is enclofed in little Cells in the Pulp, runs out in fo large a Quantity, as it were from fo many little Springs, that the Eaftern People often prefer thefe to better Melons. The *Armenians* call them *Carpous*, a Name they have borrow'd from the *Greeks*, who call all Fruits fo: And thus *Carpous* means Fruit, by way of Excellence. The beft Water-Melons are produc'd in the falt Lands between *Three-Churches* and the *Aras*. After Rains, the Sea-Salt lies in Chryftals upon the Fields, and even crackles under the Feet. Three or four Leagues from *Three-Churches*, in the way to *Teflis*, there are Pits or Quarries of Foffile Salt, which would abundantly fupply all *Perfia*, without being exhaufted. They cut the Salt into large Lumps, in the fame manner as we cut the Stone out of our Quarries; and each Buffalo carries two of them. One fometimes meets large Droves of thefe Animals in the great Roads, laden with nothing elfe but this Salt, for in the *Levant* the Buffalo's are among the Beafts of Burden.

THE People of the Eaft imagine that the Salt grows in thefe Pits, and that the fame places from whence they have once taken do in time fill up again: but who has made any accurate Obfervations on this Head? I was told the fame at *Cardona* in *Spain*, where are the beft Salt-Pits in the World. This Mountain is nothing but a Mafs of Salt, which appears like a Rock of Silver when the Sun fhines upon thofe places which are not cover'd with Earth. They who work in the Quarries of Marble have taken up

the

Lepidium Orientale Nasturtij —
Crispi folio Coroll. Inst. Rei herb. 15.

118

the fame Notion, and believe more from a Tradition among them, than upon any good Reafons, that the Stones do, by an internal Principle, actually grow like Truffles and Mufhrooms : thus the Suppofition of the Vegetation of Foffils prevails more than might perhaps be imagined, but our Notions in thefe Matters fhould be built upon Experiments and Obfervations well confirm'd, and not upon Suppofitions and Prejudices.

WE liv'd very well in the Monaftery of *Three-Churches*, where we were lodg'd to our Satisfaction : and there being not many Strangers, we had as many Chambers as we pleas'd. The Religious, who are moftly *Vertabiets*, that is *Doctors*, drink with Ice, and they gave us of it plentifully ; but they have not got the Secret of driving the Gnats from the Convent. We were forc'd in the night to leave our Chambers, and caufe our Mattreffes to be carried into the Cloifter near the Church, upon a Pavement of broad Stones that is very well kept. The Gnats were there lefs troublefome than they were within, but yet this did not prevent their fucking a great deal of our Blood ; all the morning our Faces were full of Knobs and Swellings, notwithftanding all our Precaution. The Parterres on the left of the Church are very pleafant. The *Amaranthuffes* and *Pinks* are their chief Ornaments ; but there is nothing extraordinary in thefe Flowers, to make it worth the while to bring the Seeds into this Country : on the contrary, the Curious among the *Perfians* would be much better pleas'd, if they could furnifh themfelves with the Kinds we raife in *Europe*. We gathered on the Parterres of the Convent nothing but the Seeds of that fine Species of the *Perficaria*, whofe Leaves are as large as Tobacco, and which we obferved at *Teflis* in the Prince's Garden. I have here inferted a Defcription of a fine Species of *Lepidium Orientale Nafturtii Crifpi Folio*, which grows in the Fields between the Monaftery and the River *Aras*.

THE Root runs deep in the Ground, about a foot long, and as thick as the Little-finger, hard, woody, white, a little fibrous, and produces a Stem two or three feet high, full of Branches, of a bright green colour, with Leaves at bottom, of four inches long, and two broad, very much like thofe of the *Nafturtii Crifpi Folio*, but a little more flefhy ; fmooth on both fides, of a bright green colour, divided into great pieces even to the Stalk, which is pretty long from the main Stem without Leaf. The laft piece is bigger than the others, rounded and jagged as they are which are

upon.

upon the reft of the Stalk, which however are fometimes cut deeper than this. The Leaves which grow along the Stem are cut more flender and thin. From their Knots fhoot out Branches which expand themfelves pretty much, and are adorned with Bunches of Leaves, for the moft part not divided, and very much like thofe of the *Common Iberis*. The Branches are fubdivided into fmall Sprigs with white Flowers: Each Flower has four Leaves of a line and a half long, rounded at the Point, and very fharp at their beginning. The Cup confifts likewife of four Leaves; the Pointal, which is half a line in length, fhaped like the Head of a Pike, paffes into a Fruit of the fame Form, flat, and divided into two Cells, each of the whole length, and containing a ruddy Seed inclining to a brown, half a line in length, and flatted. The whole Plant has the Tafte and Acrimony of *Garden Creffes*.

DURING our ftay at *Three-Churches* we endeavoured to procure Perfons to carry us to Mount *Ararat*, but could get none. The Carriers who came from other Parts, faid, they would not venture the lofing themfelves in the Snows; and they of the Country were employed in the Caravans, and would not harrafs their Horfes in a place which appear'd fo frightful. This famous Mountain is but two fhort days Journey from the Monaftery, and we afterwards knew it was not poffible to get on it, becaufe it is all open, and there is no paffing any farther than to the Snow. Whatever the Religious here fay, 'tis no fuch wonder that there is no coming at the top, fince almoft one half of it is covered with Snow frozen hard, and which has lain there ever fince the Flood. Thefe good Men believe, as an Article of their Faith, that the Ark refted upon this Mountain. If it be the higheft in *Armenia*, according to the Opinion of the People of this Country, it is very certain likewife that it has the greateft quantity of Snow on it of any. That which makes *Ararat* feem fo very high, is, that it ftands by it felf, in form of a Sugar-loaf, in the middle of one of the greateft Plains one can fee. We muft not judge of its heighth from the quantity of Snow which covers it, for the Snow even in the hotteft Summer lies upon the leaft Hills in *Armenia*. If the Monks of *Armenia* are asked, whether they have any Relicks of the Ark? they very gravely anfwer, That it lies ftill buried in the vaft heaps of Snow upon Mount *Ararat*.

WE

WE went *Auguſt* the 8th to *Erivan*, a conſiderable City, and the Capital of *Perſian Armenia*, three hours Journey from *Three-Churches* ; not
ſo much to ſee that Place, as, according to the Advice of the Religious
of *Three-Churches*, to pray the Patriarch to appoint us Perſons to carry us
to Mount *Ararat*, which we ſhould never have procur'd without his Order.
The City of *Erivan* is full of Vineyards and Gardens, and ſtands upon a
little Hill, which is at the end of the Plain ; and the Houſes extend
themſelves into one of the fineſt Vales of *Perſia*, conſiſting of Meadows,
intermingled with Orchards and Vineyards. The People of *Erivan* are
ſimple enough to believe that their Vines are of the ſame ſort with thoſe
which *Noah* planted there. Be that as it will, they yield a very good
Wine, which is a greater Commendation than to ſay they are deriv'd from
thoſe planted by the Patriarch. The Vale is water'd with fine Springs,
and the Country-houſes are almoſt as numerous as about *Marſeilles*. The
Tops of the Hills only, by their Drought, diſhonour this Country ; but the
Vine would do wonders here, if there were Perſons to cultivate it. The
beſt Lands are cover'd with Grain, Cotton, and Rice ; this laſt is moſtly
deſign'd for *Erzeron*. The Houſes of *Erivan* have only a Ground-floor,
without any Stories above, and are built of Mud and Dirt, after the manner of other Towns of *Perſia*. Each Houſe ſtands by itſelf, and is defended by a ſquare, angular, or round Encloſure about ſix feet high. The
Walls of the Town, tho with a double Rampart in ſome places, are hardly
above twelve feet high, and are defended only by very indifferent round
Ravelins, four or five feet thick. All theſe, together with the Walls, are
made of Mud dried in the Sun, and without any Cement. The Walls of
the Caſtle, which ſtands in the higheſt part of the Town, are not much
better, but run round it three times. The Caſtle itſelf is almoſt oval, and
contains above eight hundred Houſes, inhabited by *Mahometans* ; for the
Armenians, who go thither to work all the day, return into the Town to
lodge at Night. They aſſur'd us the Gariſon of this Caſtle conſiſted of
2500 Men, for the moſt part Tradeſmen. The Place is impregnable on
the North-ſide ; but it is Nature only that makes it ſo, by fortifying it,
not with Mud Ramparts, but with a prodigious Precipice, at the bottom
of which runs the River. The Gates of the Caſtle are plated over with
Iron. The Portcullis and Guard-houſe ſeem to be regular and well enough
contriv'd

contriv'd. The antient Town was probably a Place of greater Strength, but was deftroy'd during the Wars between the *Turks* and *Perfians*. M. *Tavernier* affirms it was given up to Sultan *Murat* by Treafon, and that the *Turks* left in it a Garifon of two and twenty thoufand Men. But notwithftanding this, *Cha-fefi* King of *Perfia* took it by main Force. He himfelf was the firft in the Attack, and the two and twenty thoufand *Turks*, who would not furrender, were cut in pieces. *Murat* reveng'd this in a barbarous manner at *Babylon*, where he put to the Sword all the *Perfians* he found, notwithftanding he had promis'd them their Lives upon their Capitulation.

TOWARD the South, upon a fmall rifing Ground, about a Mile from the Citadel, is the little Fort *Quetchycala*, cover'd with a double Wall; but thefe Works are in more danger from the Rains than from Cannon. *Quetchycala* refembles the Forts of Clay, which are fometimes built at *Paris* for the Exercifes of the Academifts. The Port-holes of all the Fortifications of *Erivan* are of a very fingular Make, jutting out beyond the Wall like a Mask, about a foot and a half, and ending in the fhape of a Cowl or Hog's Snout; which effectually fecures the Heads of the Soldiers, who are order'd to make the Difcharge, and is no ill Contrivance for Cowards; but then they are not able to obferve the Enemy, unlefs it be juft at the Gap or Opening, and they come into the only Place where they can be kill'd: and yet if the Befieg'd fuffer the Enemy to come to the Foot of the Wall, 'tis then impoffible to fire upon them.

M. *CHARDIN*, who knew *Erivan*, and the Country thereabouts, better than any of our Travellers, has defcrib'd the Rivers very exactly. The *Zengui* runs North-Weft, and the *Queurboulac* South-Weft, rifing from forty Springs, which its Name expreffes. The *Zengui* comes from the Lake of *Erivan*, two Days Journey and a half from the City, but I don't know whether this be the fame *Zengui* I mention'd before. The Lake which is very deep, and about five and twenty Leagues in Compafs, is well ftock'd with excellent Carp and Trout, which are however of no great Service to the Religious of a Monaftery built on an Ifland in the middle of the Lake, they being not permitted to eat of 'em but four times a Year; nor indeed are they fuffer'd to converfe toge-

ther,

ther, but at the fame times. The reft of the Year they keep a perpetual Lett. VII.
Silence, and eat nothing but the Herbs which their Garden produces, and
that juft as Nature prepares them, without Oil or Salt. Thefe poor Monks
are like fo many *Tantalus's*, who have conftantly in their View, and within their Reach, excellent Fruits which they dare not touch. And notwithftanding this, Ambition is not wholly banifh'd from this Place; the
Superior is not content to have the Title of Archbifhop, but likewife
takes to himfelf that of Patriarch, which he difputes with the Patriarch
of *Three-Churches.*

WE pafs'd the *Zengui* to *Erivan* upon a Bridge of three Arches, under which they have contriv'd certain Rooms or Apartments, where the
Kan or Governour of the Place fometimes comes to divert and cool
himfelf in the hot Seafons. This *Kan* raifes every Year above twenty
thoufand *Tomans* from this Province, that is, above nine hundred thoufand Livres of *French* Money, without reckoning what he gets by the
Pay of the Troops appointed to guard the Frontiers. He is oblig'd to
advife the Court of all the Caravans, and all Ambaffadors who pafs that
way. *Perfia* is the only Country I know of, where Ambaffadors are maintain'd at the Prince's Charge : And yet, nothing in my Opinion can be
more honourable for a great King. As foon as an Ambaffador or fimple Envoy, has fhewn the Governours of the Provinces, that he is charg'd
with Letters for the King of *Perfia*, they immediately give him the *Tain*,
which is an Allowance for his daily Subfiftence, of fo many Pounds
of Meat, Bread, Butter, Rice, and a certain Number of Horfes and
Camels.

AT *Erivan* there is very good living. Partridges are common, and
Fruits are brought thither in abundance. The Wine is admirable ; but
the Culture of Vines is very difficult, by reafon of the Cold and Frofts,
which oblige the Dreffers not only to cover, but even to bury 'em under
Ground at the beginning of Winter, where they are kept till the Spring.
Tho the Town be but ill built, there are however fome fine Places in it. The
Governour's Palace, which is in the Fortrefs, is confiderable for its Largenefs, and the Diftribution of its Apartments. The *Meidan* is a great
open Square, hardly lefs than four hundred Paces over. The Trees there
are as fine as in the *Bellecour* at *Lyons.* The *Bazar*, which is the Place

where they fell their Merchandife, is not difagreeable.   The Baths and
Caravanferas likewife have their Beauties, efpecially the new Caravanfera by the fide of the Fortrefs.   At entring, one feems to be going into
a Fair or Market-place, for we pafs thro a Gallery, in which are fold all
forts of Stuffs.

THE Churches of the Chriftians are fmall and half under ground.
That belonging to the Bifhop's Palace, and the other call'd *Catovique*,
were built, as they fay, in the times of the laft Kings of *Armenia*.   By
the fide of the Bifhop's Palace is an old Tower, of a very fingular kind
of Building; it would have borne fome refemblance to *Diogenes's Lant-*
horn, had it not been fo much after the Oriental Tafte.   It is flat-fided,
and its Dome has fomething very agreeable; but the People of the Place
don't know for what Ufe it was defign'd, nor when it was built.   The
Mofques of the Town have nothing particular.   M. *Chardin* fays, the *Turks*
took *Erivan* in 1582. and that they built a Fortrefs there; that the *Per-*
fians having retaken it in 1604, put it into a Condition to hold out againft
Batteries of Cannon; that it fuftain'd a Siege of four Months in 1615.
which the *Turks* were at laft oblig'd to raife; that they could not gain
the Town till after the Death of *Abas the Great*; and that the *Perfians*
retook it again in 1635, and have continued Mafters of it ever fince.

AFTER we had walk'd about the Town, we went to vifit the Patriarch of the *Armenians*, who is lodg'd in an old Monaftery out of the
Town, but not fo well by far as at *Three-Churches*.   This Patriarch, whofe
Name is *Nahabied*, was a good old Man, of a ruddy Complexion, who
out of Humility, or for his Eafe, had nothing on but a mean blue linen
Caffock.   We kifs'd his Hand, according to the way of the Country,
which, our Interpreters told us, pleas'd him much; for many *Franks* don't
fhew him that Refpect: but we would even have kifs'd his Feet, if we
had ever fo little fufpected that he requir'd it, we had fo great need of
his Intereft.   In requital, he order'd us a Treat, which was truly very
frugal.   They brought, on a wooden Salver, a Plate of Nuts between
two other Plates, one of Plumbs, the other of Raifins.   But they did not
offer us either Bread or Cake, or Bifket.   We eat a Plumb, and drank
one Glafs to the Prelate's Health, of an excellent red Wine; but who
could have drank again without a Bit of Bread?   Our Interpreters were

in

*Monument
at Athens.*

in the Entry, and had the Senſe to get ſome themſelves, but dar'd not
offer it to us, tho we ſhould willingly have pardon'd their Freedom at
that time: after the Treat, they came into the Room, and we order'd
them to requeſt the Maſter of the Houſe, to cauſe us to be furniſh'd with
good Horſes and Guides, which we would pay for, to conduct us to
Mount *Ararat*. *What Buſineſs,* ſays he, *have you at Mount* Macis? which
is the Name this Mountain bears among the *Armenians,* but the *Turks* call
it *Agrida*. We anſwer'd, *That being near a Place ſo celebrated, on which it
was ſuppos'd* Noah's *Ark had reſted, we ſhould be much blam'd at our Return home, if we did not go to ſee it.* You will find it *very difficult,* ſays
the Patriarch, *to go even ſo far as to the Snows; and as for the Ark, God
has never yet favour'd any one with the Sight of it, except only one Saint,
who was of our Order, and after fifty Years ſpent in Faſting and Prayer, was
miraculouſly carried thither ;* but the *exceſſive Cold ſeiz'd him in ſuch manner,
that he dy'd upon his Return.* Our Interpreter made him laugh, by anſwer-
ing in our name, *That after having ſpent half our Life in Faſting and Prayer,
we ſhould rather beg of God to let us ſee Paradiſe, than the Remains of* Noah's *Houſe*. At *Three-Churches* they had told us, that one of the Religious of their Order, whoſe Name was *James,* and who was afterwards Biſhop of *Niſibis,* reſolv'd to go to the top of the Mountain, or periſh in
the Attempt, accounting it a Happineſs to endeavour to find the Remains
of the Ark; that he executed his Deſign with a great deal of Difficulty ; and notwithſtanding all his Pains and Diligence to aſcend the Hill,
he always found himſelf when he awak'd in one certain Place about half
way to the Top; that this good Man perceiv'd in a few Days, that all
his Attempts to get higher were vain ; and that in his Trouble an Angel appear'd to him, and brought him a piece of the Ark. *James* return'd
to the Convent with his choice Burden ; but before the Angel left him,
he told him that God would not ſuffer Men to pull in pieces a Veſſel
which had ſav'd ſo many Creatures. Thus the *Armenians* amuſe Strangers with ſuch like Stories.

T H E Patriarch ask'd us whether we had ſeen the Pope; and was much
diſpleas'd with us when we anſwer'd, we intended to ſee him in our Return home. *What,* ſays he, *do you come ſo far to ſee me, and han't yet
ſeen your own Patriarch ?* We dared not tell him we were come into *Arme-*

nia to fearch for Plants. *What think you*, continues he, *of my Church at* Itchmiadzin, *have you any fo fine in* France? We anfwered him, *That every Country had its own manner of building ; that our Churches were of a quite different Tafte, and that we did not fee the Skill of the Workmen, but only in the Candlefticks, Lamps, and other Veffels.* Thefe were certainly not made in *Armenia.* While this Venerable Prelate (who would pafs with us for a good Country-Schoolmafter) gave his Orders, we defired to fee the Chappel, and we put three Crowns into the Bafon to pay for our Treat : we do thefe kind of Charities more out of Decency than Devotion. At our return, they offered us another Glafs, which however we at firft refus'd, there being yet no Bread come; but we were obliged to drink, to return our Thanks to the Patriarch who drank our Health : and all this paffed very agreeably. After the ordinary Civilities, he fent one of the Houfe with us to carry a Letter of Recommendation to the Religious in the Road to Mount *Ararat* : fo we went to lie this Night two hours from *Erivan* in a Convent of *Armenian* Monks at the Village *Nocquevit.* We drank there excellent Claret inclining to an Orange-colour, and as good as that of *Candy :* but for fear we might want Bread, we made our Interpreters give them to underftand, that we would deal honourably with them : this anfwered our Expectations, for we were handfomely treated ; and we were as good as our Word the next Morning before we went away.

THE Country about *Nocquevit* is admirably fine, all manner of Fruits are there in great plenty, and they neglect fuch Melons as would be in great efteem at *Paris.* The Buildings in thefe Parts are only of fquare pieces of Mud and Clay dried in the Sun, for want of Timber.

AUGUST the 9th, we fet out at four in the morning, with our Faces ftrangely disfigured by the prickings of the Gnats, who attack'd us very furioufly fome nights ago. We continued our Journey over a large and fine Plain which led to Mount *Ararat.* About eight a clock we reached *Corvirap* or *Couervirab*, which, as they told us, fignifies in the *Armenian* Tongue, *The Church of the Well. Corvirap* is another Monaftery of the *Armenians*, whofe Church is built by a Well, into which they affirm St. *Gregory* was caft and miraculoufly fed, as *Daniel* was in the Lion's Den. The Monaftery looks like a fmall Fort on the top of a little Hill which commands the whole Plain, and it was from this Eminence that we firft

faw

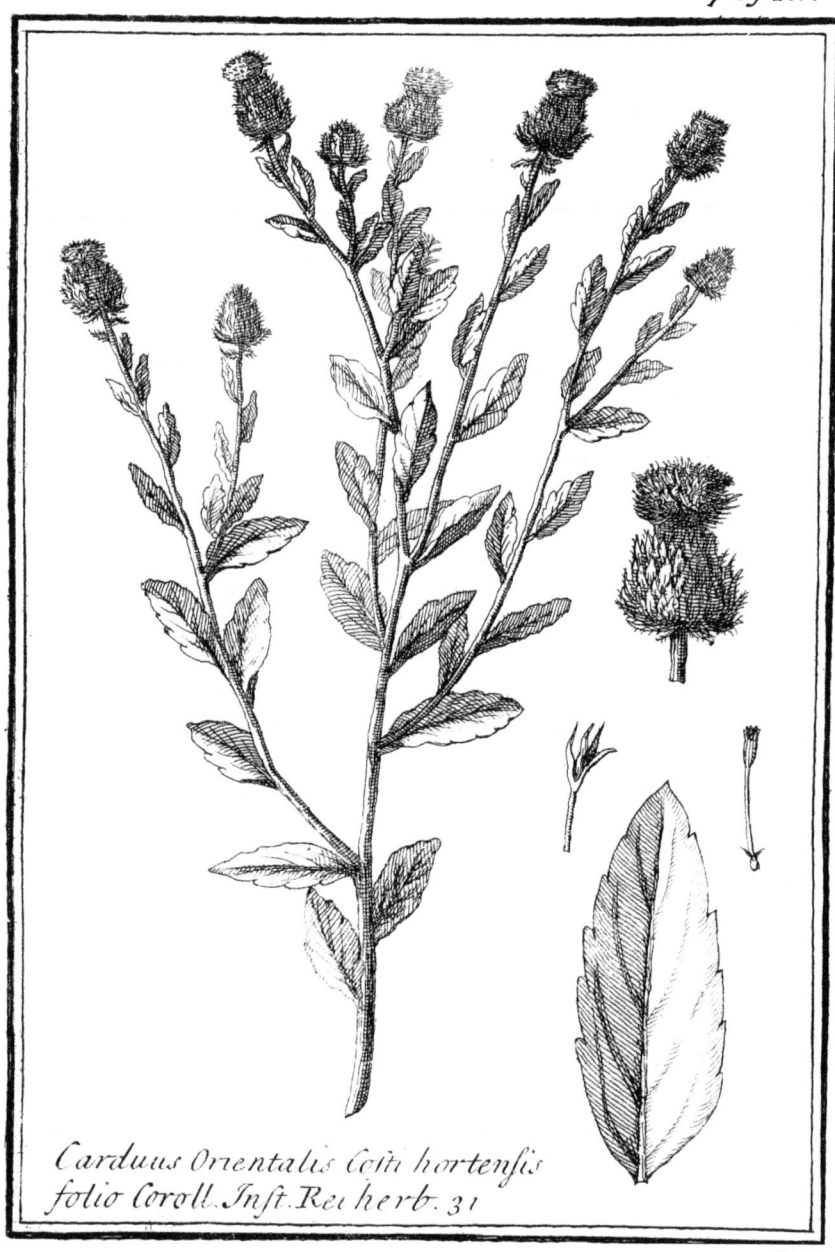

Carduus Orientalis Coin hortensis folio Coroll. Inst. Rei herb. 31

faw the River *Aras*, fo well known formerly by the name of *Araxes*; it runs Lett. VII.
along about four Leagues from Mount *Ararat*. We were oblig'd to repofe
and refrefh our felves in this Monaftery, for we had had very bad Nights
by reafon of the Gnats, and the Heat was intolerable in the day. And
this was the Life we led, even from the time of our leaving *Teflis*; but all
our Fatigue was abundantly recompens'd with the fight of *Araxes* and
Mount *Ararat*. From *Corvirap* we could diftinctly difcern the two Tops
of this famous Mountain. The fmaller one, which is moft fharp and pointed of the two, was not covered with Snow; but the greater one was pro-
digioufly laden with it. The Plants we found in this Monaftery, while our
Guides repofed themfelves, were thefe.

CARDUUS Orientalis Cofti Hortenfis Folio. Coroll. Inft. Rei Herb.
Pag. 31.

T H E Root of this Plant is about one foot long, hard, woody, white, at
the upper end about the thicknefs of the Little-finger, furnifhed with many Fibres, and covered with a reddifh Skin; it fends out a Stalk of two
or three feet long, branched from the beginning, hard, firm, whitifh, two
inches thick, with Leaves about three inches long, and one and a half
broad, a little jagged about the Edges, like the Leaves of that fort of *Tanfy*
which the *French* call *le Coq*, which word to me feems to be a Corruption of *Coftus Hortenfis*. The Leaves of this *Carduus* are lefs and lefs
as they grow nearer the top of the Plant, and lofe their Indentings or
Jags, but end in a fmall foft Point or Prickle. From their Knots fhoot
out Branches all along the Stalk, each of which ends in a yellow Flower.
The Leaves which grow along the Branches are flender, and fometimes
fmall as Threds. The Calix or Cup of the Flower is eight or nine lines
high, and almoft as thick. 'Tis like a Pear confifting of feveral Scales
which are whitifh, pointed, firm, prickly, and fometimes inclining to a
purple colour at the Extremities. The Prickles about the edges are fofter,
and grow out like the Hair on the Eye-lid. Each Flower confifts of fmaller yellow Flowers or Fleurons, which run out beyond the Cup about five
or fix lines, divided into as many fmall Points, out of the middle of
which grows a Sheath with a very fine Thred at the top. The Fleurons
in little Bags, or Embrio's, bear the Seed of about two lines long, and one
broad, with a white Tuft on it. They which are not untimely, become

Seeds

Seeds of three lines in length. The Flowers have no Smell that we could perceive, but the Leaves are very bitter.

WE had the Pleasure this day to find a Plant of a new Kind, and we gave it the name of one of the moſt learned Men of this Age, equally eſteemed for his Modeſty and Integrity; I mean Mr. *Dodart* of the Royal Academy of Sciences, Phyſician to her Royal Highneſs the Princeſs Dowager of *Conti.*

THIS Plant ſends out Stalks of a foot and a half high, ſtrait, firm, ſmooth, woody, of a bright green, two lines thick, branch'd from the bottom, round like a Buſh, and furniſh'd with Leaves of an inch or fifteen lines long, and two or three wide, a little fleſhy, jagged on the ſides, eſpecially towards the bottom of the Plant, for higher they are ſtraiter, and leſs indented; ſome of them are even as ſmall as the common *Linaria* or *Flaxweed.* The top of the Branches is adorn'd with Flowers growing out of the Knots of the Leaves. Each Flower is a Head of a deep Violet-colour, of eight or nine lines long; the bottom is a Pipe of one line diameter, opening into two Lips, the uppermoſt of which is in the ſhape of the Bowl of a Spoon, the Convex-ſide being turn'd up, and about one line and a half long, cleft in two parts, pretty much pointed; the lower Lip is three lines long, rounding, but divided into three parts, the middlemoſt of which is the ſmalleſt, and moſt pointed; this Lip is rais'd toward the middle with a ſort of white Hair or Down. The Calix is a ſmooth Cup of two lines high, divided into five Points; it ſends out a Pointal that is ſpherical and near a line in diameter, which is inſerted in the Pipe of the Flower, as it were by *Gomphoſis,* and has at the top a very fine Thred, and paſſes into a ſpherical Cod of three lines diameter, ending in a Point. This Cod is reddiſh, hard, divided into two Cells by a middle Partition, which are furniſhed with each a fleſhy *Placenta* or Cake, divided into little hollows, which hold a ſmall brown Seed.

ALL along the Plain, by the ſide of the *Araxes,* grows abundance of *Liquoriſh* and *Dodder.* The Liquoriſh is in all reſpects like the common ſort, except only that the Husks or Cods are longer, and full of Prickles. The *Dodder* grows ſo faſt upon the Stalks of the *Liquoriſh,* that it ſeems to be part of the ſame Plant; when it is plucked off, one ſees certain Tuber-

cules

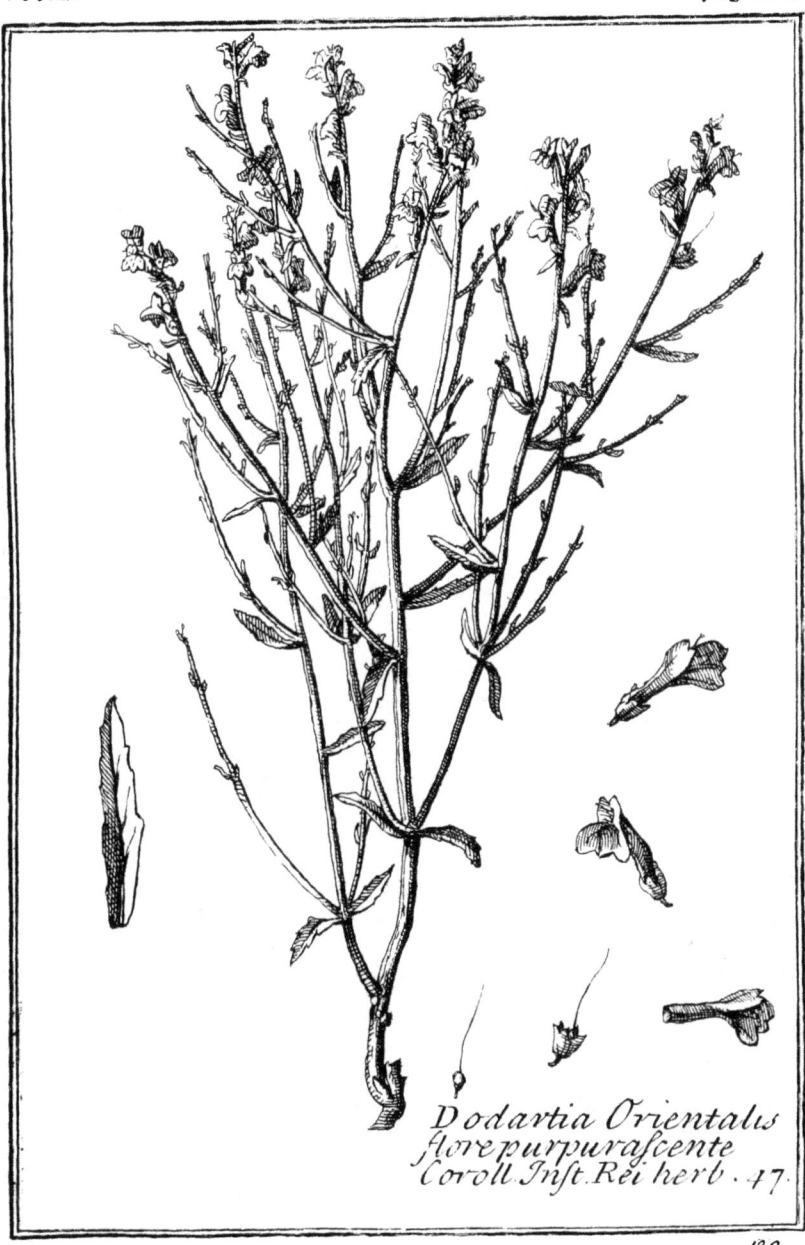

Dodartia Orientalis
flore purpurascente
Coroll. Inst. Rei herb. 47.

cules of about one line thick, which are like fo many Nails or Pegs fticking into the Plant on which it grows. Thefe Stalks are one line in thicknefs, and fometimes more. We at firft took them for Stalks of a Species
of *Bindweed,* whofe Leaves were gone. One cannot compare the Leaves
of this *Dodder* to any thing better than to Cat's-Gut, about the bignefs of
a Packthred; but they are firm, hard to break, bitter, a little aromatick,
of a pale green colour, divided into feveral branches twifted about the
neighbouring Plants, from whence they fuck the Juice for their Nourifhment by the Tubercules before-mention'd. Thefe Tubercules are
commonly plac'd obliquely at the diftance of a line from one another; and
in different places there are no Roots, no more than to other Species of
the fame kind, when the Tubercules can furnifh Juice enough for its Nourifhment. Its Flowers grow in Bunches like a Head, of a pale gridelin, two
lines in height, and one and a half in diameter. They are Cups divided
into five obtufe Points, which are bored through at bottom, to receive a
Pointal from a Cup of two lines high, divided alfo into five parts. This
Pointal paffes into a Fruit like that of the great white *Bindweed,* four lines
long, and three in the diameter, membranous, of a pale green, afterwards reddifh, ending in a fmall Point, and confifting of two pieces, the
uppermoft of which is a kind of Cap: it contains generally four Seeds as
big as thofe of the *Bindweed* juft mention'd. Thefe Seeds are roundifh on
the back, and on the other fide corner'd, a line and a half long, and one
line thick, and as it were divided into two Lobes by a very thin Mem-
brane, hollow below, and fticking to a fpungy and clammy *Placenta.*

THESE Seeds are nothing elfe but membranous Bladders, in each of
which is rolled fpirally, or wrapt up like a Snail, a young Plant; which
is a Twift or String of a bright green colour, half an inch long, and a
quarter of a line broad at the beginning, but growing narrower and fharper
towards the end, faftned at the broad end to a fpungy and clammy *Placenta,* which is partly in the Seed-Veffel and partly in the Cup. The Creator of all things feems to have defign'd by this Plant to fhew us, that the
Embrio's of Plants are contain'd in fmall in the Bud of the Seeds; and that
fo the Seeds are as fo many Bladders in which the young Plants lie entire,
waiting only a proper nutritive Juice to make their parts fwell, and become
vifible. There are many things in Nature which would difcover to us the

Structure of things unknown, if we gave but due Attention. *Malpighi* had a wonderful Talent this way : and indeed our Notions and Syftems ought not to be form'd nor eftablifh'd, but upon a great Number of Obfervations. For example, in the Month of *October*, in the Body of a Tulip-Root, we have obferv'd an entire Tulip, on whofe Stalk, tho not three lines high, might be feen the Flower, which was not to appear till the *April* following : we could plainly difcern the fix Leaves of the Flower, their Chieves, their Tops, the Pointal or young Fruit, the Seed-Veffels, and the Seeds they contain'd. And after all this, who can refufe to believe that all thefe Parts were fhut up even in a yet narrower Space, and are render'd more or lefs vifible, in proportion as the nuititive Juice has fwell'd and dilated the fmaller Parts ?

THE Birds we faw in thefe Plains, which extend themfelves even to the River, would poffibly have furnifh'd us with fome ufeful Anatomical Obfervations, if we had been provided with a Gun to fhoot any of them. We faw there a fort of *Herns*, whofe Bodies were not bigger than a Pidgeon, tho their Legs were a foot and a half high. The *Egrets* are common enough there; but nothing comes near the Beauty of an admirable Bird, the Skin of which I keep in my Cabinet, and whofe Figure I have feen in the Book of Birds, painted for the King. It is as big as a Raven, the Wings are black, the Feathers of the Back towards the Rump of a purple Colour, and they towards the Neck very fharp-pointed, and of an admirable fhining golden green ; they towards the middle of the Neck are of a bright Flame-Colour, and they which cover the reft of the Neck and the Head of a dazling green. Upon the Head is a Tuft of the fame Colour, about four inches high, the longeft of which are like a Battledore with a long Handle. The Bill of this Bird is brown, like that of a Raven. One may with more reafon call this the *King of the Ravens*, than that which they brought from *Mexico* to *Verfailles* ; feeing that *American* Bird, tho it be a very fine one, has nothing in which it agrees with our common Ravens.

IT troubles me very much that we pafs'd by *Corvirap*, without going to *Ardachat*. Till I came to *Paris*, and read M. *Chardin*'s Voyage, I did not know that *Ardachat* was, according to the Tradition of the *Armenians*, the Remains of the old *Artaxata*. *The People of the Place*, fays

this

this Author, *call this Town* Ardachat, *from the Name of* Artaxerxes, *whom in the East they call* Ardechier. *They assure us, that among the Ruins one may see those of the Palace of* Tiridates, *which was built* 1300 *Years ago.* They likewise say, *there is one Front of the Palace which is but half ruin'd; that there remain four Ranks of Columns of black Marble; that these Columns surround a large piece of wrought Marble, and that they are so thick that three Men can't encompass them with their Arms.* This Heap of Ruins is call'd Tact-Tardat, *that is to say,* the Throne of *Tiridates.*

T A V E R N I E R also mentions the Ruins of *Artaxata* between *Erivan* and Mount *Ararat*, but says nothing more. The Situation of *Artaxata* is so well describ'd by *Strabo*, that we cannot mistake it, if we observe the Course of the *Araxes. Artaxata*, says this Prince of antient Geographers, *was built upon the Design which* Hannibal *gave to King* Artaxes, *who made it the Capital of* Armenia. *This Town is situate,* continues he, *upon an Elbow of the River* Araxes, *which forms a kind of Peninsula, and so is encompass'd by the River as with a Wall, except on the side of the Isthmus; but this Isthmus is secur'd by a Rampart and a good Ditch.* The Country about is called the Artaxan Lands.

THIS Description of *Strabo* increases my Vexation, for we might have seen whether *Ardachat* is in a Peninsula, or perhaps we might have found it a little higher or lower; but our Guides observing we busied our selves so much in the Search of Plants, believ'd we had no Regard to any thing else. Who can imagine that *Hannibal* came from *Africa* to *Araxes*, to be Engineer to a King of *Armenia? Plutarch* however confirms it, and says that this famous *African*, after the Defeat of *Antiochus* by *Scipio Asiaticus*, fled into *Armenia*, where he gave a great deal of good Counsel to *Artaxes*, and among other things advis'd him to build *Artaxata* in the most advantageous Situation in his Kingdom. *Lucullus* made as if he intended to besiege this Place, in order to draw *Tigranes* his Successor to a Battle; but the King of *Armenia* came to encamp upon the River *Arsamias*, to dispute the Passage of the *Romans*; and, according to this Observation, *Arsamias* can be no other than the River of *Erivan*. The *Armenians* were beaten at this Passage, and afterwards in a second Rencounter. But our Historian says that *Lucullus* thought it most proper to make towards *Iberia*; and therefore *Artaxata* was not taken. *Pompey,* who had the Command of the Army after him, prefs'd *Tigranes* fo hard, that he was oblig'd to deliver up his Capital without ftriking a Stroke. *Corbulon*, the *Roman* General under the Emperor *Nero*, forc'd King *Tiridates* to yield up *Artaxata*; and far from fparing it, as *Pompey* did, he caus'd it to be entirely ruin'd. But *Tiridates* came to *Rome*, and made his peace with the Emperor, who not only return'd the Diadem upon his Head, but likewife gave him liberty to take Workmen with him from *Rome*, to rebuild *Artaxata*; which, by way of Acknowledgment, the King of *Armenia* call'd *Neronia*, from his Benefactor. 'Tis furprizing that none of the Authors who fpeak of this Place, have ever given us the Name by which they then call'd Mount *Ararat*, which we were now about to afcend.

THE 10th of *Auguft* we departed from *Corvirap*, and travelled feven Hours to find the Ford of *Araxes*, which is but a Mile from the Monaftery. Tho the Stream be very rapid, the Ford is fo large and wide, that one of our Guides ventur'd to pafs it upon an Afs, tho indeed he had Difficulty enough to get over. We arriv'd about eleven a Clock at the Foot of the Mountain; and we din'd, according to the Cuftom of the Country, in the Church of a Convent, in the Village *Acourlou*: this Convent, which is ruin'd, was formerly call'd *Araxil-vane*, that is to fay, *the Monaftery of the Apoftles*. All the Plain beyond the *Araxes* is full of fine Plants. We obferv'd one of a very fingular kind, to which I gave the Name *Polygonoides*, becaufe it was very like the *Ephedra*, which was formerly call'd *Polygonum Maritimum*.

'TIS a Shrub of three or four feet long, very bufhy and fpreading, its Trunk is crooked, bowing in and out, hard, brittle, as thick as an Arm, cover'd with a reddifh Bark, and divided into Branches, which are crooked in like manner, and fubdivided into leffer, on which, inftead of Leaves, grow Cylindrical Slips or Sprigs, about half a line thick, of a Sea-green Colour, an inch or fifteen lines long, compos'd of feveral Pieces pointed together End to End, fo very like the Leaves of the *Ephedra*, that 'tis impoffible to diftinguifh them without feeing the Flowers. From the Articulation of thefe Sprigs proceed others jointed in the fame manner, and thefe laft put out all along their whole Length certain Flowers of three lines diameter. They are a fort of Bafins cut into five Parts to

the

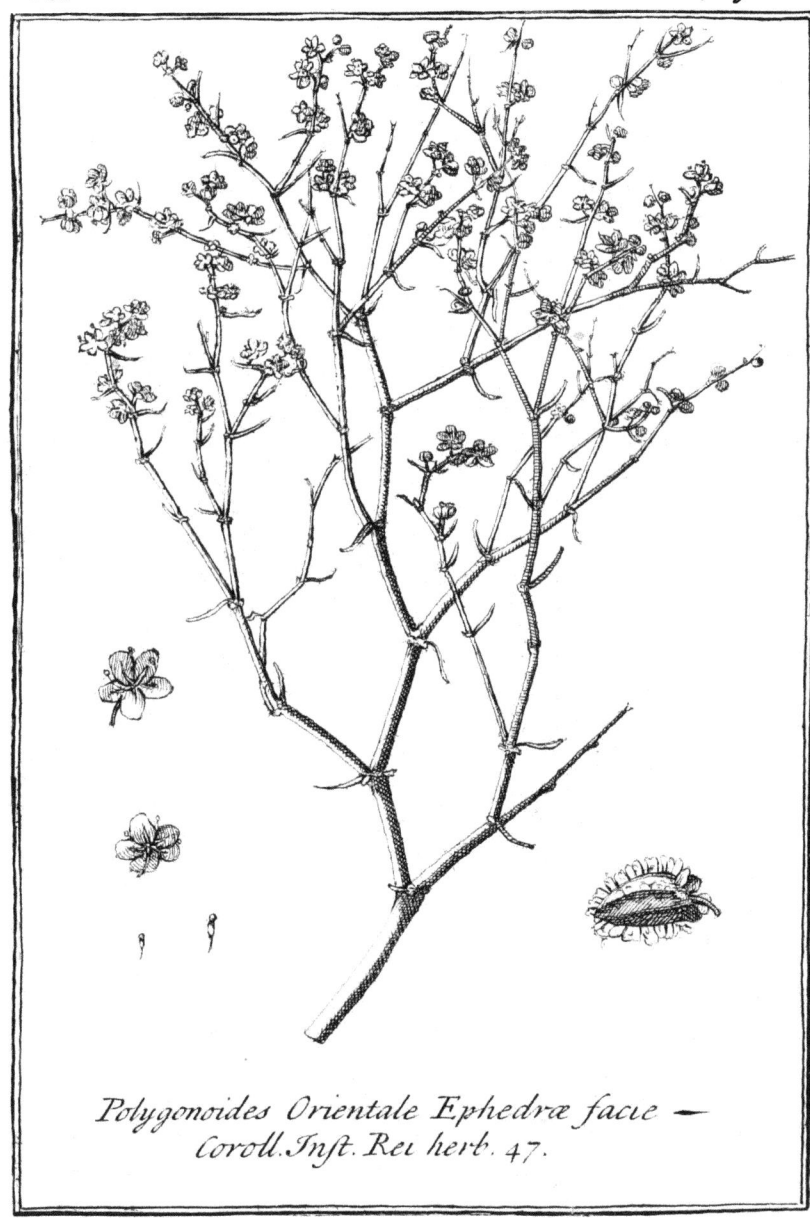

Polygonoides Orientale Ephedræ facie ——
Coroll. Inſt. Rei herb. 47.

the very Center, of a pale-green Colour in the middle, but every where Lett. VII.
elſe white.   From the bottom of each Baſin riſes a Piſtile, one line and 〰
a half long, corner'd, rais'd with little Beards, and encompaſs'd with
white Chieves, but of a purple Colour at the top.   Each Flower is ſupported by a very fine and ſhort ſmall Stalk.   The Piſtile becomes a Fruit
half an inch long, four lines broad, of a conical Figure, and deeply furrow'd thro the whole Length.   Theſe Furrows are ſometimes ſtrait and
ſometimes ſpiral.   The Beards are terminated with Wings divided into
very ſmall Fringes.

When the Fruit is cut croſs-wiſe, the pithy Part appears, which is
white and corner'd.   The Flowers ſmell like thoſe of the *Linden-Tree*,
wither but ſlowly, and remain at the bottom of the Fruit like a kind of
Roſe.   The Leaves are of an herbiſh Taſte, but are ſtiptick.

W E begun this Day to go up Mount *Ararat* about two a Clock in the
Afternoon, but not without difficulty : We were forc'd to climb up in
looſe Sand, where we ſaw nothing but ſome *Juniper* and *Goats-Thorn*.
This Mountain, which lies between South and South-South-Eaſt from
Three-Churches, is one of the moſt ſad and diſagreeable Sights upon Earth.
There are neither Trees nor Shrubs, nor any Convents of Religious, either *Armenians* or *Franks*.   M. *Struys* would have done us a particular
Favour, if he had told us where the *Anchorites*, he mentions, reſided ; for
the People of the Country don't remember to have heard that there
ever were in this Mountain either *Armenian* Monks or *Carmelites* : All
the Monaſteries are in the Plain.   I don't believe the Place is inhabitable
in any other Part, becauſe the whole Soil of *Ararat* is looſe or cover'd
with Snow.   It ſeems too as if this Mountain waſted continually.

F R O M the top of a great Abyſs, which is a dreadful Hole, if ever
there was any, and which is oppoſite to the Village from whence we
came, there continually fall down Rocks of a blackiſh hard Stone, which
make a terrible Noiſe.   There are no living Animals but at the bottom
and towards the middle of the Mountain : they who occupy the firſt
Region, are poor Shepherds and ſcabby Flocks, among which one finds
ſome Partridges : the ſecond Region is poſſeſs'd by Tygers and Crows.
All the reſt of the Mountain, that is, the half of it, has been cover'd
with Snow ever ſince the Ark reſted there, and theſe Snows are cover'd

half the Year with very thick Clouds. The Tygers we faw gave us no fmall Fear, tho they were not lefs than two hundred Paces from us, and we were affur'd they did not ufe to moleft the Paffengers; they were feeking Water to drink, and undoubtedly were not hungry that Day. However, we laid our felves along upon the Sand, and let them pafs by very refpectfully. They fometimes kill fome of them with a Gun, but the chief way of taking them is with Traps or Nets, by the help whereof they take young Tygers, which they tame, and afterwards lead about in the principal Towns of *Perfia*.

T H A T which is yet more inconvenient and troublefome in this Mountain, is, that the Snow which is melted, runs into the Abyfs by a vaft Number of Sources which one can't come at, and which are as foul as the Waters of a Land-flood in the greateft Storm. All thefe Sources form the Stream which runs by *Acourlou*, which never becomes clear. They drink Mud there all the Year ; but we found even this Mud more delicious than the beft Wine : 'tis always cold as Ice, and has no muddy Tafte. Notwithftanding the Amazement this frightful Solitude caft us into, we endeavour'd to find the pretended Monaftery, and inquir'd whether there were any Religious fhut up in Caverns. The Notion they have in the Country that the Ark refted here, and the Veneration all the *Armenians* have for this Mountain, have made many imagine that it muft be fill'd with Religious; and *Struys* is not the only Perfon who has told the Publick fo. However, they affur'd us there was only one forfaken Convent at the Foot of the Gulph, whither they us'd to fend one Monk every Year from *Acourlou*, to gather in fome Sacks of Corn which grows in the Country about it. We were oblig'd to go thither the next day for Water to drink, for we foon confum'd the Water our Guides, by the Advice of the Shepherds, had furnifh'd themfelves with. Thefe Shepherds are more devout than others, and indeed all the *Armenians* kifs the Earth as foon as they fee *Ararat*, and repeat certain Prayers, after having made the Sign of the Crofs.

W E encamp'd this day juft by the Shepherds Cottages, which are very forry Huts; they move from place to place as they have occafion, for they can't continue there but in good Weather. Thefe poor Shepherds, who had never feen any *Franks*, efpecially *Botanifts*, were almoft

*

as much afraid of us as we were of the Tygers : However, it was neceffary they fhould become more familiar with us; and we began to fhew
them fome Marks of our Friendfhip for them, and gave them fome Cups
of good Wine.   In all the Mountains in the World, one may gain upon
the Shepherds with this Liquor, which they are much fonder of than of
the Milk they live on.   Two of them were fick, and in vain reached
feveral times to vomit : We affifted them, and gave them Eafe immediately; which procured us great Efteem with their Companions.

A S we continually purfued our Defign, to inform our felves of the
Particulars of this Mountain, we caus'd a great many Queftions to be put
to them : But every thing being well weigh'd and examin'd, they advis'd
us to return back, rather than venture to advance farther up to the
Snow.   They inform'd us there was no Fountain throughout the whole
Mount, only the Stream of the Abyfs, which we could not come at to
drink but near the forfaken Convent, before mention'd; and that we
could not go in a whole Day to the Snow, and down again to the
Bottom of the Abyfs; but muft be like Camels, who drink once in the
Morning for the whole Day, it being impoffible to carry Water with
us, and climb fo horrible a Mountain, where they themfelves often loft
their Way : That we might judge what a miferable Place it was, from
the Neceffity they were under to dig the Earth from time to time to
find a Spring of Water for themfelves and their Flocks : And that it
would be to no purpofe to afcend higher in fearch of Plants, becaufe we
fhould only find Rocks hanging over our Heads, and heap'd one upon
another.   And, in fhort, that it would be Folly to proceed on our Way;
for our Legs would fail us : And that, for their parts, they would not
accompany us for all the Treafures of the King of *Perfia.*

T H I S Day we met with fome Plants, which were handfom enough :
But we expected to find fomething more extraordinary the next Day,
notwithftanding what the Shepherds had faid to us.   And the very Name of
Ararat would raife any one's Curiofity.   Who would not expect to find
fome of the moft extraordinary Plants upon a Mountain which ferv'd, as
I may fay, for a Ladder to *Noah,* whereby he and all other Creatures
came down from Heaven to inhabit the Earth? And yet we were vex'd
to meet with *Cotonafter folio rotundo* J. B.   *Conyza acris, cærulea* C. B.
Hieracium

Hieracium fruticofum, angufti folium, majus C.B. *Jacobæa, Sencionis Folio,* Strawberry Plants, *Orpin, Eye-bright,* and I know not how many of the moft common Plants, intermingled with fome others that were more rare, which we had already feen in feveral Places. But two we found which feemed wholly new

LYCHNIS Orientalis maxima, Bugloffi folio undulato. Coroll. Inft. Rei Herb. 23.

THE Root of this Plant is a foot and half long, whitifh, divided into large Fibres, pretty hairy, at the Neck about an inch thick, divided into feveral Heads, from whence fpring Stalks three feet high, ftrait, firm, four lines thick, gutter'd, of a pale green, hairy, clammy, adorn'd with Leaves two by two, about five inches long, and one broad, like thofe of *Buglofs,* wav'd, jagged at the Edge, rais'd at bottom with a pretty thick Rib, which fends feveral Veffels through the whole length of the Leaves. They leffen confiderably towards the middle of the Stalk, and from their Knots fhoot out on every fide Branches or Sprigs divided generally into three Foot-ftalks, each of which bears a Flower; and fo all the Flowers feem to be difpos'd into Stories. Each Flower confifts of five white Leaves, about two inches long, half an inch thick at the top, deeply hollow'd, and ending at bottom with a greenifh Tail. Out of the middle of thefe Flowers proceeds a Tuft of Chieves of the fame Colour, very fmall, but much longer than the Leaves, and having Tops which are Sea-green. The Cup is a Pipe of one inch long, and three lines thick, whitifh, ftrip'd with green, cut into Points; at the bottom of which is a Piftile of four lines long, and one thick, of a pale green, furmounted by three white Threds as long as the Chieves.

GEUM Orientale, Cymbalariæ folio molli & glabro, flore magno albo. Coroll. Inft. Rei Herb. 18.

THIS fine Species of *Geum* grows out of the Cracks of very fteep Rocks. Its Root is fibrous, whitifh, four or five inches long, hairy. Its Leaves grow in bunches, fo like the *Cymblaria communis,* that they are eafily miftaken for it; only they are more firm. For the moft part, they are nine or ten lines broad, and feven or eight lines long, cut into large indentings like Gothick Arches, fhining, ftanding upon a Foot-ftalk of an inch, or two inches and half long. The Stalks are a fpan
long,

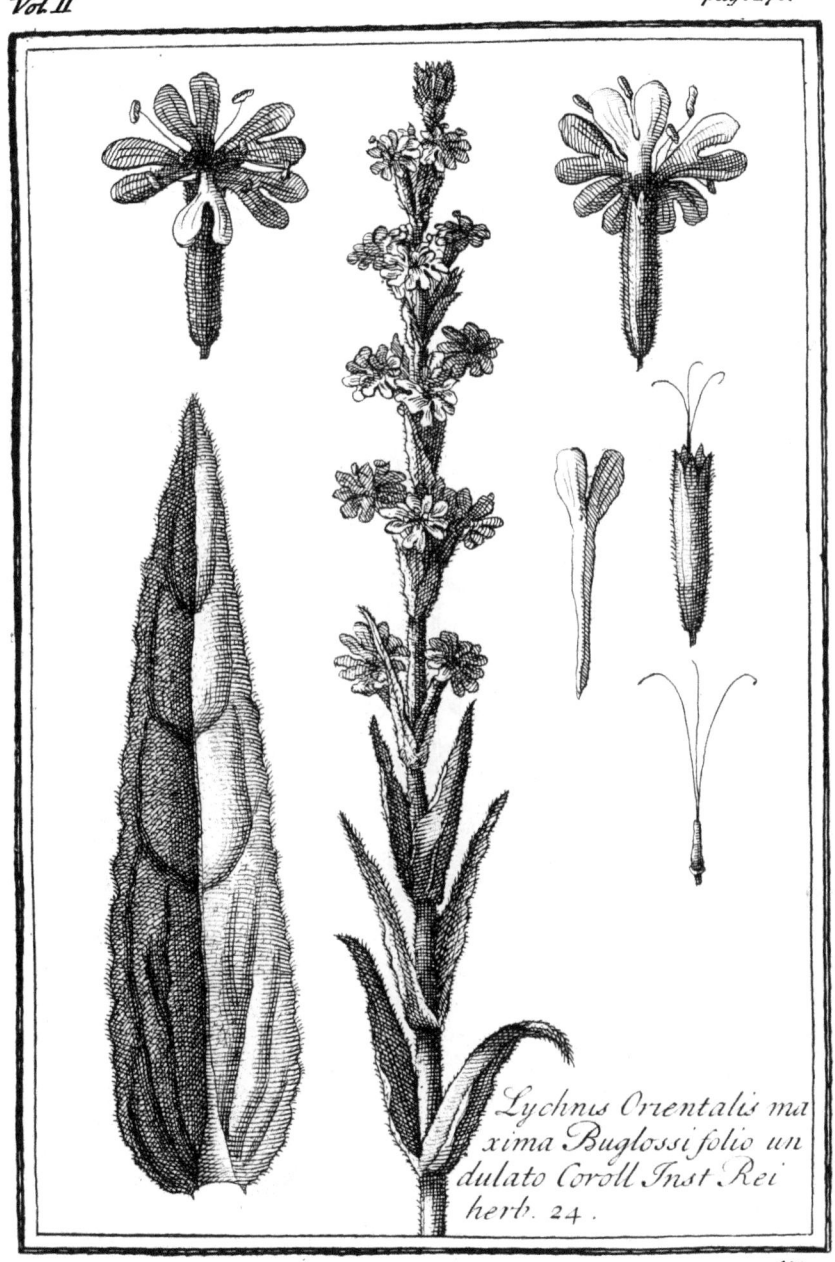

*Lychnis Orientalis ma
xima Buglossi folio un
dulato Coroll Inst Rei
herb. 24.*

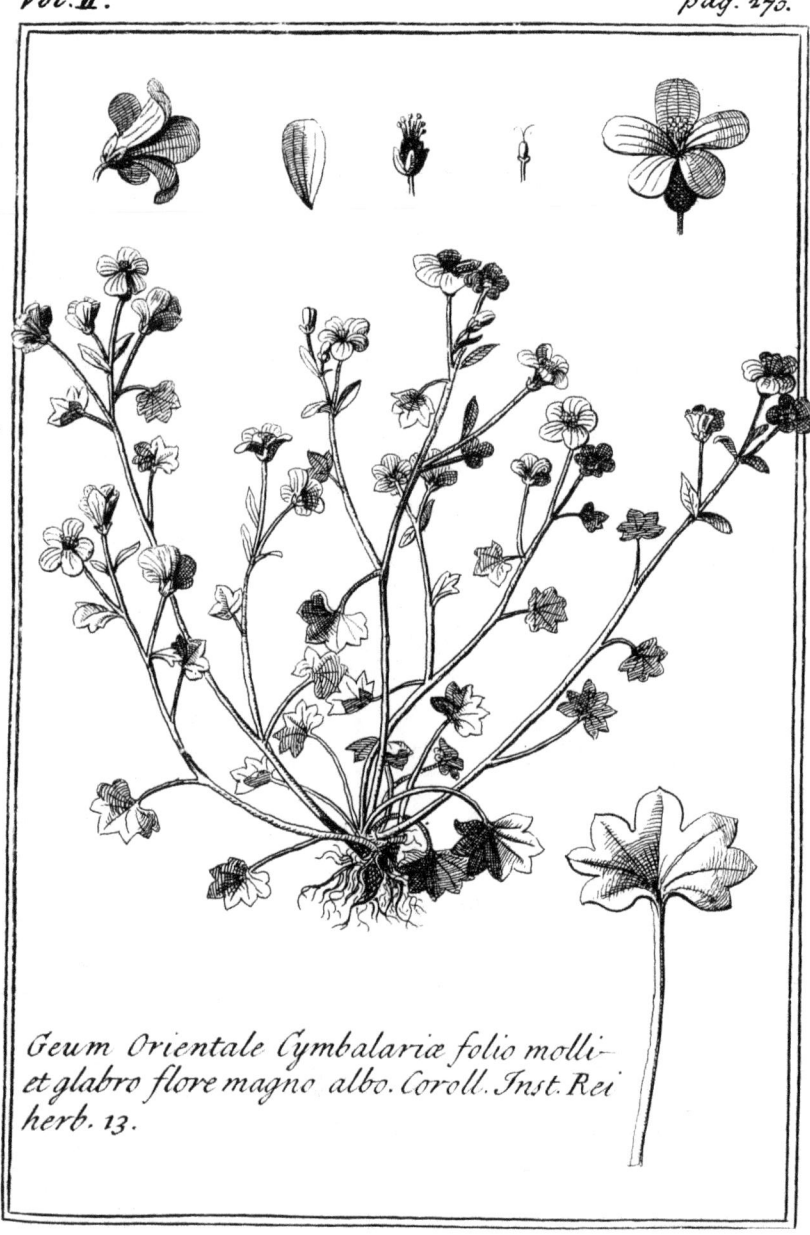

*Geum Orientale Cymbalariæ folio molli-
et glabro flore magno albo. Coroll. Inst. Rei
herb. 13.*

long, and hardly one third of a line thick, weak, almoſt lying upon the Lett. VII
Rocks, afterwards rais'd, and having a few Leaves, whoſe Indentings
are more pointed than in the lower Leaves. All along the Stalk and
Branches it is hairy, and charg'd with Flowers of five Leaves, half an inch
long, and about three lines wide at the Extremity, white, vein'd with
green at the Baſe. The Chieves, which ariſe out of the middle of theſe
Leaves are white, and not above two lines long, with very ſmall green-
iſh Heads. The Cup is cut to the Center into five Parts ſtrait and hairy.
The Piſtile is a pale-green, pretty round at bottom, in ſhape like an Ewer,
with two Lips, as in the other Species of the ſame Kind. It becomes a
Coffin of the ſame Shape, membranous, brown, divided into two Apartments, three lines high, in each of which there is a ſpungy Placenta, charg'd
with very ſmall and blackiſh Seeds. The Leaves of this Plant have a herbiſh Taſte, a little ſaltiſh. The Flowers have no Smell, the Roots are ſwee-
tiſh, and a little ſtiptick.

AFTER we had writ our Journal fair, we three at Table held a
Council, to conſider what Route to take the next day. We ran no hazard of being underſtood, becauſe we talk'd *French* ; and who is there
upon Mount *Ararat* who can boaſt he underſtands *French* ? Not even *Noah*
himſelf, if he was to come thither again with his Ark. We conſider'd
what the Shepherds had ſaid, which we look'd upon as very material,
eſpecially that inſuperable Difficulty of the want of Drink ; for we
reckon'd it nothing to ſcale a Mountain they repreſented ſo frightful.
How vexatious is it, ſaid we, to have come ſo far, to have gone up one
quarter part of the Mountain, to have found but three or four rare Plants,
and turn back again without going any farther ! We advis'd with our
Guides : they, good Men, unwilling to expoſe themſelves to the danger of
dying for Thirſt, and having no Curioſity, at the expence of their Legs, to
meaſure the Height of the Mountain, were at firſt of the ſame Sentiments
with the Shepherds, but afterwards concluded we might go to certain
Rocks, which ſtood out farther than the reſt, and ſo return to reſt at
Night in the ſame Place we were now in. This Expedient ſeem'd very
reaſonable, and with this Reſolution we went to Bed ; but who could
ſleep under the Inquietude in which we were ? In the Night the Love of
Plants overcame all other Difficulties ; and we three by ourſelves conclu-

ded

ded it was for our Honour to afcend the Mountain up to the Snow, and venture being devour'd by Tygers. As foon as it was Day, for fear we fhould die of Thirft in our Journey, we began to drink plentifully, and put our felves to a fort of voluntary Torture. The Shepherds, who were become a little fociable, laugh'd heartily, and took us for Perfons who were endeavouring to deftroy our felves. After this Precaution it was neceffary to dine, and it was no lefs Punifhment to eat without being hungry than it was to drink without Thirft: but it was abfolutely neceffary; for there was no Conveniency upon the way, and we were fo far from being able to carry Provifions with us, that it was with difficulty we could carry even our Clothes thro fuch bad Ways. We order'd two of our Guides to go with our Horfes, and wait for us at the abandon'd Convent, at the bottom of the Abyfs: we are forc'd to defcribe it thus, to diftinguifh it from that other abandon'd Convent at *Acourlou*, which ferves only for a Retreat for Paffengers.

A F T E R this, we began to travel towards the firft Range of Rocks, with one Bottle of Water, which to eafe ourfelves we carried by turns; but notwithftanding we had made Pitchers of our Bellies, in two hours time they were quite dry'd up; and Water fhook in a Bottle is a very difagreeable fort of Drink: our only Hope therefore was to come at the Snow, and eat fome of it to quench our Thirft. The Pleafure of Simpling is, that one may, under pretext of feeking Plants, ramble as much as one pleafes out of the direct Road, and fo tire ourfelves lefs than if we were forc'd to afcend right up: Moreover, 'tis a very agreeable Amufement, efpecially when we difcover any new Plants. However, tho we did not meet with many Novelties, yet the Hope of a good Harveft made us advance briskly. It muft be acknowledg'd that the Sight is very much deceiv'd, when we ftand at the Bottom, and guefs at the Height of a Mountain, and efpecially when it muft be afcended thro Sand as troublefome as the *Syrtes* of *Africa*. It is impoffible to take one firm Step upon the Sand of Mount *Ararat*, and in good Philofophy one lofes a great deal more Motion than when one walks on firm Ground. What a Feaft was it for thofe who had no Water but what was in their Bellies, to fink every Step up to the Ancle in Sand! In many Places, inftead of afcending, we were oblig'd to go back again down to the middle of the Mountain; and in order to continue
nue

nue our Courſe, to wind ſometimes to the right, and ſometimes to the Lett. VII.
left: when we met with any Mouſe-ear, it made our Boots as ſmooth as
Glaſs, and ſo ſlippery that we were forc'd to ſtand ſtill. However, this
time was not wholly loſt, for we employ'd it in diſcharging the Water
we had drank: but in truth we were two or three times about to have
given up our Déſign. And it had been better we had, than in vain to
ſtrive againſt ſuch a horrible Sand, and a Mouſe-ear ſo ſhort, that the
moſt hungry Sheep could not brouze on it. However, the Reflection
that we had not ſeen all, would have given us Uneaſineſs afterwards, and
we ſhould have been apt to fancy we had neglected the beſt Places. 'Tis
natural to flatter our ſelves in theſe ſorts of Enquiries, and to believe that
we only want a lucky Minute to find ſomething extraordinary, which
would make amends for all our Pains. Beſides, the Snow which was always in our View, and which ſeem'd to draw nearer to us, tho indeed it
was a great way off, attracted us very powerfully, and bewitch'd our
Eyes continually; and yet the nearer we approach'd it, the fewer Plants
we found.

TO avoid the Sand, which fatigued us intolerably, we took our way
to the great Rocks heap'd on one another, like *Oſſa* upon *Pelion*, to
ſpeak in the Language of *Ovid*. We paſs under them as thro Caverns,
wherein we are ſhelter'd from all the Injuries of the Weather, except the
Cold, which we felt there very ſenſibly, and ſerv'd a little to allay our
Thirſt. We were oblig'd to leave this Place quickly, left we ſhould get
a Pleuriſy; and came into a very troubleſome way, full of Stones, much
like the Stones us'd at *Paris* by the Maſons; and we were forc'd to leap
from one Stone to another. This Exerciſe we found very tireſome, and
we could not but laugh to ſee our ſelves forc'd to take ſuch Methods, tho
in truth it was but from the Teeth outwards. For my part, being quite
tired out, and not being able to go any farther, I firſt began to repoſe my
ſelf, which was an Excuſe for the reſt of the Company to do the like.

AS the Converſation is commonly renew'd when we are ſat down,
one talk'd of the Tygers which walk'd about very quietly, or play'd at
a good reaſonable diſtance from us. Another complain'd that his Waters
did not paſs off well, and that he could not breathe: and for my own
part, I never was more afraid that ſome lymphatick Veſſel was broken

in my Body. In fine, amidſt all theſe little Paſſages with which we endeavour'd to amuſe our ſelves, and which ſeem'd to give us new Strength, we came about Noon to a place more pleaſing, for it ſeem'd as if we were ready to take hold of the Snow with our Teeth. But our Joy laſted not long; for what we had taken for Snow was only a Chalk Rock, which hid from our Sight a Tract of Land above two hours Journey diſtant from the Snow, and which ſeem'd to us to have a new kind of Pavement, not of little Flints, but ſmall pieces of Stone broken off by the Froſt, and whoſe Edges cut like Flints. Our Guides told us their Feet were quite bare, and that ours would quickly be ſo too; that it grew late, and we ſhould certainly loſe our ſelves in the Night, or break our Necks in the Dark, unleſs we choſe to ſit our ſelves down to become a Prey to the Tygers, who ordinarily make their chief Attempts in the Night. All which ſeem'd very probable; however, our Boots were not bad yet. After having look'd on our Watches, which we kept in very good Order, we aſſured our Guides that we would go no farther than a Heap of Snow which we ſhew'd them, and which did appear to be hardly bigger than a Cake: But when we came to it, we found more than we had need of; for the Heap was above thirty Paces in diameter. We every one eat more or leſs, as we had a mind; and by Agreement reſolv'd to advance no further. This Snow was above four Feet thick; and being frozen hard, we took a great Piece to fill our Bottle. It can't be imagin'd how much the eating of Snow revives and fortifies: Some time after we felt a glowing Heat in our Stomachs, like that in the Hands, after having held Snow in them half a quarter of an Hour; and far from cauſing griping Pains, as moſt imagine it muſt, it was very comfortable to our inward Parts. We deſcended therefore from the Snow with a wonderful Vigour, much pleas'd that we accompliſhed our Deſire, and that we had now nothing farther to do but to retire to the Monaſtery.

AS one good Fortune is generally followed by another, by chance I perceived a ſmall green Plat, which glitter'd among the ruinous Fragments of Stone. We ran thither as to a Treaſure, and were highly pleaſed with the Diſcovery. It was an admirable Species of *Veronica Telephii folio:* But we did not ſtay there long, our Thoughts being now much taken up with our Return. And our pretended Vigour was not of long

Dura-

Duration . For we came to Sands which lay behind the Abyfs, and were full as troublefome as the former. When we endeavour'd to flide along, half our Bodies were buried : Befides, we could not keep the direct Way, but were oblig'd to go to the Left to come to the Edge of the Abyfs, of which we had a mind to take a nearer View. And indeed it is a moft frightful Sight : *David* might well fay, fuch fort of Places fhew the Grandeur of the Lord. One can't but tremble to behold it ; and to look on the horrible Precipices ever fo little, will make the Head turn round. The Noife made by a vaft Number of Crows, who are continually flying from one Side to the other, has fomething in it very frightful. To form any Idea of this Place, you muft imagine one of the higheft Mountains in the World opening its Bofom, only to fhew the moft horrible Spectacle that can be thought of. All the Precipices are perpendicular, and the Extremities are rough and blackifh, as if a Smoke came out of the Sides, and fmutted them. About Six a clock after Noon we found our felves quite tir'd out, and fpent ; and were not able to put one Foot before another, but were forc'd to make a Virtue of Neceffity, and merit the Name of *Martyrs to Botany.*

W E at length obferv'd a Place cover'd with Moufe-ear, whofe Declivity feem'd to favour our Defcent, that is to fay, the Way *Noah* took to the Bottom of the Mountain. We ran thither in hafte, and then fat down to reft our felves ; and found there more Plants than we had all the Journey befide : And what pleas'd us mighty well, was, that our Guides fhew'd us from thence, but at a great diftance from us, the Monaftery whither we were to go to quench our Thirft. I leave it to be guefs'd what Method *Noah* made ufe of to defcend from this Place, who might have rid upon fo many Sorts of Animals which were all at his Command. We laid our felves on our Backs, and flid down for an Hour together upon this green Plat, and fo pafs'd on very agreeably, and much fafter than we could have gone on our Legs. The Night and our Thirft were a kind of Spurs to us, and caus'd us to make the greater fpeed. We continued therefore to flide in this manner as long as the Way would fuffer us ; and when we met with fmall Flints which hurt our Shoulders, we turn'd, and flid on our Bellies, or went backwards on all four. Thus by degrees we gain'd the Monaftery ; but fo diforder'd and fatigu'd by our manner of travelling, that we were not able to move Hand or Foot. We found some good Company in the Monaſtery, the Gates of which are open to every body for want of Faſtnings. The People of the Town had taken a Walk thither, and were juſt going away as we came; but to our great misfortune had neither Wine nor Water. We were therefore forc'd to ſend to the River; but had no Veſſel beſide our Leathern Bottle, which held not above a Quart. And what a Puniſhment was it for the Guide on whom the Lot fell, to go the River, and fill it? He had the Happineſs indeed to be the firſt who drank; but no body envied him: For he paid dear enough for it; the Deſcent from the Monaſtery to the River was near a quarter of a League down-right, and the Way very rugged: One may gueſs how pleaſant his Journey was back again. It took up half an Hour to go and come; and the firſt Bottle was almoſt drank out at one Draught. The Water ſeem'd like Nectar; but we were forc'd to wait another Half-hour for a ſecond Bottle, which was Miſery enough We took Horſe that Night for the Town, to get ſome Bread and Wine; for after all the Pains we had taken, we found our Bellies very empty. We did not reach the Town till about Midnight; and he that kept the Key of the Church, in which we were to lodge, was ſleeping at his Eaſe at the other End of the Town. We were very happy now in having found ſome Bread and Wine. After this light Supper we got into a good ſound Sleep, without being diſturb'd by Dreams, any Uneaſineſs, or Indigeſtion, or ſo much as in the leaſt feeling the Sting of the Gnats.

THE Day following, being the 12th of *Auguſt*, we departed from *Acourlou* at Six in the Morning to return to *Three-Churches*, where we arriv'd the 13th, after having forded the *Araxes*; which loſt us much Time, for this River is known to be very unmanageable ever ſince the Time of *Auguſtus*. 'Tis too rapid to have any Bridge laid over it; and it did formerly carry away thoſe which the Maſters of the World build over it. This *Araxes*, on whoſe Banks have appear'd the moſt famous Warriors of Antiquity, *Xerxes, Alexander, Lucullus, Pompey, Mithridates, Anthony*; I ſay, this *Araxes* ſeparated *Armenia* from the Country of the *Medes*, and therefore *Three-Churches* and *Erivan* are in *Media*. Antient Authors, with good reaſon, make this River to come from thoſe famous Mountains in which are the Springs of the *Euphrates*; for we found it at *Aſſancala* near

to

to *Erzeron,* not far from whence lies the *Euphrates,* as was obferv'd above. Lett. VII.
Thofe Geographers who fay the *Araxes* comes out of Mount *Ararat,* are greatly miftaken; and muft have taken the River near *Acourlon* for the *Araxes,* which is larger between *Ararat* and *Erivan* than the *Sein* is at *Paris.*

THE 14th of *Auguft* we ftaid at *Three-Churches,* waiting for fix Horfes we had fent for to *Erivan,* in order to return to *Cars.* We had the misfortune to fet out without Company, for all the Caravans which were at *Three-Churches* were bound for *Tauris.* So civil as the *Perfians* were, we did not care to come near their Frontiers, efpecially in the Neighbourhood of *Cars.* There fell this Day fo much Snow upon Mount *Ararat,* that its fmaller Top was all white with it. We gave Thanks to God that we were fafe return'd; for we might have been loft there, or died of Hunger upon the Mountain. We fet out next Day at Six a clock, and travell'd till Noon upon a very dry Plain, cover'd with different Kinds of *Saltwort,* *Harmala,* that Kind of *Ptarmica* which *Zanoni* took for the firft Kind of *Southern-wood* of *Diofcorides.* The *Alhagi Maurorum* of *Bauvolf,* which furnifhes the *Perfian Manna,* was every where to be feen. I have before given a Defcription of it. We encamped this Day upon the Banks of a River, near a Village, render'd very agreeable by the fine Greens thereabout. We ftaid there but about an Hour; and ftill leaving Mount *Ararat* on the Left, went towards the Weft to come to *Cars.* We continued our Journey till Six a Clock in the Evening, but over Plains full of Flints and Rocks.

I imagine the Country which *Procopius* calls *Dubios,* can't be far from Mount *Ararat.* 'Tis a Province, fays he, not only very fruitful, but likewife extremely convenient and pleafant for the Goodnefs of the Climate and its Waters, about eight Days Journey from *Theodofiopolis.* One fees here nothing but large Plains, on which are feveral Villages not far from one another, inhabited by Factors, who have fettled there to facilitate the Commerce of *Georgia,* *Perfia,* the *Indies,* and *Europe;* the Merchandize of thofe Countries being brought thither as to the Centre of Trade. The Patriarch of the Chriftians in this Country is called *Catholick,* becaufe he is generally own'd as the Head of their Religion. It is plain from hence, that the Trade between the *Perfians* and *Indians* is not

new.

new. Perhaps this *Dubios* is the Plain of *Three-Churches*, and that the *Romans* carried their Merchandizes thither as to the greateſt Fair in the World. There is no Place more proper for a common Mart for the Nations of *Europe* and *Aſia*.

THE 16th of *Auguſt* we ſet out at Three a clock in the Morning, without Convoy or Caravan. Our Guides made us travel till about Seven in dry, ſtony, uncultivated, and very diſagreeable Plains. We got on horſeback about Noon, and put on for *Cochavan*, the laſt Town in *Perſia*. Fear began to ſeize us, upon our approaching to this Frontier: But I was not aware of any Danger I was expoſed to in paſſing the River of *Arpajo*, or *Arpaſou*. Some one or other is drown'd there every Year, according to Report; and I was in great danger of being one of thoſe who pay that Tribute. The Ford is not only dangerous, becauſe of its Depth, but beſides this, the River brings down from time to time great Pieces of Stones which roll down from the Mountains, and cannot be diſcern'd in the Bottom of the River, and avoided. The Horſes can't ſet their Feet firm upon the Bottom: They often ſtumble, and even break their Legs when they get in between theſe Stones. We paſs'd over two and two together: My Horſe in his place, after having ſtumbled, raiſed himſelf up again without any Hurt, but not without putting me into a very great Fright. I then gave my ſelf up to his ſage Conduct, or rather to my good Fortune, and let him go as he would, ſpurring him with the Heels of my Boots, which had a Piece of Iron ſticking out very little, in form of a Semicircle; for they have no Spurs in the *Levant*. My poor Beaſt ſunk a ſecond time into a Hole, leaving only his Head above Water, out of which he could not recover himſelf but after a great deal of Struggling, during which I was in a very bad Condition. The Outcries, not to ſay the Roarings of our Guides, increaſ'd inſtead of leſſening my Fear. I did not underſtand any thing they ſaid to me, and my Companions could give me no manner of Aſſiſtance. But my Hour was not yet come: The Lord would have me return to herborize again in *France*; and I eſcap'd with no other Damage but the Trouble of drying my Clothes and Papers, which, according to the Cuſtom of that Country, I carried in my Boſom; for we had left our Baggage at *Erzeron*, and travell'd with as little Luggage as poſſible.

THIS

THIS Washing was the more inconvenient, becaufe we dared not Lett. VII.
go into the Town of *Chout-louc*, in the *Turkish* Dominions. Our Guides,
who were of *Erivan*, and expected they fhould be obliged to pay the
Capitation in *Turky*, tho the *Perfians* don't exact it of the *Turks* who come
into their Country, would ftop upon the Banks of a River about a
quarter of a League from this Town. The Air of this River did not
warm me much, and contributed lefs to dry my Clothes. We were
therefore oblig'd to pafs the Night without Fire or any hot Victuals;
nay, we had not fo much as any Wine left. And to compleat the Mif-
fortune, my Half-bathing, which I had no Inclination to, had given me
a Diforder, which caufed me to rife oftner than I could have wifh'd. We
fhould, however, have remain'd tolerably content under thefe Misfortunes, had not a Man of thofe Parts, I don't know of what Religion,
took it into his Head to make us an unpleafant Vifit, notwithftanding
all the Care our Guides had taken to lie concealed. He pretended to
come only very charitably to advife us we were not fafe in that Place;
that it would be very happy for us if we were not plunder'd in the
Night; that he thought even our Lives in danger; that we would do well
to retire into the Town, the *Sous-Bachi* whereof is a fworn Enemy to
the Robbers; but that he could not fecure us from the Robbers in the
Country, into whofe Hands we fhould probably fall the next Day in our
Way to *Cars*. We order'd our Guides to faddle our Horfes, that we
might go into the Town not only for greater Security, but that I might
there dry my felf: But thefe Wretches, notwithftanding all the Inftances
we could make to them, would not ftir, and treated our Advifer as a filly
whimfical Fellow. We were angry with them in vain; they would not
ftir an Inch: The five Crowns Capitation-Tax was of more Confideration with them than our Lives. I promis'd them to pay the Tax for
them, if the *Sous-Bachi* fhou'd demand it: But that was nothing; they
look'd upon it only as an Artifice of mine to prevail with them to go.
One of them, to recommend himfelf to us, had taken a great deal of
pains to pick up an Armful of Sticks, which he brought to me to dry
my Clothes. But our Advifer, whofe Kindnefs we wonder'd at, advifed
us not to make a Fire, left we fhould by that means difcover our felves
to any ill Men who might be wandering about: Nay, he even affured

us,

us, that if the *Sous-Bachi* knew our Intention, he would oblige us to lodge in the Town : That fure we had in Charge all the Diamonds of the Kingdom of *Golconda*, feeing we avoided every body with fo much, Precaution. All this fignified nothing to our *Perfians*; they thought of nothing but the Capitation : But we were fully revenged on them the next Day, when they were taken by the Throat at the Gates of *Cars*, and obliged to pay the Tax.

T H E Y might glory as long as they would in being Subjects of the King of *Perfia*, and of the good Ufage the Subjects of the Grand Signior found in their Country ; all was in vain : The *Turks* of *Cars* were hard-hearted ; and they were forc'd to pay five Crowns each, and take a *Carack*, which is a kind of Acquittance, to fecure them from being obliged to pay a fecond time. They were foolifh enough to propofe it to us to repay them this Tax, becaufe 'twas in our Service they had fuffer'd this Oppreffion. We anfwer'd, we had not agreed to any fuch Article in our Bargain with them ; but that neverthelefs we would have paid it voluntarily, if they would have gone to lodge in the Town, inftead of forcing us to lie all Night in the open Fields, at the mercy of Robbers and Wolves.

A N D in truth we had a very ill Night by the River : And it feem-ed much longer after our Advifer went away ; for the good Man, when he faw all his Rhetorick could not prevail, left us. We could not tell but he was come as a Spy to obferve us, and might inform his Companions that we had befides our Baggage certain Merchandife : But this which to him might have feem'd to be Merchandife, was only a *Collection of dry Plants* in two *Turkifh* Coffers. Our Advifer did not fail to feel the Weight of them while he was giving us his Advice, and admir'd they were fo light. To fpeak freely, I believe our apparent Poverty fav'd us ; for all our Baggage was not worth their coming from the Town to fetch. Neverthelefs the Nights being very cold in the *Levant*, and this being much more cold to me than any of the Company, becaufe my Clothes were not dry, I was in a very great Perplexity. The Way we were to go to *Cars* added to my Uneafinefs : They talk'd of nothing but Robbers ; and we had no Letters to *Cars* to be fupplied with Money, if we fhould be robbed.

W E

WE had likewise the Diſſatisfaction to come away without ſeeing Lett. VII. the Ruins of *Anicavac*, or *Anicagué*, that is to ſay, the City of *Ani*, which is the Name of a certain King of *Armenia*. Theſe Ruins are in the *Perſian* Dominions, half a League out of the Road we had paſs'd; but our Guides did not obſerve to ſay any thing of it to us, till we were come to our Lodging. I don't believe there is any thing curious to be ſeen by Travellers among theſe Ruins : There are nothing but the Remains of antient *Greek* Towns which deſerve to be ſeen ; becauſe one often meets with Inſcriptions, which frequently help very much to remove ſeveral Difficulties in antient Geography.

WE departed hence the 17th of *Auguſt* at Four a clock in the Morning, and travell'd till Seven without meeting with any body on the way. The Clearneſs of the Day reviv'd us much; and as the Danger I was in of being drown'd had brought me under an Inconveniency, which often obliged me to diſmount from my Horſe, I propoſed to the Company to ſtop a while to repoſe our ſelves. The Place was very agreeable, and we ſpread our Cloth, and eat up the Proviſions we had left. After this Repaſt, we continued our Journey in a plain Low-Country, very pleaſant, and well cultivated. We diſcover'd three or four conſiderable Towns, and perceiv'd we drew near to one of the principal Cities in thoſe Parts. We found charming Paſture at the foot of a ſmall Hill, which was very agreeable; and the Shepherds, who were not far out of the great Road, look'd like a very good ſort of People.

WE arriv'd at *Cars* about Four of the clock, and ſtaid there till the 22d of *Auguſt*, waiting for Company. A great Party of *Curdes* had encamp'd themſelves upon the Mountains, two Days Journey from *Cars* in the Road to *Erzeron*; and as we had no *Armenian* Biſhop to intercede for us, we judg'd it would be very imprudent to run the hazard of paſſing without the Caravan. While we waited for one, we viſited ſeveral ſick Perſons with Succeſs, that is, as to their Health; for all our Viſits procur'd us nothing more than ſome Plates of Fruit, or Meaſures of Milk. The Country about *Cars* is very fit for herborizing; and we walk'd about very freely, by the Favour of ſome Friends we had gain'd by coming from *Erzeron*. The Aga, who had a *Fiſtula in ano*, tho he

Vol. II.   O o   had

had no advantage by our Remedies, came to give us Thanks, and affured us he would not let us depart thence without a good Guard. Another Gentleman, whom we had done fome Service to, who had been miferably afflicted with the *Hæmorrhoids*, would accompany us in Perfon, with three or four of his Family, till we fhould be out of danger : So certain it is, that there are many good People every where ; and that a Box of Medicines well chofen and prepar'd, and properly ufed, is a good Paffport. There's no Part of the World where one can't raife one's felf Friends by the help of Phyfick. The greateft Lawyer in *France* would be taken for a very ufelefs Perfon in *Afia*, in *Africa*, and in *Armenia :* The moft profound and zealous Divines would not be more efteem'd, unlefs the Lord would efficacioufly touch the Hearts of the Infidels : But the Fear of Death prevailing in all Places, they are every where glad of Phyficians, and pay them a great deal of Refpect. The greateft Commendation that can be given the Gentlemen of our Profeffion, is the general Acknowledgment that they are neceffary ; for God has given Phyfick for the Comfort of Mankind. I beg your Lordfhip to pardon this fhort Digreffion in favour of my Profeffion.

H E R E is a Defcription of fome fine Plants which grow in the Neighbourhood of *Cars*.

CAMPANVLA Orientalis, foliorum crenis amplioribus & crifpis, flore patulo fubcæruleo. Coroll. Inft. Rei Herb. 3.

T H E Root of this Plant, which fhoots down into the Clefts of the Rocks, is about a foot long, and about an inch thick at the Neck, parted into feveral Heads, pretty flefhy, and divided in thick hairy Fibres, white within, but drawing to a yellowifh towards the Heart. The Rind is brown and reddifh. The Stalks, of a foot and half or two feet high, come out in Bunches feven or eight together, about two or three lines thick, firm, full of white Pith, fmooth, pale-green, furnifh'd at bottom with Leaves pretty firm, four inches long taking in the Stalk. They are not unlike thofe of the *Nettle*, fmooth, bright green, deeply notch'd with large Dents pointed and unequal, which are again cut or notch'd, jagg'd, and even, divided toward the bottom into certain fmall unequal Pieces. Thefe Leaves grow lefs all along the Stalk, and quite lofe their

*

Foot-

Campanula Orientalis, foliorum.
crenis amplioribus et crispis, —
flore patulo, subcæruleo. Coroll. Inst.
Rei. Herbar. 3.

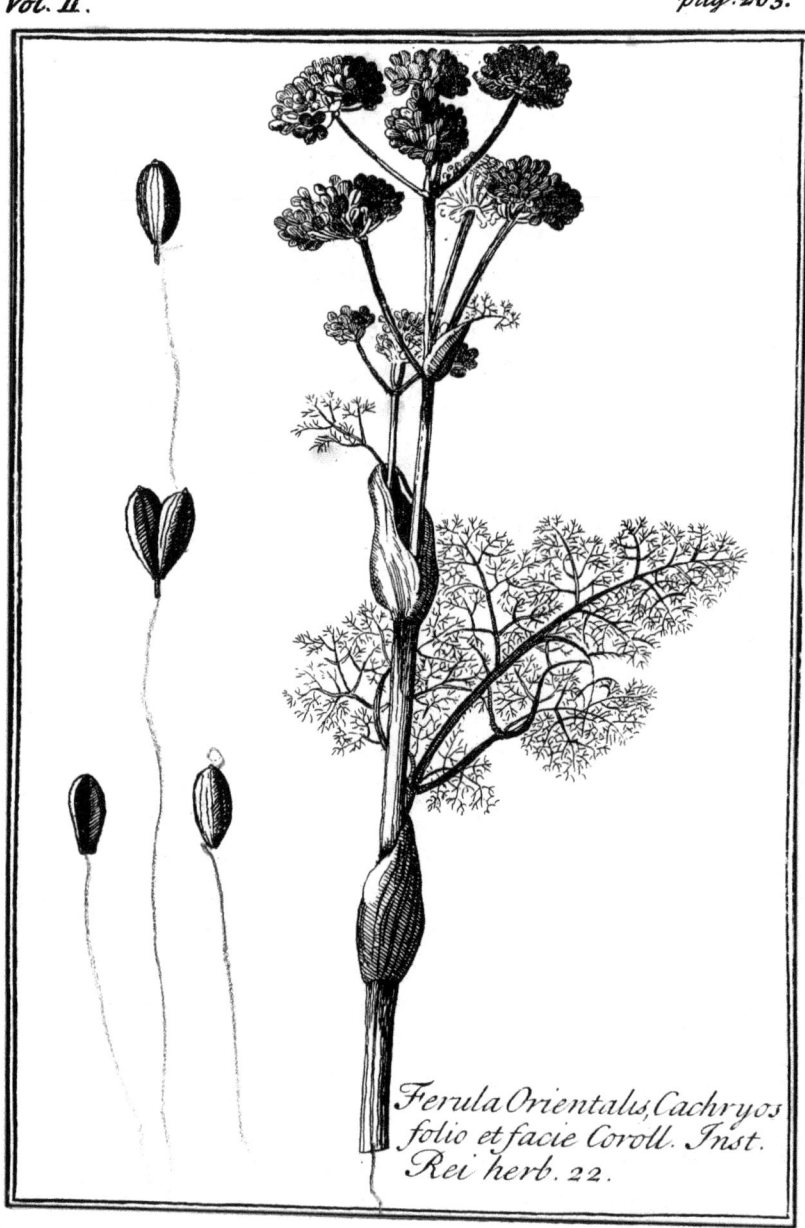

Ferula Orientalis, Cachryos folio et facie Coroll. Inst. Rei herb. 22.

Foot-ftalk or Tail toward the Top, where they refemble the Leaves of Lett. VII. the Herb call'd *Golden-Rod*; but they always are jagged. From the Knots ∿ fpring, even from the bottom, Flowers upon very fhort Foot-ftalks, which widen into a Bafon of more than an inch diameter, and half an inch deep, cut into five Parts. From the bottom of this Bafon proceed fo many Chieves or Threds with yellow Tops or Heads. The Piftile is as long as the Flowers, and ends in the fhape of an Anchor with three Arms. The Cup is another fort of Bafon, of about five lines high, pale-green, fplit into five Points. When this Plant is bitten off, as frequently happens near *Cars*, it puts forth Branches from the bottom. We faw fome whofe Flowers were very white, and others with bluifh Flowers. The Leaves are of a herbifh Tafte, and pretty ftrong. The Root is very much of a fweetifh Tafte, the Flowers are without Smell. The whole Plant yields a Milk which is pretty fweet, but which fmells like *Opium*.

FERULA Orientalis, Cachryos folio & facie. Coroll. Inft. Rei Herb. 22.

THE Root is as thick as an Arm, and two feet and a half long, branched, a little hairy, white, cover'd with a yellowifh Peel, and yields a Milk of the fame Colour. The Stalk rifes to three feet high, is half an inch thick, fmooth, firm, reddifh, full of a white Pith, furnifh'd with Leaves like the *Fennel*, of a foot and a half or two feet long, the fides of which divide and fubdivide themfelves into Slips as fmall as the Leaves of the *Cachrys, ferulæ folio, femine fungofo lævi*, of *Morifon*; which this Plant fo much refembles, that one might eafily be deceiv'd, were it not for the Seeds. The Leaves which accompany the Stalks are fome much lefs than others, and their Diftances are unequal. They begin by a Thred of three inches long, and two thick, fmooth, reddifh, terminated by a Leaf of about two inches long, cut as fmall as the others. Above the middle of the Stalk come out many Branches from the Knots of the Leaves, which are not much above a fpan long, and bear fmall *Umbellæ*, charg'd with yellow Flowers from five to feven or eight leaves apiece, half a line long. The Seeds are very like thofe of the *Ferula communis*, about half an inch long, and two lines and a half broad, thin toward the Edges, reddifh, and a little ftrip'd on the Back, bitter and oily.

LYCHNIS Orientalis, Buplevri folio. Coroll. Inft. Rei Herb. **24**.

THE Stalk of this Plant is three feet high, two lines thick, hard, firm, ftreight, knotty, fmooth, cover'd with a white Powder like that on the Stalks of *Pinks,* accompanied at bottom with Leaves four inches long, and four lines broad, fea-green, pointed like thofe of the *Bupleurum an-guſtifolium, Herbariorum Lob.* rais'd on one fide, for otherwife they are not vein'd. Thofe at the firft Knot of the Stalk are longer, but not a-bove four or five lines broad; the reft are more ftrait, the laft are like thofe of the *Pink.* From their Knots, all along the Stalk from the middle upward, grow out Branches half a foot in length, with very fmall Leaves: thefe Branches bear each three or four Flowers, whofe Cup is a Pipe or Tube of an inch or fifteen lines long, one line thick toward the bottom, and two lines at the top, where it is divided into five Points, fea-green and fmooth. From the bottom of the Tube come out five Leaves which reach over about half an inch, hollow'd into two parts very round, white be-low, but of a yellowifh green upwards, each rais'd by two white parts, which ferve to form the Crown of the Flower. The Chieves are white, with yellowifh tops. The Piftile, which is of a pale-green, oblong, hav-ing at the end two white Tufts, becomes a Fruit but half an inch long, and three lines in diameter, upon a Foot-ftalk of three lines high. This Fruit is a hard Shell, oval, reddifh, opening at the point into five or fix Parts, and yields a greyifh kind of Seeds, much like thofe of Henbane. The whole Plant is of a herbifh Tafte, and very mucilaginous.

THE 23d of *Auguſt* we left *Cars,* with a fmall Caravan, defign'd to guard a Sum of Money the *Carachi-Bachi,* or *Receiver of the Capitation,* fent to *Erzeron.* They were all chofen Men, well arm'd, and refolv'd to fight; whereas the Merchants Caravans are made up of fuch as would chufe to fleep in a whole Skin, as we fay, and had rather be ranfom'd than come to Blows. All things confider'd, this is the beft way for them, for a Mer-chant makes a very good Market, when he faves his Life and Merchan-dize by a handful of Crowns. We travell'd but four Hours this day, and encamp'd near *Benecliamet,* a Town in a large Plain, where we met a frefh Guard of *Turks,* confifting of refolute well-made Fellows.

*

THE

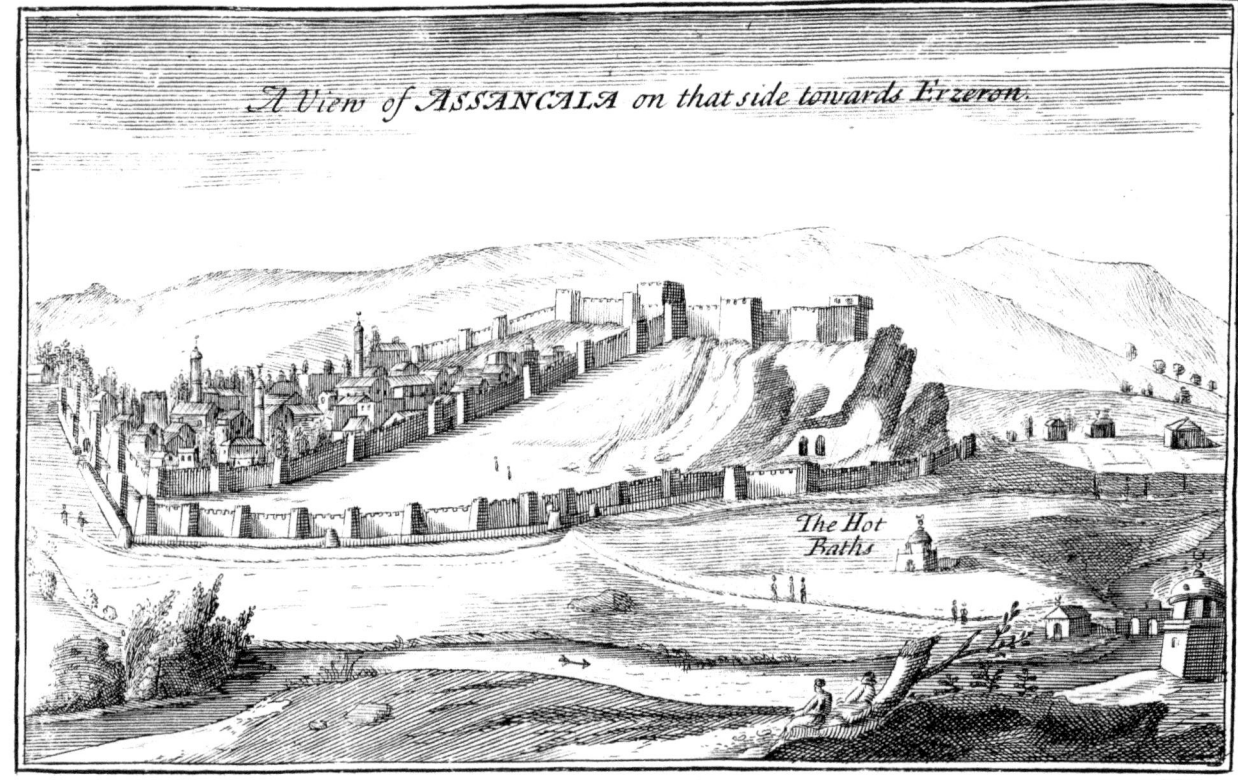

A View of ASSANCALA on that side towards Erzeron.

The Hot Baths

T H E 24th of *Auguſt* the *Carachi-Bachi*, who had an Order from the Baſſa of *Cars*, to take out of the Towns in his way, as many Men as he ſhould think needful to ſecure his Charge of Money, took from the Mountains about thirty Perſons well arm'd, who gave us a great deal of Diverſion, for it was rumour'd that the *Curdes* would attack them for their Booty.   This new Guard was reliev'd the next day by another of equal Strength.   A Caravan of ſixty *Turks* will face two hundred *Curdes*, theſe being only arm'd with Lances, while the *Turks* have good Guns and Piſtols.   We did not ſet out this day till nine a-clock, with deſign to lodge at *Kekez*, a Town ſituate in the ſame Plain, at about three Hours diſtance.   We were join'd by a Recruit of ſeven or eight Perſons, who carried Rice to *Erzeron* ; but they added no great Strength to us.

W E went but four Leagues the next day : we travell'd all Night by Moon-light among Mountains, where there were ſeveral dangerous Paſſes, and a few Men might eaſily attack us ; but the Darkneſs favour'd our March, while the *Curdes* ſlept at their Eaſe.   We reſted our ſelves the 26th till nine of the Clock in the Morning, and then went only upon one of the higheſt Mountains in that Country, cover'd with *Pines*, *Black Poplars*, and *Aſpines*.   Apprehending ſome Ambuſcade, we detach'd ſome of the *Turks* to view the Paſſes, and they brought to the *Carachi-Bachi* four Peaſants, who aſſur'd him the Robbers were behind us, and that we were a great way out of their reach.   Upon this News we ſtopt about Three of the clock after Noon near a ſmall River, where we had ſtopt before in our way to *Cars*, along which we found a beautiful kind of *Valerian*, whoſe Roots are very like thoſe of the *Great Garden Valerian*, as thick, and aromatick.   The Leaves are more ſtreight ; but as the *Great Valerian* is not, that I know of, to be found in the Champain, I perſuade myſelf 'tis only this which has been now ſome Ages cultivated in Gardens.

T H E 27th of *Auguſt* we travell'd near ſix Hours, and ſtopt at *Lavander*, an inconſiderable Village.   The 28th, after a good long Journey, we arriv'd at the Baths of *Aſſancala*, built very neatly on the Banks of the *Araxes*, a ſmall Day's Journey from *Erzeron*.   They are warm, and much frequented.   The *Araxes*, which comes from the Mountains, wherein are the Springs of the *Euphrates*, is not large at *Aſſancala* : the Plain is
more

more fruitful than that at *Erzeron*, and produces better Wheat. In general, all forts of Corn are but indifferent in *Armenia*: for the moft part it produces but fourfold, efpecially about *Erzeron*; but then there is a vaft quantity, which makes amends. If they had not the Conveniency of watering their Lands, they would be almoft barren.

IN the middle of the Plain of *Affancala* arifes a horrible fteep Rock, upon which they have built a Town and Fort which threatens all the Neighbourhood, and where they are more in danger of Famine than of Cannon. There are not above three hundred Men in the Garifon, tho it requires five hundred to defend it. The Walls are built in a fpiral line all round the Rock, and ftrengthen'd with fquare Towers, whofe Cannon, if they were well furnifh'd and mann'd, would hinder any Approaches, for thefe Towers are not rais'd higher than the Walls, and appear only like Platforms. The Ditches are not above two Fathom over, and not fo deep, cut into a very hard Rock. If this Place was upon the Frontier, it might be made impregnable with fmall Charge. The Merchandize carried from *Erzeron* to *Erivan* by way of *Affancala*, pays half a Piafter whether by Horfe or Camel, tho the Difference of Weights is very great. They who come from *Erivan* to *Erzeron*, pay but half as much. Our dry'd Plants paid nothing; the *Turks* and *Perfians* don't much efteem that fort of Merchandize, which however we valued more than the fineft Silks in the *Levant*.

THE Way from *Affancala* to *Erzeron* is very fine. We travell'd it in fix Hours time, and run the fame day to embrace Mr. *Prefcot* the *Englifh* Conful, our very good Friend, who would have taken the charge of our Clothes, Money, and dry'd Plants. We went next day to pay our refpects to the Beglerbey *Cuperli* our Protector, who afk'd us a thoufand Queftions concerning what we had feen in our Route, and efpecially of the Difference we found between *Turky* and *Perfia*. After having return'd him thanks for his Recommendation to the Baffa of *Cars*, we related to him part of our Adventures: we prais'd much the good Temper of the *Perfians*, and the good Reception they give the *Franks*. Among other things, he faid to us, that the Patriarch of *Three-Churches* was a good Oil-Merchant, alluding to the Proceedings between him and the *Ar-*

menian

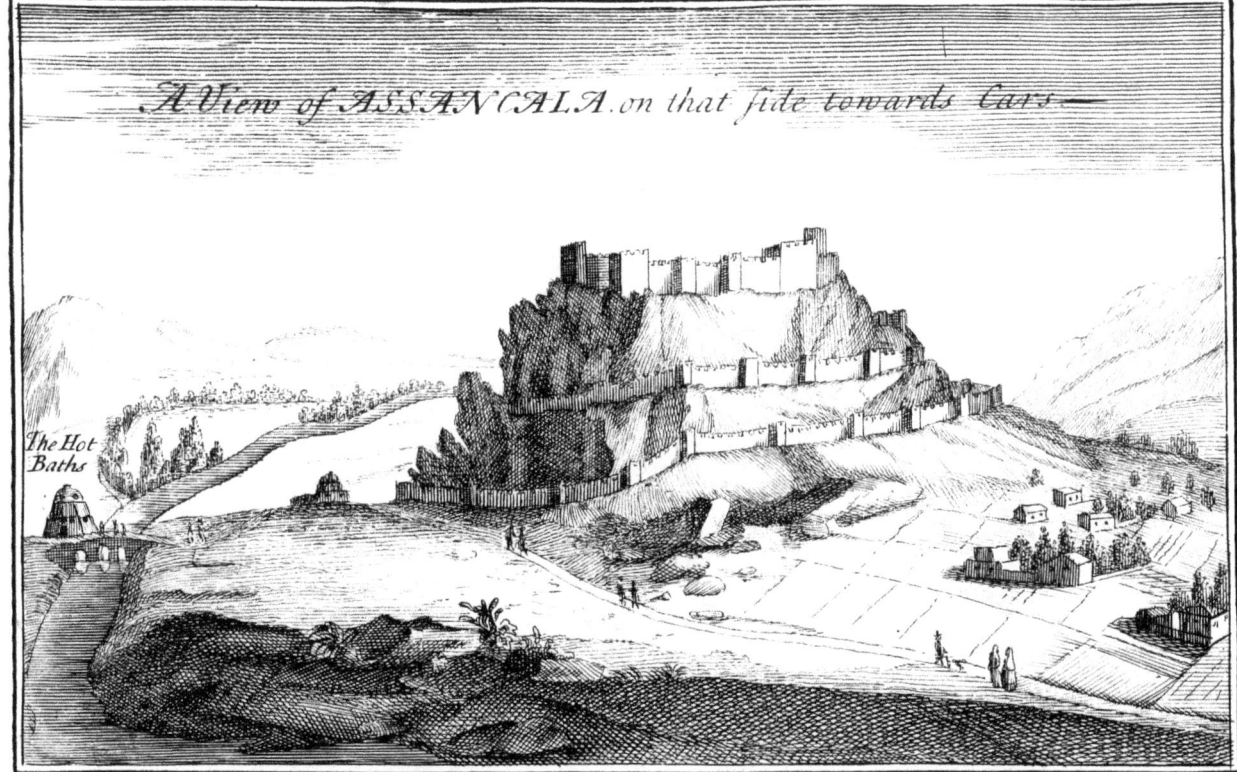

A View of ASSANCALA on that side towards Cars.

The Hot
Baths

Artemisia Orientalis
Tanaceti folio in odo-
ra Coroll. Inst Rei
herb. 34.

menian Patriarch of *Jerusalem*, for the Sale of the Holy Oil us'd in the Lett. VII.
Adminiſtration of the Sacraments among the *Armenians*.

W E went to viſit the Country, after we had tir'd our ſelves in the
Town, and run all over the fine Valley of the *Forty Mills*, where we had
left too many rare Plants in Flower, not to go and get the Seeds. With the
ſame Deſign we ſpent the firſt of *September* at the *Red Monaſtery* of the
Armenians, from whence we went up towards the Head of the *Euphrates*, to continue our Harveſt. The *Curdes*, thank God, had left theſe
Mountains, ſo that our ſecond Crop was got in with much more Tranquillity the firſt. This Crop was of Seeds of Plants we had before
ſeen, rather than of any thing new; but theſe Seeds were not the
leaſt Advantage of our Journey. By their means it is that *Armenian*
Plants are ſpread over the King's Garden, and the moſt famous Gardens
of *Europe*, to the Directors of which we had communicated a great part.
In this manner we employ'd our ſelves about *Erzeron*, ſometimes on one
ſide, and ſometimes on another, and glean'd to very good purpoſe. Here
is the Deſcription of a very fine ſort of *Mugwort*, of which I believe no body yet has made any mention. It is found in the Churchyard
of the *Armenians*, and in ſome Places about the City, where it blows only in *Autumn*.

T H E Root of this Plant is about a foot long, hard, woody, as thick
as the little Finger, furniſh'd with hairy Fibres, white within, cover'd
with a reddiſh Rind. The Stalks grow in Bunches, about two feet high,
ſtreight, firm, ſmooth, pale-green, reddiſh in ſome Places, brittle, accompany'd with Leaves exactly like thoſe of *Tanſy*, but inſipid, and without Smell: the biggeſt are about three inches long, and two thick, dark-
green, ſmooth, cut deeply even to the Rib, and again cut into very ſmall
Dents: they grow leſs and leſs to the very top, without changing their
Figure. From their Knots grow out Branches but half a foot long, ſub-
divided into many Sprigs, all charg'd with Flowers very cloſe, and rais'd
high; they are a ſort of Buttons or Buds, like thoſe of the common Mugwort, compos'd of certain Demi-fleurons very ſmall and purpurine, en-
clos'd in a Calix or Cup made up of ſmall Scales, of a deep-green Colour.
Each Fleuron or ſmall Flower bears an Embryo of Grain, which becomes

a

a very fmall Seed, reddifh, half a line long. We perceiv'd neither Smell nor Tafte in this Plant. It loves a fat, frefh, moift Earth.

TO the South Eaft of *Erzeron* lies the Vale of *Caracaia,* which is full of fine Plants. We obferv'd there, among other things, the true *Monks-hood,* as reprefented by the Figure *Clufius* has given of it. The *Caryo-phyllata aquatica, nutante flore,* C. B. is common there. Nothing was pleafanter to us than from time to time to meet with the Plants of the *Alps* and *Pyrenees.*

WHILE we waited for the Caravan from *Tocat,* of which we were to take the advantage to go to *Smyrna,* we went to converfe in the Caravan-feras to learn News. We found there a Company of thofe who go into *Perfia,* and the Dominions of the Great Mogul, to fetch Drugs into *Turky.* They inform'd us the People of the Country made their chief Ma-gazine at *Machat,* a City in *Perfia*; but we did not learn much of them, for neither they who fill the Warehoufes, nor they who go into the Places and Villages, whither the Peafants bring the Drugs out of the Country, know any thing of them. Nothing is more difficult than to write a good Hifto-ry of Drugs, that is, to defcribe not only all that which conftitutes the *Materia medica,* but to give a Defcription of the Plants, Animals, and Minerals from whence they are taken. One muft not only go into *Per-fia,* but likewife into the Empire of the Mogul, which is the richeft in the World, and where Strangers are mighty well receiv'd, efpecially fuch as have a great deal of Silver and Gold. Every thing is bought there for ready Money, and it is not permitted to carry out any thing but Goods ; fo that all foreign Money is kept in the Country, and new-coin'd. But what a trouble would it be, even in that Kingdom, to go a-bout to inform one's felf of what concerns the Nature of Drugs? One fhould be obliged to go to the feveral Places where the Drugs are found or prepar'd, in order to defcribe the Plants from which they are produc'd ; and to how many Inconveniences would this expofe one? A Man's whole Life would fcarce be enough to examine thofe only which are produc'd in *Afia :* one muft go thro *Perfia,* the Empire of the Mogul, the Ifles of *Ceylon, Sumatra, Ternate,* and I know not how many other Coun-tries, in which it will be more difficult to travel, than in the Empire of

the

the Mogul. *Rhubarb* alone would require a Voyage to *China* or *Tarta-*
ry. Afterwards one muſt go down into *Arabia*, *Egypt*, and *Ethiopia.* I
ſay nothing of the Drugs, which are only found in *America*, and which
are not leſs valuable than thoſe brought from other Parts of the World.
In going to *America*, one ought to ſtop at the *Canaries*, to deſcribe *Dra-
gons-Blood.*

AFTER this, I am not at all ſurpriz'd if they who attempt to write
the Hiſtory of Drugs make ſo many Miſtakes, and myſelf particularly.
They only relate uncertain Facts, and give imperfect Deſcriptions. It is
more ſhameful that we don't know thoſe Drugs that are prepared in
France. Where can one find an exact Account of *Vermilion*, *Turnſel*,
Verdigreaſe, *Pitch*, *Turpentine*, the *Fir-Tree*, the *Balm*, *Agarick*, and our
Vitriols?

IN our Converſation in the Caravanſeras of *Erzeron*, we learn'd from
thoſe of the Caravan of *Wan*, a *Turkiſh* Town on the Frontiers of
Perſia, eight days Journey from *Erzeron*, that they carefully lay up in
heaps the Dirt of the great Roads, which are frequented by Caravans of
Camels. This Earth they waſh, and every Year get out of it above a
hundred Quintals of *Nitre*, which is diſpos'd of chiefly in *Curdiſtan* to
make Powder. They aſſur'd us that the Fields near the Roads from
Wan yield no *Nitre.* However, it muſt contain ſomething proper to
become *Nitre*, by being mix'd with the Urine of Camels.

POWDER for Cannon is not worth fifteen Sols the Oque at *Erzeron*; 'tis only fit to charge, but 'tis neceſſary to have better for
Prime. They all uſe a Cartouch to charge withal; and nothing is
better contriv'd to make a quick Shot with our Fuſees. Thoſe M. *de la
Chaumete* has invented are much better, and give better Fire than thoſe
they uſe. They were never carried to the Perfection they now are
by M. *de la Chaumete.* The Pouches us'd in the *Levant* are made of
Tubes of Cane, commonly in a double Row, much like the antient
Flutes of *Pan*, or to uſe a more intelligible Compariſon, like thoſe
Whiſtles us'd by the Colliers who travel from Province to Province
in ſearch of Work. The Pouch us'd in the Eaſt is light, curve, and
fits eaſy to the Side. Its Tubes are four or five inches deep, and

Vol. II.   P p   cover'd cover'd with a very convenient Skin : Each Tube holds its Charge, which is a Tube or Pipe of Paper fill'd with a proper Quantity of Powder and Ball for one Shot. When they would charge their Fuſee, they take one of theſe Tubes out of the Pouch, and bite a Hole in that part where the Powder is, and pour it into the Barrel of the Fuſee, letting the Lead follow, which is in the other part of the Paper-Tube. They ram it down with a Gun-ſtick ; and the ſame Paper which held the Powder and Ball, ſerves for the Wadding.

I am, MY LORD, *&c.*

LET-

LETTER VIII.

To *Monfeigneur the Count* de Pontchartrain, *Secretary of State*, &c.

My Lord,

A S we us'd every Evening, during our Stay at *Erzeron*, to fet down what we had learn'd that Day in our Converfation with the *Armenians*, efpecially in the Convent where we lodg'd, we found at length that our Remarks, together with thofe we had made in other Convents, as we pafs'd, would furnifh an entire Letter concerning the Genius, Manners, Religion, and Commerce of that Nation. I therefore pray your Lordfhip to accept of the Fruits of our Converfations.

Of the Manners, Religion, and Commerce of the Armenians.

T H E *Armenians* are the beft People in the World, civil, polite, and full of good Senfe and Probity. I fhould account them happy in not underftanding the Ufe of Arms, if it were not by the Corruption of Mankind become neceffary to ufe them fometimes, purely to defend our felves againft the Violence of others. But the *Armenians* trouble themfelves with nothing but Trade, which they follow with the utmoft Attention and Application. They are not only Mafters of the Trade in the *Levant*, but have a large Share in that of the moft confiderable Places in *Europe*. They come from the fartheft Parts of *Perfia* to *Leghorn*. Not long fince they fettled at *Marfeilles*. There are many in *Holland* and *England*. They travel into the Dominions of the *Mogul*, to *Siam*, *Java*, the *Philippine* Iflands, and throughout all the Eaft, except *China*.

T H E Center of the *Armenian* Merchants is not in *Armenia*, but at *Julfa*, the famous Suburb of *Ifpahan*, defcrib'd by all Travellers. This

Suburb, which deferves rather to be called a City, feeing it contains above thirty thoufand Inhabitants, is a Colony of *Armenians*, which the Great King of *Perfia*, *Cha-Abbas*, the firft of that Name, fettled at firft in *Ifpahan*, and was remov'd a little after to the other Side of the River *Zenderou*, to feparate them from the Mahometans, who defpis'd them on the fcore of their Religion. 'Tis faid this Alteration happen'd under the Leffer *Cha-Abbas*; others fay 'twas much fooner. 'Tis certain, however, that the firft Eftablifhment of this Colony was by the Great *Cha-Abbas*, Cotemporary with *Henry* IV. to whom he fent Father *Jufte*, a Capuchin, in Quality of Ambaffador: But he did not arrive till after the Death of the King. *Cha-Abbas* fuccefsfully aim'd at two Things, for the Good of his Kingdom. He fecur'd it from being attack'd by the *Turks*; and greatly enrich'd it by eftablifhing Commerce. To hinder the *Turks*, whom the *Perfians* call *Ofmalins*, from penetrating into his Countries, he judg'd it proper to take from them the Means of maintaining a numerous Army upon his Frontiers: And as *Armenia* is the principal Place where the *Turks* ordinarily make their Attempts, he unpeopled it as much as he thought proper for his Defign. The Lot fell upon the City of *Julfa*, the greateft and moft powerful of the Country; the Ruins whereof are ftill to be feen upon the *Araxes*, between *Erivan* and *Tauris*. The Inhabitants of *Julfa* had Orders to come to *Ifpahan*; and from that time, the City they abandon'd was called the *Antient Julfa*. The People of *Nacfivan*, and the Neighbourhood of *Erivan*, were difpers'd into feveral Parts of the Kingdom. They fay this Prince caus'd above twenty thoufand Families of *Armenians* to be tranfplanted into the fingle Province of *Guilan*; from whence come the fineft Silks of *Perfia*.

AS *Cha-Abbas* had nothing in view but to enrich his Kingdom, and was convinc'd that it could not be done but by Commerce, he caft his Eyes upon the Silk Trade as the moft valuable, and the *Armenians* as the propereft Perfons to carry it on; having no opinion of the Diligence of his other Subjects, who he knew to have no Genius for Trade. The Frugality of the *Armenians*, their Oeconomy, their Credit, their Vigour in undertaking and performing great Voyages, appear'd to him very neceffary for the carrying on his Defign. Their Profeffion of the Chriftian Religion, which facilitated their Commerce with all the *Euro*

pean Nations, feem'd likewife farther to favour his Intentions. And in Lett. VIII.
fhort, he made the *Armenians,* who were Husbandmen, to become Merchants; and now they are the greateft Traders upon Earth.

THUS this Prince, who had a wonderful Genius for the Affairs
of War or Civil Government, improv'd the Talents of the People, and
the Trade of his Kingdom. For the better eftablifhing and fettling the
Trade, he entrufted the *Armenians* of *New Julfa* with a certain Quantity
of Bales of Silk, to carry by Caravans into foreign Countries, and
throughout *Europe,* on condition that they fhould themfelves go with
them, and at their return fhould pay for each Bale a certain Price, fettled by Perfons of Judgment before they went. To encourage them to
pufh the Trade, he let them enjoy all that could be made of the Silk
above the Price agreed on. And the Succefs anfwer'd the Hopes of the
Prince and his Merchants. Though the Silk be ftill the beft Commodity in *Perfia,* it was then of much more Value. There were then hardly any Mulberries in *Europe:* And Gold and Silver, at that time very
fcarce in *Perfia,* began to fhine there by the Return of the Caravans; to
which the Riches of the Kingdom are even at this day owing. The
Armenians brought back alfo the Cloths of *England* and *Holland,* Brocades, *Venice* Glafs, Cochineal, Watches, and every thing they thought
fit for their own Country, or the *Indies.* Could there be a finer Eftablifhment? To how many Manufactures has it given rife in *Europe* and
Afia? *Abbas* the Great chang'd the Face of the whole Earth: All the
Commodities of the Eaft were made known in the Weft, and thofe of
the Weft ferve as new Ornaments for the Eaft.

NEW JULFA foon ftretch'd it felf upon the River *Zenderou.* It
feem'd by the Magnificence of their Houfes, and the Beauty of their
Gardens, that the Inhabitants had taken their Tafte from the beft Cities
in *Europe.* In the midft of *Perfia* is now feen every thing that is curious
throughout all the Countries where the Merchants have extended their
Correfpondence. The King does now no longer affift them with a Stock
to carry on the Trade: The Inhabitants of *Julfa,* by their Agents and
Factors, carry on this vaft Trade themfelves, and diftribute throughout
the World all that's curious or valuable in the Eaft. Thefe Agents are
Armenians, who, in confideration of a certain Profit allow'd them, un-

* dertake

dertake to go with the Merchandize in the Caravan, and fell the Goods in the beft manner they can for thofe who employ them.

THE *Armenians,* whether they act for themfelves, or for the Merchants of *Julfa,* are indefatigable in their Journeys or Voyages, regarding no Weather in the moft rigorous Seafons. We have feen feveral, and even of the very rich ones, pafs great Rivers on Foot up to the Neck in Water, to help up their Horfes when fallen, and fave their Bales of Silk, or their Friends: But the *Turkifh* Carriers give themfelves very little Trouble with the Goods they carry, and are not anfwerable for any thing that may happen. The *Armenians,* in paffing a River, lead their Horfes; and nothing's more inftructive than to fee with what Charity they affift one another, or even thofe of any other Nations in the Caravan. Thefe good People are very conftant and regular in their Way, always equal, and fhun Strangers who are turbulent and troublefome, as much as they efteem thofe that are peaceable; but fuch they entertain very civilly and freely. When we did any Service to any of their Sick, the whole Caravan return'd us their Thanks. If they hear at any place that a Caravan is coming that way, they will go two or three Days Journey to meet their Brethren with Refrefhments, and with the beft Wine; which they don't only offer to the *Franks* likewife, but by their Civility force them to accept it, and drink their Health. They are unjuftly accus'd of drinking too much; we never faw them abufe themfelves that Way: On the contrary, it muft be allow'd they are the moft fober, and thrifty, and modeft of all Travellers. If, when they fet out on a Journey, they carry a great quantity of Provifions with them, they often bring a good part of it back again. The Provifions coft them nothing for Carriage; becaufe generally when they hire fix Camels, they have a feventh allow'd them above the Agreement, to carry their Baggage, Clothes, *&c.* The Provifions they furnifh themfelves with, before they fet out, are Meal, Bisket, Smoak'd-Meat, Potted-Butter, Wine, *Aquavitæ,* and Dried Fruits.

WHEN they ftay in Towns, they lodge feveral together, and live at fmall Expence. They never go without Nets: They fifh on their Journey; and they made us often eat with them of very excellent Fifh. They exchang'd Spices for frefh Meats when they had opportunity, or for

other

other Commodities they had a mind to.   In *Asia* they sell the Wares of Lett.VIII.
Venice, France, Germany; small Looking-Glasses, Rings, Necklaces, Enamels, little Knives, Scissars, Buckles, Needles, are more enquir'd after
in the Villages than good Money.   In *Europe* they carry Musk and
Spices.   Whatever Fatigues they go through, they as carefully observe
the Fast of the Church, as if they were at repose in a City; and know
nothing of Dispensations, not even in Sickness.   The only thing to
be blam'd in the *Armenians*, in relation to Trade, is, that if they succeed
ill in any foreign Country where they are trading, they never return home
again; they say they have not the Face to shew themselves after they are
become Bankrupts: But their Creditors obtain no Satisfaction by this.
However, this Justice must be done them, to own there are very few
Bankrupts among them.

T H E Merchants of *Julfa* have made a Treaty with the Great Duke
of *Muscovy*, whereby they are permitted to import into his Dominions
whatever they think proper; while no *European* Merchant of any Nation
is allowed to go any farther than *Aftracan*, a strong Town, possessed by
the *Muscovites* ever since the Year 1554.   'Tis situate on the other side
of the *Caspian* Sea, upon the Frontiers of *Asia* and *Europe*.   The Great
Duke encourages this Trade as much as possible: The of *Julfa* pay
Custom for every thing they import into *Muscovy*; but they pay no Duties for what they export from *Muscovy* into *Persia*.   The Way they go
and come, is this: From *Ispahan* they carry their Merchandize to *Tauris*,
Schamakee, and *Nosava*, a Port of the *Caspian* Sea, three Days Journey
from *Schamakee*:   At *Nosava* they ship the Silk, and other Commodities
of *Persia*, and the Empire of the *Mogul*, for *Aftracan*:   From *Aftracan* they
are carried by Land to *Moscow*; and thence to *Archangel*, the farther most
Port of *Muscovy* on the North-Sea.   The *English* and *Dutch* carry on a
great Trade to that Port: There they ship Goods for *Stockholm*, and
from thence by the *Straits* of *Elfinore* they are carried into *England* and
Holland.

F R E D E R I C K, Duke of *Holftein*, according to *Olearius*, built *Frede-*
rickftad in the Dutchy of *Holftein*, to settle there a Trade for Silk more
considerable than any in *Europe*.   To this purpose he resolv'd to hold a
Correspondence with the King of *Persia*, in order to facilitate the Carriage

riage by Land. But this not being practicable without the Permiſſion of the Great Duke of *Muſcovy,* he thought fit in the Year 1633 to ſend him a ſolemn Embaſſy; to which he nam'd *Cruſius,* one of his Counſellors of State, and *Brugman,* a Merchant of *Hamburgh.* This laſt by his ill Management, together with the Dangers they were to run through in paſſing among the *Tartars* of *Dageſthan,* cauſed the Deſign to miſcarry. He was afterwards convicted of Male-Adminiſtration, and condemn'd to die, and was accordingly executed at *Gottorp, May* 5. 1640. The *Dutch,* who have ſince attempted to make themſelves Maſters of the Silk of *Perſia,* which comes from *Aſtracan,* are oblig'd to take a great Quantity every Year; for which reaſon they don't get much by this Trade, becauſe the *Armenians* make them take the good and bad together. Mr. *Preſcot* aſſur'd us, that the *Engliſh* loaded a great deal of Goods of *Aſia* at *Archangel*; and that there was the beſt *Caviar* that could be eaten. That which they ſell in *Turky,* comes from the *Black-Sea:* It is very ſlovenly, and put up in Skins; but the *Caviar* of the *Caſpian* Sea is manag'd with a great deal of Care, and they put it up very cleanly. We eat Sturgeon's Spawn at Mr. *Preſcot*'s which had been ſalted in the Neighbourhood of the *Caſpian* Sea, and *Caviar* ſalted in the ſame Places, which was very excellent: The Sauſages made at *Marſeilles* are not better.

WE could not but laugh to ſee the way of Trafficking among the *Armenians* in the Caravanſeras of *Erzeron.* They begin by putting Money upon a Table, as among the *Turks*; after that they haggle a great while, and add one Piece after another, but not without a great deal of Noiſe. We believ'd by their way of talking they were ready to cut one another's Throats; but they meant nothing like it. After having puſhed one another backward and forward with a great deal of Violence, the Brokers or Mediators ſqueeze the Hands of the Seller ſo very hard as to make them cry out, and don't let them go till they agree that the Buyer ſhall not pay above ſo much as they think a reaſonable Price: After that, every one laughs. They ſay, with reaſon, that the Sight of the Money makes them ſooner agree.

AS to their Religion, every body knows the *Armenians* are Chriſtians; and they would be very good Chriſtians, were it not for the Schiſm whereby they ſeparate from us. They are ſaid to be *Eutychians,* that is to ſay,

that

that they own but one Nature in Jefus Chrift, or rather two Natures fo con-Lett.VIII.
founded together, that tho they admit the Properties of each Nature in
particular, they neverthelefs allow but of one Nature. Their moft able Bifhops would clear themfelves of this Herefy, and pretend that the Miftake arifes from the Barrennefs of their Language ; which not furnifhing
them with proper Terms, is the Caufe that they often confound the Words
Nature and Perfon. When they fpeak of the *Hypoftatical Union*, they
think they exprefs it fufficiently by confeffing that Jefus Chrift is perfect
God and perfect Man, without Mixture, Change, or Confufion. The
truth is, they don't all explain themfelves in the fame manner; and the
greateft Part of them have a great Veneration for two famous *Eutychians, Diofcorus* and *Barfuma.* When they are reproached with having excommunicated the Fathers of the Council of *Chalcedon* for having condemned the firft of thefe Hereticks, they avow that tho it appears ridiculous
to excommunicate the Dead, the Cuftom was introduc'd among them to
revenge themfelves on the *Greeks,* who in all their Feafts excommunicate
the *Armenian* Church : That their Defign was not merely to excommunicate the Fathers of the Council of *Chalcedon,* who had condemned *Diofcorus,* Patriarch of *Alexandria,* without having duly examined the Caufe;
but that their Intention was to excommunicate the prefent *Greek* Bifhops,
as the Succeffors of the Prelates of the moft famous Affembly which was
ever held in *Greece :* That the *Greek* Fathers had dealt very unjuftly by
Diofcorus, in confounding his Sentiments with thofe of *Eutychius,* feeing
Diofcorus always maintain'd that the Word Incarnate was perfect God and
perfect Man. The Source of the irreconcilable Enmity between the
Armenians and the *Greeks* is from that Council : And the Enmity is fo
great, that if a *Greek* comes into an *Armenian* Church, or an *Armenian*
into a *Greek* Church, they think the Church to be defiled, and confecrate
it a-new.

WHEN one examines into their Opinions, one finds a great many
Articles of Schifm which are not to be attributed to the *Armenian*
Church, but to particular Perfons. For example; It is not true that
they three times a Year excommunicate the *Latin* Church : The good Peo-
ple never think on it; and there is nothing like it to be found in their
Rituals : Tho at the fame time it is very true, that fome of the more vio-

lent Bifhops, or *Vertabiets*, who have declared againft the *Latin* Church, have, or even do ftill practife it: For in an ill-govern'd Church, oftentimes every one does as he pleafes.   The Patriarch *Ozuietfi*, fworn Enemy of the *Latins*, may perhaps have added to this Excommunication the Name of Pope *Leo*, becaufe he confirm'd the Condemnation of *Diofcorus*. How great foever their Efteem be for their great Doctor *Altenafi*, 'tis entirely wrong to attribute to the whole *Armenian* Church the feveral Injuries which this Fanatick has vomited out againft the *Roman* Church.

ONLY the moft filly and ignorant of the *Armenians* believe the Little Gofpel.   This Little Gofpel is ftuffed with Fables and Extravagancies concerning the Infancy of our Lord.   For example; *That the Virgin being big with him,* Salome, *her Sifter, accufed her of having proftituted her felf to fomebody: The Virgin anfwered her, that fhe need only lay her Hand upon her Belly, and fhe would know how fhe was with Child.*   Salome *accordingly put her Hand upon the Virgin's Belly, and a Fire came out, which confum'd half her Arm.   She acknowledged her Fault, and drew back her Hand, and her Arm was perfectly healed, after having by order of the Virgin put it upon the fame Place.*   They pretend that *the Son of God had done himfelf wrong to pafs thro the Womb of a Woman ; that he only feem'd to do fo ; and that the Jews fubftituted fome other Perfon in his ftead.*   They have borrow'd from the *Mahometans* this laft idle Fancy.   They fay alfo, that *Jefus Chrift being at School to learn the* Armenian *Tongue, would never pronounce the firft Letter of their Alphabet, unlefs the Mafter would give him a Reafon why it reprefented an* ɯ *inverted.*   The good Man, not knowing the Infant Jefus, gave him a Box on the Ear.   *Well,* faid Jefus, without any Emotion, *fince you don't know, I will tell you : This Letter reprefents the Trinity by its three Legs.*   The Mafter of the School admir'd his Knowledge, and fent him to his Mother, confeffing that the Child was wifer than himfelf.   M. *Thevenot*, who alfo mentions this Story, affirms it is in an *Armenian* Manufcript in the King's Library, which gives an account of the Hiftory and Inventors of their Characters; but it does not carry back the Invention above four hundred Years.   They probably ufed the *Greek* Characters.

THE *Armenians* relate that *Jefus Chrift being a fowling with* St. Bartholomew *and* St. Thaddeus, *he kill'd five Partridges on the Banks*

of

of the Araxes, *and that a great many People came about him to hear him preach; but that Night coming on, the two Apostles put him in mind of dis-missing them.* Jesus answer'd them, *that after having fed their Souls with necessary Food, he ought to take care of their Bodies, and for that purpose they should boil the six Partridges with an Oque of Rice.* The whole Company were fill'd with them; and it being not Day-light, every one thought he had a whole Partridge. The King of *Armenia,* who took great delight in that Game, was very angry at this, and order'd the Apostles and their Master to be kill'd: Jesus sav'd himself in the Ark on the top of Mount *Macis;* but St. *Bartholomew* and St. *Thaddeus* paid for the whole.

THE pleasantest Story they tell, is that of *Judas: This Wretch,* as they say, *repenting that he had betray'd his Master, thought there was no other way to save his Soul, but to hang himself, and go to the Limbo, whither he knew Jesus Christ would descend to deliver the Souls; but the Devil, who resolv'd to carry him to Hell, play'd him a sly Trick in his way, and kept him up by the Feet, hanging as he was, till Jesus Christ had made his Visit to the Limbo: after which, he let him fall, and so dragg'd him away among all the Devils.* The *Georgians* tell a thousand ridiculous Stories of this kind, taken out of their Little Gospel. I believe these two Pieces were made by the same Hand.

THO the *Armenians* won't hear Purgatory mention'd, they pray over the Tombs, and say Masses for the Dead; it is perhaps owing to the Avarice of their Priests, that their Opinions being chang'd, they still continue the Use of so profitable a Ceremony. According to the greatest part of their Priests, there is neither Paradise nor Hell: they believe Hell was destroyed after Jesus Christ took thence the Souls of the Saints, as well as of the Damned. As to the Creation of the Soul, they hold *Origen's* Sentiments, without knowing there ever was an *Origen* in the World; for they imagine that all the Souls were created in the beginning of the World. There are *Millenarians* among them, who know nothing of *Papias* or St. *Irenæus.* They believe that after the universal Judgment, Jesus Christ shall remain a thousand Years upon Earth with the Predestinated, to make them enjoy Happiness. The greatest part of the *Armenian* Doctors are of opinion, that the Souls wait the univerfal Judgment in a Place between Heaven and Earth, where they flatter themfelves they fhall enjoy a day of Glory, tho they are under fears of being condemn'd to eternal Punifhment.

ST. *NICON*, who was of the *Leffer Armenia*, and pafs'd fome Years of his Life in Miffions in the *Greater Armenia* in the tenth Century, has left us a Treatife in *Greek* concerning *the Errors of the* Armenians: the Original is in the King's Library, and *Cortelerius* has tranflated it into *Latin*. St. *Nicon* mentions fome very fingular things concerning the Creed of this People; and does not only accufe them of being Difciples of *Eutychius*, *Dioſcorus*, *Peter* the *Armenian*, and *Mantacunez*, but likewife of being in the Herefy of the *Monothelites*. He mentions fome of the Fables which are ftill in their Little Gofpel.

HOWEVER, this People were favour'd with two Apoftles our Lord fent them foon after his Paffion. *Baronius* affirms, that St. *Bartholomew* and St. *Thaddeus* fuffer'd Martyrdom in *Armenia* forty four Years after the Death of Jefus Chrift, in recompence for the Faith they had preach'd there. Unhappily it made no great Progrefs there; for *Eufebius* tells us, that a holy Bifhop call'd *Meruzanes* fow'd the good Seed there in the Reign of *Decius*, and God fpread his Bleffings to fuch a degree among this People, that there were none but Chriftians among them in the time of *Dioclefian*. *Maximinian* fet himfelf to deftroy them, but the *Armenians* took Arms in defence of their Faith; and this, as *Eufebius* fays, was the firft War undertaken for Religion. In fine, God went on to open the Eyes of this People by the Miniftry of St. *Gregory the Illuminator*, an *Armenian* by Birth, but brought up at *Cefarea* in *Cappadocia*, where he was confecrated by St. *Leontius*. St. *Gregory* return'd into his own Country in the Reign of *Conſtantine the Great*, converted *Tiridates* King of *Armenia* by a very fingular Miracle; and this Prince, who at firft caus'd him to be ill us'd, was fo touch'd with it, that he by an Edict oblig'd all his Subjects to embrace the Chriftian Religion. The Saint compleated by his Doctrine, by his Example, and by his Miracles, what the King could only command and order. A Slave, who became a Chriftian at *Conſtantinople* at the fame time, contributed not a little by his Miracles to propagate the Chriftian Religion in the fame Country.

WE

WE muſt not confound St. *Gregory the Illuminator,* firſt Patriarch of Lett. VIII. the *Armenians,* with another Saint of the ſame Country and Name, who in the tenth Century dy'd in *France,* ſhut up in a Solitude near *Pluviers* in *Beauce,* in the Dioceſe of *Orleans.* He ſpent ſeven Years in this Hermitage, faſting according to the Cuſtom of his Country, that is to ſay, in a manner which thoſe in the Weſt dare hardly imitate. He eat nothing at all on Monday, Wedneſday, Friday, and Saturday ; and if he broke his faſt Tueſday and Friday after the Sun-ſet, he eat only three Ounces of Barley-Bread, ſome raw Herbs, a handful of Lentils ſoak'd in Water, and ſhot in the Sun : on Feaſt-days and Sundays he fed a little better, but he never eat Meat.

THE Clergy of *Armenia* conſiſts of a Patriarch, Archbiſhops, Biſhops, *Vertabiets* or Doctors, ſecular Prieſts, and Monks. The Patriarch has borne the Name of *Catholicos* a great while ; for *Procopius* obſerves, that the *Armenians* borrow'd this Term of the *Greeks.* The *Armenians* have many Patriarchs in the Dominions of the King of *Perſia,* and the Grand Signior. Beſides him of *Itchmiadzin,* who is the chief of 'em all, they reckon in *Perſia* him of *Schamakee* near the *Caſpian* Sea, and him of *Nacſivan,* whom the *Armenian* Roman Catholicks own for their Patriarch next the Pope. In *Turky* there are two Prelates, who have made themſelves Patriarchs by the Grand Viſier, who would give this Title to all the Prelates, if they would buy it of him, as the Biſhop of *Cis* near *Tarſus* in *Cilicia,* and the *Armenian* Biſhop of *Jeruſalem* have done, who by Preſents obtain their Miſſion and Authority from the Port. The *Armenians* have another Patriarch at *Caminiec* in *Poland* : for Father *Pidou,* Religious Theatin of *Paris,* and Apoſtolical Miſſionary, knew ſo well how to manage the *Armenians* of *Poland,* and eſpecially their Archbiſhop, that he brought 'em back to their Mother the Church of *Rome* in the Year 1666. They purg'd their Books of all the Errors which ſeparate Schiſmaticks from us. The Patriarch acknowledg'd the Pope for Head of the true Church, and carried the Sacrament thro the Streets in a general Proceſſion, which was made to return thanks to God in the more ſolemn manner.

THE Patriarch of *Itchmiadzin* is the richeſt of all in one Senſe, for they aſſur'd us he has near ſix hundred thouſand Crowns Revenue. All the

the *Armenians* who acknowledge him, and are above the Age of fifteen Years, pay him five Sols a Year. Men of Subſtance give him to three or four Crowns. But notwithſtanding this, he is poor in another Senſe, and truly poor; for he is oblig'd to pay the Capitation, to keep thoſe in his Flock, who are not themſelves able to pay this Tax. Often he expends his whole Revenue this way, and part of what he had laid up. The Archbiſhops and Biſhops ſend him every Year the State of the poor Families in their Dioceſes, which are threatned with being ſold or forc'd to change their Religion, when they don't pay the Capitation. This Patriarch is cloth'd as plainly as the other Prieſts; he lives very frugally, and has but a few Domeſticks: but he is the moſt conſiderable Prelate in the World, in regard to the Authority he has over his Nation, which tremble at the leaſt Threat of Excommunication from him. They ſay there are fourſcore thouſand Villages which own him. To keep his Place, he is oblig'd to make many Preſents to the Governour of *Erivan*, and the powerful Men at Court. A Man muſt be a great Slave to Ambition, to buy ſuch kind of Poſts.

HE was formerly the only Patriarch among the *Armenians*, who had Power to make the *Holy Chriſm* or *Mieron*, from the *Greek Myron*, a liquid Compoſition or perfum'd Oil. He furniſh'd all Parts of *Perſia* and *Turky*; even the *Greeks* too bought it with great Veneration, and they ſaid commonly that a Fountain of Holy Oil flow'd from *Three-Churches*, which water'd the whole Eaſt. The Patriarch ſent it to the Archbiſhops and Biſhops of the *Armenians* to diſperſe it, and to uſe it in Baptiſm and the extreme Unction: but above forty years ſince *Jacob*, a *Vertabiet* and *Armenian* Biſhop, who reſided at *Jeruſalem*, took upon him to erect himſelf into a Patriarch under the Influence of the Grand Viſier, and refus'd to take the *Mieron* from the Patriarch of *Three-Churches*. As Oil is a very cheap Commodity in *Paleſtine*, and this Liquor does not corrupt, he made more than could be us'd for Anointings among all the *Armenians* in *Turky* for many Years: and this was the Foundation of a great Schiſm among them. The Patriarchs excommunicated one another; he of *Three-Churches* commenc'd a great Suit at the Porte againſt him of *Jeruſalem*. The *Turks* are too wiſe to decide the Queſtion, and content themſelves with receiving the Preſents both Parties make,

as

as they revive the Suit; and each goes on to fell his Oil as well as he can.

IT is prepar'd between the Vespers on *Palm-Sunday* and the Mass on *Holy-Thursday*, which is celebrated on this day on a great Vessel in which is kept this Liquor. They use neither Wood nor common Coals to boil the Kettle wherein it is prepar'd, and this Kettle is bigger than that in use among the Invalids. They boil it with Wood that has been bless'd, and with any thing that has been us'd in Churches, old Images, worn-out and decay'd Ornaments, torn Books; all is kept for this Ceremony. This Fire can't smell very well; but the Oil is perfum'd with Herbs and odoriferous Drugs, which are mix'd with it. They are not ordinary Clerks who are employ'd in making this wonderful Composition; 'tis the Patriarch himself cloth'd in his Pontifical Vestments, and attended at least by three Prelates in their Pontifical Habits, who all together recite certain Prayers during the whole Ceremony. The People are more struck with this than with the real Presence of Jesus Christ; so true is it, that Men are not so susceptible of any thing as what is sensible.

THERE is nothing particular to be mention'd concerning the Arch-bishops and Bishops of the *Armenians*, but that there are many of them who have no Diocese, and who lodge in Monasteries, of which they are the Abbots. All the Prelates are subject to the Patriarch, as in other Christian Churches. It were only to be wish'd they discharg'd their Duty; but they have no Zeal, and are sunk into the most wretched Ignorance, and are often less esteem'd than the *Vertabiets*. Sometimes they are Bishops and *Vertabiets* at the same time, that is to say, Bishops and Doctors. These *Vertabiets*, who make such a noise among the *Armenians*, are not in reality great Doctors; but they are the most considerable Men of the Country, or at least pass for such. To be receiv'd to this eminent Degree, it is not necessary to study Theology for many Years: 'tis enough to understand the literal *Armenian* Tongue, and to learn by heart some Sermon of their great Master *Gregory Altenafi*, who shew'd all his Eloquence in the Blasphemies he vomited out against the Church of *Rome*. The literal Language is among them the learned Language, and they pretend it has no Affinity with the other Eastern Languages, which renders it so difficult. They say it is very expressive, and enrich'd

with

with all Terms of Religion, and Arts and Sciences; which shews that the *Armenians* were formerly Men of much greater Learning than they are at present. In short, it is a great Accomplishment among them to understand this Language; it is only to be found in their best Manuscripts. The *Vertabiets* are consecrated, but they seldom say the Mass, and are properly appointed to preach. Their Sermons turn upon very ill-contriv'd Parables, upon Passages of Scripture ill understood and ill explain'd; and upon some Stories true or false, which they have receiv'd by Tradition: however, they pronounce them with a great deal of Gravity; and these Discourses give them almost as much Authority as the Patriarch: they above all things assume that of excommunicating. After having exercis'd themselves some time in some Villages, an antient *Vertabiet* receives them Doctors with abundance of Ceremonies, and puts into their Hands the Pastoral Staff. This Ceremony does not pass without Simony; for the Degree of Doctor being look'd upon among them as a Sacred Order, they make no scruple to sell it, as they do the other Orders. These Doctors have the Privilege of sitting when they preach, and holding in their Hands a Pastoral Staff; while the Bishops, who are not Doctors, preach standing. The *Vertabiets* live on the Collection that's made for them after the Sermon, which is considerable, especially in the Places where the Caravans stop. These Preachers observe Celibacy, and fast very rigorously three quarters of a Year, when they neither eat Eggs, nor Fish, nor any thing made of Milk. Tho they speak in their Sermons half the literal and half the vulgar Language, they often preach in the vulgar Language entirely, to be the better understood; but the Mass, the Singing in the Church, the Lives of the Saints, and the Words us'd in the Administration of the Sacraments, are in the literal Tongue.

THE Curates and Secular Priests marry, as do the Papas among the *Greeks*, but can't marry a second time; and therefore they chuse Lasses, whose Complexions promise a long Life and good Health. They employ themselves in any Trade or Occupation, to get a Livelihood, and maintain their Families; which engages them so much, that they have hardly time to perform their Ecclesiastical Functions. To approach

the

the Altar with the greater Purity, they are oblig'd to lie in the Churches the Vigil of thofe Days in which they are to officiate.

THE Religious *Armenians* are either Schifmaticks or Catholicks. The Schifmaticks follow the Rule of St. *Bafil*; the Catholicks that of St. *Dominic*. Their Provincial is nam'd by the General of the *Dominicans* at *Rome*. About the Year 1320, Father *Barthelemy*, a Dominican, re-united many of the *Armenians* to the Church of *Rome*, which Pope *John* XXII. then govern'd; and this great Miffionary eftablifh'd there a great many Convents of his Order: there are ftill fome in the Province of *Nacfivan*, between *Tauris* and *Erivan*. M. *Tavernier* reckon'd ten about the Town of *Nacfivan* and the antient *Julfa*, which are but a Day's Journey diftant: all the Monafteries are governed by *Armenian* Domini-cans. To make good Subjects, they fend from time to time fome of the young Children of this Nation to *Rome*, to be brought up in the Sci-ences, and in the Spirit of the Order of St. *Dominick*. Each Monaftery is in a Borough; and they reckon in this Quarter about fix thoufand Catholicks. Their Archbifhop, who takes the Title of Patriarch, goes to *Rome*, to be confirm'd after his Election; and they follow in his Di-ocefe the *Roman* Ritual in every thing, except the Mafs and the Divine Service, which they fing in the *Armenian* Tongue, that the People may underftand it. This little Flock lives holily, is well taught, and there are not better Chriftians in all the Eaft.

THE *Armenian* Schifmaticks are much to be pitied: they faft like the Religious of *la Trappe*; and all this would fignify nothing, if they did not take care to be Orthodox. They fare very hardly two Days in a Week, Wednefday and Friday; and they eat neither Fifh, nor Eggs, nor Oil, nor any thing made of Milk. The Lents of the *Greeks* are times of Plenty and Good Cheer, in comparifon of thofe of the *Armenians*: befides the extraordinary Length, they are not permitted thro the whole to eat any thing but Roots, nor fo much of them as is needful to fatisfy the Appetite. The Ufe of Shell fifh, Oil, and Wine is forbidden them, except on the Holy Saturday; on that Day they begin again to eat But-ter, Cheefe, and Eggs. On Eafter-day they eat Meat, but that only which was kill'd on that Day, not on any of the foregoing. During the great Lent they eat no Fifh, nor hear Mafs but on Sunday. 'Tis faid at Noon, and they call it *Low-Mafs*, becaufe they place a great Hurdle before the Altar ; and the Prieft, who is not feen, pronounces only the Gofpel and Creed aloud. The Faithful communicate only on Holy Thurfday at the Mafs, which is faid at Noon ; but that of Holy Saturday is celebrated at Five or Six a clock in the Evening, when alfo they give the Communion. After that they break Lent, in the manner juft now mention'd, by eating Fifh, Butter, or Oil. Befides the Great Lent, there are four others in the Year, confifting each of eight Days ; they are inftituted to prepare for the four great Feafts of the *Nativity*, of the *Afcenfion*, of the *Annunciation*, and of St. *George*. Thefe Lents are as rigoroufly obferv'd as the great one ; they muft not fo much as fpeak of Eggs, or Fifh, or even of Oil or Butter ; fome take no manner of Nourifhment for three Days together.

THE *Armenians* have feven Sacraments, as we have ; *Baptifm*, *Confirmation*, *Penance*, the *Eucharift*, *Extreme Unction*, *Orders*, and *Matrimony*.

BAPTISM is adminifter'd among them by Immerfion, as among the *Greeks* ; and the Prieft pronounces the fame Words, *I baptize thee in the Name of the Father, of the Son, and of the Holy Ghoft* ; and plunges the Child three times in the Water, in memory of the Holy Trinity. Tho our Miffionaries fhew'd them their Miftake, in repeating all the Words at each Immerfion, there are ftill many Priefts who do it thro mere Ignorance. While the Curate recites certain Prayers of his Ritual, he makes a fmall Cord or String, one half of white Cotton, the other of red Silk, the Threds whereof he has himfelf twifted feparately. After having put it on the Neck of the Infant, he makes the Holy Unction on the Forehead, the Chin, Stomach, Arm-pits, Hands and Feet, by making the Sign of the Crofs on each Part. The Ceremony of the String is, they fay, in memory of the Blood and Water, which came from our Saviour's Side, when he receiv'd the Stroke of the Lance upon the Crofs. They baptize only on Sundays, if the Child be not in danger of Death ; and the Prieft gives it always the Name of the Saint of the Day, or of him whofe Feaft is to be the Day following, if there be no Saint for the Day on which the Baptifm is celebrated. The Midwife carries the Child to Church, but the Godfather carries it home to

the

Armenian Priests in their Sacerdotal Habits. 130.

the Mother, with the Sound of Drums and Trumpets, and other In- Lett.VIII.
ſtruments of the Country.   The Mother falls proſtrate to receive her
Child, and the Godfather at that time kiſſes the hinder part of the Mother's Head ; after that, they ſit down to Table with the Parents and
Friends, and the Clergy.   The Clergy muſt be at the Feaſt, becauſe the
Armenians believe that none but the Prieſts can-adminiſter valid Baptiſm
on any occaſion whatever.   I myſelf have heard ſay, there are Prieſts
who baptize dead Children; and I make no difficulty of believing it, ſince
they give the Extreme Unction only to thoſe who are dead.

THE Baptiſms which are adminiſter'd on Chriſtmaſs-day are the moſt
magnificent, and they put off to this Day the Baptiſms of ſuch Children
whoſe ſtate of Health will permit it.   The moſt famous Feaſts are principally celebrated in Places where there is a Pond or River.   For this
purpoſe they prepare an Altar in a Boat cover'd with fine Carpets : thither the Clergy repair as ſoon as the Sun riſes, accompanied by their Parents, Friends, and Neighbours; for whom they provide Boats fitted and
adorned in the ſame manner.   Be the Seaſon ever ſo ſevere, after the
ordinary Prayers, the Prieſt plunges the Child three times into the Water, and performs the Unctions.   The Fathers are not diſmiſs'd with a
ſmall Charge, for the Feſtival is carried on with Feaſtings and Preſents;
and therefore many Parents avoid the waiting till the Feaſt of the Nativity, and pretend their Children are in danger of dying.   And, in reality, what Folly is it, without any manner of neceſſity, to run one's
ſelf into Inconveniences? The Governours of Provinces are often preſent, and even the King himſelf ſometimes comes to *Julfa* to ſee theſe
ſort of Feaſts.   They muſt then make abundance of Preſents, beſides the
Entertainments and Collations.   Women go not to Church till forty
Days after their Delivery : they obſerve many Jewiſh Ceremonies.

IT appears by what we have ſaid, that the *Armenians* confer two
Sacraments at one time, Baptiſm and Confirmation, ſeeing they give the
Holy Chriſm to Infants.   They believe that all Prieſts can adminiſter this
Sacrament, but they think the Patriarch only can bleſs the Holy
Chriſm.

FOR the Communion, the Prieſts give the Faithful a piece of the
conſecrated Hoſt ſoak'd in conſecrated Wine; but it is ſcandalous that

they

they give it to Infants at the Age of two or three Months in their Mothers Arms, becauſe they frequently throw the conſecrated Elements out
of their Mouths. The *Armenian* Prieſts conſecrate Bread without Le-
ven, and make the Hoſts themſelves the Vigil of the Day in which they
are to offer: they are like thoſe we uſe, only they are three or four times
as thick. The Prieſt, before he begins Maſs, takes care to put the Hoſt
upon a Patin, and the Wine pure and unmix'd in a Chalice. Jeſus Chriſt,
ſay they, made the Supper with Wine, and Baptiſm with Water. The
Prieſt covers the Elements with a great Veil, and ſhuts them up in a
Cupboard near the Altar, on the ſide of the Goſpel. At the Offertory,
he goes to take the Chalice and Patin with Ceremony, that is to ſay,
follow'd by his Deacons and Subdeacons, ſome carrying Flambeaux, and
others Plates of Copper faſtned on pretty long Sticks, furniſh'd with little
Bells, which they roll about in a very harmonious manner. The Prieſt,
having a Cenſer carried before him, and being in the midſt of the Flambeaux and theſe muſical Inſtruments, carries the Elements in Proceſſion
round the Sanctuary. Then the People, miſinform'd, fall down and adore
the Elements, not yet conſecrated. The Clergy, yet more to be blam'd,
on their Knees ſing a Song which begins thus, *The Body of our Lord is*
preſent among us. The *Armenians* ſeem to have taken this abominable Cuſtom from the *Greeks*; for the *Greeks*, as we have already obſerv'd, by an inexcuſable Ignorance, do alſo adore the Elements before
their Conſecration. Their Error comes from hence, that formerly they
thought they might not celebrate this Sacrament, but on Holy Thurſ-
day; and conſecrated that Day as many Hoſts as they ſhould want
throughout the Year: theſe they kept in a Cup-board by the ſide of the
Goſpel; and the People were in the right to adore them, when the
Prieſt carried them from the Cupboard to the Altar. After this little
Proceſſion, the Prieſt puts the Elements upon the Altar, and pronounces the Sacramental Words: turning himſelf to the People, who proſtrate themſelves, kiſs the Earth, and beat their Breaſts, he ſhews them
the Hoſt and the Chalice, ſaying, *Behold the Body and the Blood of Jeſus*
Chriſt, which was given for us. After that, he turns himſelf to the Altar,
and communicates by eating the Hoſt ſoak'd in Wine. When he gives
the Communion to the Faithful, he repeats the following Words three
times,

An Armenian Deacon & Subdeacon

times, to make the Force of them be the better perceiv'd and felt; *I* Lett. VIII.
firmly believe this is the Body and the Blood of the Son of God, who took
away the Sins of the World, and who is not only my proper Salvation, but
likwife of all Men. This the People repeat very low after him word for
word.

NOTWITHSTANDING this holy Precaution, the *Armenian*
Schifmaticks don't appear to have any Senfe of the Grandeur of this
adorable Myftery. They for the moft part come to the Communion
without any Preparation, and they give it to Children of fifteen or fixteen Years old, without Confeffion, notwithftanding at this Age they are
not fo innocent as People may imagine. The *Armenians* rarely communicate in the Country, becaufe oftentimes the People have not where-
withal to have Mafs faid; and the Priefts perfuade them that a Mafs not
well paid for, is of no great efficacy.

OUR Miffionaries are to be admir'd for their Knowledge, for their
Zeal, and for their Generofity; but thefe Schifmaticks, by their Money,
deftroy all that thofe Apoftolical Men have built up in the moft folid
manner. The moft flourifhing Miffions muft fink and come to nothing,
unlefs God change the Hearts of the Schifmaticks. Thefe Wretches,
who apprehend nothing fo much as the holy Progreffes of our Priefts,
fet the Civil Powers againft them, and don't ceafe to reprefent to them,
how dangerous it would be to fuffer the *Latins* to encreafe among them;
that they are a fort of People who entertain ill Defigns againft the Government, and are devoted to the Pope and Chriftian Princes; that they
are to be look'd on as fo many Spies, who, under pretence of Religion, come to obferve the Strength of the Country; that they infpire thofe
of their Perfuafion with a Spirit of Sedition and Rebellion; that the
moft powerful Princes of *Europe* would not trouble themfelves with them,
were it not that they are a proper kind of Emiffaries, who may ferve
one day to extend their Conquefts. All thefe falfe Reafonings, accompanied with the force of Money, open the Eyes of the *Mahometans*; and
notwithftanding all the Recommendation in the World, our Miffionaries
are forc'd to withdraw themfelves. Neverthelefs, thefe Apoftles are not
difcourag'd; we every day fee in the *Levant* new Capuchins, Dominicans,
Carmelites, Jefuits, Priefts of the foreign Miffions of *Paris*. They inftruct

ftruck fuch as offer themſelves; they baptize; they bring back to the Flock, Sheep that have ſtray'd; and open the Gates of Heaven to the Elect.

WHAT a pity is it that the *Armenians* won't open their Eyes, for they are otherwiſe of a good natural Diſpoſition, and much enclined to Devotion? Their Churches are made very neat, ſince they have ſeen ours: There is in each Church but one Altar, plac'd at the Bottom of the Nave of the Church in the Sanctuary, to which they mount by five or ſix Steps. They are at conſiderable Charge to adorn this Place. No ſecular Perſon is permitted to enter it, of what Quality ſoever he may be. One may ſee by the Richneſs of this Place, that the *Armenians* handle more Crowns than the *Greeks* do Doubles. Poverty ſhews it ſelf among the *Greeks,* even in the Things they hold the moſt ſacred: They have ſcarce two ſmall Wax-Candles to ſay Maſs withal. On the contrary, among the *Armenians* one ſees fine Illuminations, and large Torches. Their Singing is alſo much more agreeable; and the Symphony of the little Bells, faſten'd to the Inſtruments above mention'd, whereof here is a Figure, inſpires an inexpreſſible Tenderneſs of Heart. They play'd on them at reading the Goſpel, and when they mov'd the Elements.

THE *Armenians* don't make more Preparation for Confeſſion than they do for the Communion. One may juſtly ſay that their Confeſſions are for the moſt part ſo many Sacrileges. The Prieſts don't underſtand the Nature of this Sacrament; and the Penitents, who are very great Sinners, as well as we, don't know how to diſtinguiſh Sin from what is not. Unhappily, neither the one nor the other are capable of a good Act of Contrition. The Declarations of the Sins are vague and indeterminate: Without dwelling upon thoſe they have committed, ſome of them confeſs three times more than they have committed, and recite by Heart a Catalogue of enormous Crimes, which has been formerly made for a Rule or Model by which to examine themſelves. If they confeſs they have robb'd or murder'd, the Confeſſor often anſwers that God is full of Mercy: But there is no Forgiveneſs among them for one who has not obſerv'd their Faſts, or for having eaten Butter on a *Wedneſday* or *Friday;* for their Prieſts, who make their Religion to conſiſt in great Abſtinences, impoſe monſtrous Penances for ſuch Faults: They will ſometimes enjoin whole

*

Months

Celtis Orientalis minor, foliis minoribus, et crassioribus, fructu flavo Coroll. Inst Rei herb. 42.

Months of Penance on thofe who confefs they have fmoked, kill'd a Cat, or a Moufe, or a Bird.

I SHOULD here give an Account of the Extreme Unction us'd among the *Armenians,* feeing they reckon it among their Sacraments : But there is nothing more abfurd than their Practice in this Particular ; for they never give it till after Death, and then almoft only to facred Perfons, others being denied the ufe of it.

THEY have particular Rules and Cuftoms in relation to Marriage. A Widower can marry but one Woman ; and amongft them none may contract a third Marriage, which would be accounted Fornication : And in like manner a Widow can't marry a Batchelor. There is no great harm hitherto. Nay, perhaps Marriages would be better and more agreeably manag'd thus among them, than they are among thofe of other Religions, if the Perfons were permitted to know one another before the Marriage : But among them they know nothing of making Love. Marriages are wholly manag'd according to the Pleafure of the Mothers, who generally confult only their own Husbands. After having agreed upon the Articles, the Mother of the young Man comes to the Houfe where the young Woman dwells, accompanied with a Prieft, and two old Women : She prefents her with a Ring in behalf of her Son. The young Man fhews himfelf at the fame time, keeping his Gravity as much as may be ; for he is not permitted to laugh at the firft Interview. 'Tis true, this Interview is very indifferent ; becaufe the fair one, or ugly one, does not fo much as fhew even her Eyes, fhe is fo veil'd. They make the Curate drink, who makes the Betrothings. 'Tis not cuftomary to publifh the Banns. The Day before the Nuptials the Bridegroom fends Suits of Clothes ; and fome Hours after goes himfelf, to receive the Prefent his Bride is to make him. The next Day they mount their Horfes ; and take a great deal of care to have very fine ones. The Bridegroom, coming out from the Houfe of his Bride, goes firft, having his Head covered with a Coronet or Garland of Gold or Silver, or with a Gawfe Veil of a Flefh-colour, according to his Quality : This Veil hangs half way down his Body. In his Right-hand he holds one End of a Girdle, which his Bride, who follows him on horfeback, cover'd with a white Veil, holds by the other End : This Veil hangs down even to the Horfe's

Legs

Legs. Two Men walk by the fide of the Bride's Horfe, to hold the Reins. The Parents, Friends, the Flower of the Youth, on horfeback and on foot, accompany them to Church with the Sound of Inftruments of Mufick, in Proceffion, Tapers in their Hands, and without any Diforder. They alight from their Horfes at the Church-door, and the young Couple walk to the Steps of the Sanctuary, holding the Girdle by the Ends all the way they go. There they ftand together a-breaft; and the Prieft having put the Bible on their Heads, asks them if they will take one another for Husband and Wife; and they bow their Heads to fignify their Confent. Then the Prieft pronounces the Sacramental Words, he performs the Ceremony of the Rings, and fays Mafs. After that, they return to the Bride's Houfe in the fame Order they came. The Husband goes to bed firft, the Wife pulling off his Shoes and Stockings, who is alfo left to put out the Candle, and does not put off her Veil till fhe gets into bed. Thus the Marriages are celebrated; and thefe are the Ceremonies obferv'd by the new-married People among the *Armenians:*

> *Concealment thus abates the Husband's Flame,*
> *And hides the Blufhes of the willing Dame.*

But after all, this is no better than, as we fay in *Englifh, buying a Pig in a Poke.* They fay there are *Armenians* who would not know their Wives, if they fhould find them lying with other Men. Every Night they put the Candle out before they take off their Veil; and the greateft part of them never fhew their Faces all the Day. An *Armenian* returning from a long Journey, could not be affured that he had the fame Wife in bed with him, and that fome other Woman had not, for the fake of his Subftance, taken the place of his dead Wife.

WHEN the Daughters lofe their Mothers before their Marriage, commonly the next Relation takes the Care of the Marriage. Sometimes the Mothers betroth their Children at two or three Years of Age. There are fome Mothers, who, even while they are with Child, agree together to marry the Children they go with, if one be a Boy, and the other a Girl: And this is one of the greateft Marks of Efteem and Friendfhip which Perfons of Figure can give one another. They betroth them as foon as they are born; and after the Betrothing, to the Confummation Lett. VIII. of the Marriage, the young Man, on *Eafter-Day*, every Year fends his Miftrefs a Suit of Clothes. I fay nothing of the Feafts and Rejoicings at the Marriage. The Feaft lafts three Days; and the Men are not mix'd with the Women: They fay they drink much on both fides. Thefe good Women unveil among themfelves, talk merrily, and to be fure do not fpare the Liquor.

THE *Armenians* don't ufe many Ceremonies at prefent in conferring Holy Orders. He that defigns for the Ecclefiaftical State, offers himfelf to the Curate, accompanied with his Father and Mother, who confirm the Declaration their Son makes of his Defire to dedicate himfelf to God. The Curate well inform'd of his Defign, without taking the pains to reprefent to him the Weight of the Burden he is taking upon him, without exhorting him to beg of God the neceffary Graces for perfevering in fo holy a State, without requiring of him the Practice of fuch Virtues as are infeparable from the Miniftry, contents himfelf with putting a Cope on him, and repeating fome Prayers. This is the firft Ceremony. They repeat it fix times, Year after Year, without obferving any Rules between the Times; but when the Ecclefiaftick attains the Age of eighteen Years, he may be confecrated: thefe Impofitions of the Cope, accompanied with certain particular Prayers, being only fufficient for the other Orders, which are the Clerkfhip, Subdeaconfhip, and Deaconfhip. In the mean time, if the Prieft intends to marry, which is the conftant Practice among them, after the fourth Ceremony, they caufe him to marry the Woman he has a mind to. After the Impofition of the Cope, he addreffes himfelf to a Bifhop or Archbifhop, who puts on him all the Sacerdotal Habits. This Ceremony cofts much more than the former; for they pay dearer in proportion as they advance in Orders. Formerly the *Armenian* Priefts could not marry a fecond time after the Death of their Wives, and they are not entirely free as to this Point at prefent; but they are not permitted to fay Mafs if they marry a fecond Wife, as tho their Character was effac'd by this fecond Marriage. The new Priefts are oblig'd to continue in the Church a whole Year, to perform Divine Service: After which time likewife, the moft part lie in the Church the Eve of the Day in which they are to celebrate.

Vol. II.   S f

brate. Some remain there five Days, without going to their Houſes, and eat nothing but hard Eggs, and Rice boiled in Water and Salt. The Biſhops eat no Meat or Fiſh but four times a Year. The Archbiſhops live on Pulſe. As they make the Perfection of their Religion to conſiſt in their Faſts and Abſtinences, they encreaſe them in proportion as they advance in Dignity: Upon this foot the Patriarchs muſt almoſt ſtarve themſelves to Death. Our Miſſionaries are oblig'd to comply a little with their Uſages and Manners; for one cannot merit their Eſteem by any thing ſo much as by extravagant Faſtings.

THE Prelates prepare Holy Water but once a Year: And this Ceremony they call the *Baptiſm of the Croſs,* becauſe on the Day of *Epiphany* they plunge a Croſs into Water, after having recited divers Prayers. And after the Holy Water is made, every one fills his Pot, and carries it home. The Prieſts, and eſpecially the Prelates, draw a very conſiderable Advantage from this Ceremony.

I am, My Lord, *&c.*

LET-

LETTER IX.

To *Monseigneur the Count* de Pontchartrain, *Secretary of State,* &c.

My Lord,

 E began to turn our backs upon the *Levant* in good earnest  the 12th of *September*; and tho we were at the bottom of Natolia, we seem'd to see the tops of the Steeples in *France,* when we had resolv'd to make towards the *Mediterranean.* We went, however, that Day but one Mile from *Erzeron* with part of the Caravan, which was going for *Tocat.* We set out the next Day, being the 13th of *September,* for the *Baths* of *Elijah,* where the rest of the Merchants were assembled. These Waters seem'd to us to be warmer than those at *Assancala,* and than those in the Neighbourhood of the great Monastery of *Erzeron.*

THE 14th of *September* we travell'd from Five in the Morning till Noon in a flat Country, so dry and burnt up, that we found no Plants nor Grain there. Our Caravan consisted of not above three hundred Persons, almost all *Armenians,* who carried Silk to *Tocat, Smyrna,* and *Constantinople.* We set out the 15th, at half an Hour after Five, and about Noon encamp'd on that Branch of the *Euphrates,* which runs through the Plain of *Erzeron* under *Elijah*'s Bridge. We had all along kept on the Left-side of it : But the Country seem'd much more rugged than the Day before : They are Rocks which confine the *Euphrates* in its Course toward the West. The Banks of this River are cover'd with a fine Species of *Barberry-tree,* taller than ours, and which is distinguish'd by its

Fruit.

Fruit. 'Tis a Bunch confifting of feven or eight cylindrical Berries, about four lines long, and two thick, black, cover'd with a Flower like that on Plumbs frefh gather'd, full of a violet-colour'd Juice, not fo fharp, and much more agreeable than that of the *Barberry-tree.* The Shrub we are fpeaking of has Leaves about two inches long, and near ten lines broad, a little fharp, and indented. The Wood of it is yellow, furnifh'd with hard Thorns, fome fingle, and fome with two or three Points. This Plant was rais'd from the Seed in the King's Garden.

THE 16th of *September* we travell'd from half an Hour after Four in the Morning till One after Noon, in a narrow Valley, difagreeable, un-cultivated; wherein we found but one Caravanfera : and the *Euphrates,* which runs continually towards the Weft, makes divers Windings. We were oblig'd to pafs this River twice, having learn'd of a Caravan, con-fifting of about twenty four Camels, that the Road to *Tocat* was full of Robbers. Upon this News we affembled together, to advife what might be beft to do ; and it was refolv'd to put our felves into the beft Pof-ture we could. In the Center we plac'd all the Horfes laden with *Silk;* and we were fometimes among them, and fometimes in the Rear. We arriv'd about Eleven of the clock at the Entrance of a Valley, much narrower than the former : And while we entrenched our felves upon the Brow of a little Hill, at the fight of this dangerous Place, we de-tach'd three Fufiliers to go and reconnoitre the Paffage. Happily they brought us word that they faw but three or four arm'd Horfemen, who were making to the Mountains ; and fo we pafs'd the Defile without fpeaking a Word, and with all the fpeed we could. In this place the *Euphrates* makes a confiderable Elbow, bending towards the South to ap-proach another of its Branches, which goes to *Mammacoutum.* We con-tinu'd our Route towards the South-weft, and were oblig'd to encamp half an Hour from this Paffage, almoft half way up the fide of a rug-ged Mountain, in a frightful Solitude, where we could fee neither Village nor Caravanfera : We had a great deal of Difficulty to find Cow-dung enough to boil our Kettle.

THE 17th of *September* our Route was fhort, but very troublefome : We pafs'd over a very bare Mountain ; at the foot of which we enter'd into a well-cultivated Valley, where we encamp'd, after four Hours travel,

vel, near *Caraboulac* a very pretty Village. This Day we were join'd by
a Caravan of Silk-Merchants, as numerous as our own. It came from
Erzeron two Days after us; but it had made more haste, upon a Rumour
which was spread, that one *Pacha Manfoul* had put himself at the Head
of the Robbers. This Recruit pleased us much; and we together left
Caraboulac about Five in the Morning to go to *Acpounar*, another Village,
where we arriv'd about One a clock after Noon. The Route would be
pleasant enough, were it not that we are forc'd to pass a very high open
Mountain.

THE 18th of *September* we set out at Four of the clock in the
Morning, to go, however, not very far; for we encamp'd about three
quarters past Eight near a Brook, which runs towards the West. It is
true, we pass'd a Mountain cover'd with Pines, the Descent of which is
very rugged, and leads to a Valley narrow and winding; on the Left of
which one sees the Remains of an antient Aqueduct with round Arches,
which seem pretty antient. This Day we pass'd the River which runs
into the *Black-Sea* at *Vatiza.* This River comes from the South; whereas in our Maps it's made to run from the East.

THE 19th of *September* we continu'd our Journey to the North-
west, in another very narrow Valley: After which we enter'd upon a
fine Plain to the West, in which runs an agreeable Rivulet, on the
Edge of which stands the Village *Sukmè.* A little on this side the Village, to the Right of the main Road, are seen two Pieces of antique Columns; upon the least of which are very antient *Greek* Characters, which
we could not stay to examine, for fear of the Robbers; and besides, the
Inscription appear'd to be much decay'd. Perhaps it mentions the Name
of some antient Town, upon the Ruins of which *Sukmè* is built. After
a Route of five Hours and a half, we encamped near another Village,
called *Kermeri.*

OUR Journey the 20th of *September* was of seven Hours; and we
rested at *Sarvoular,* another Village, built in the same manner as *Kermeri,* that is to say, very poorly. At the Descent of a Mountain, and
the Entrance of a dangerous Place, we discover'd five or six Robbers on
horseback; who retir'd from us, upon our threatening to fire on them.
We alit from our Horses, and took in our hands our Fusees, or Pistols,

or

or Sabres, or Lances; for we had in our Company fuch as were arm'd with all thefe different Weapons: But there were few who had Refolution enough to ufe them. For my part, I freely own that I did not find I had a Soul for War at that time. The Bales of Silk were in the middle of our Troop, and thofe of our Horfemen who were the moft fprightly and active, were plac'd fome in the Van, and fome in the Rear. Certain Robbers appear'd a quarter of a League from us, upon fome neighbouring Hills: But notwithstanding, we enter'd upon a fmall Plain, terminated by a little Dale, at the Entrance of which were pofted fifteen or twenty of thefe Robbers, who feeing us move forward in good Order, thought fit to retire. Thefe poor Wretches are Mountaineers, who rob thofe to whom they find themfelves much fuperior; but have not the Senfe to underftand one another, and form their Parties well. 'Tis certain, if they had attack'd us with Refolution, they might have carried off half the Bales of Silk. Some Robbers, who mingled themfelves with us, in the Morning, when we were loading our Bales of Silk, had more Management and Cunning; for they drove off two Mules with their Burdens, and we heard no more of them. The Mountains over which we pafs'd are cover'd with Copices of *Yoke-Elm,* among which grow *Pines, Savine,* and *Juniper.* The Water-Melons are excellent in all thefe Parts: The beft have a pale-red Flefh, and reddifh Seeds, inclining to black; the others have a yellowifh Flefh, and black Seed: The lefs fweet have a white Flefh.

THE 21ft of *September* we fet out at Five in the Morning, and pafs'd over the higheft, rougheft, and moft fatiguing and troublefome Mountain in the Country, always on our Guard, for fear of Robbers. The Sight of an infinite Number of rare Plants, was a great Confolation to us in our Dangers. Thefe Plants grow among common *Oaks, Willows, Lote-Trees, Tamarisk, Pines, Barberries with black Fruit.*

THE 22d of *September,* from Five in the Morning till Noon we faw nothing but very rugged Rocks, all of white Marble, or red and white Jafper; among which the River *Carmili* runs with Rapidity from Eaft to Weft. We had for our Inn a very bad Caravanfera, or rather a Barn, wherein we found a Bank rais'd three Feet high, on which every one laid his Bedding. The *Turks* carry only a Carpet for their ufe in the Night.

Night. This Place receives Light only by Openings, which are lefs than the Windows of the Capuchins Chambers. We were happy, however, in finding this Retreat; for befides that it had rain'd almoft all Day, it hail'd the whole Night. We obferv'd this Day fome wild *Almond-Trees*, which are much lefs than the common *Almond-Trees*; but their Branches don't terminate in a fharp Point, like the wild *Almond* of *Candia*. The Leaves of this Kind we fpeak of, are not above five or fix lines broad, and an inch and a half long, of the fame Colour and Contexture with thofe of our Almond-Trees. The Fruit of the wild *Almond-Tree* is hardly eight or nine lines long, and feven or eight thick, but very hard. The Kernel is not fo bitter as our Bitter-Almonds, and fmells like the Kernel of a Peach-ftone. We faw here in thefe Parts likewife a kind of *Micocoulier*, or *Lote-Tree*, which was very remarkable.

THIS Tree grows hardly any higher than a Plumb-Tree, but is more bufhy: Its Branches are of a white Wood, cover'd with brown-green Bark: Its Leaves are ftiffer and firmer than thofe of our *Lote-Tree*, fmaller, thicker, lefs pointed, ordinarily of an inch and a half long, much like thofe of an Apple-Tree, but of the Contexture of thofe of the *Micocoulier* or *Lote-Tree*; they are a brown-green above, a whitifh green underneath, of a herbifh Tafte, indented on the Edges, and one of the Ears of the Bafe is fmaller and lower than the other. The Fruit grows out of the Knots of the Leaves, four lines long, almoft oval, yellow, inclining to a brown when they are thorough ripe. Their Flefh is yellowifh, fweet, but ftiptick: The Kernel is green, and includes a pithy Seed, like the common Kind.

THE 23d of *September* our Journey was eight Hours and a half long. We found at going out of the Caravanfera a very high Mountain, very rugged, and bare: But we afterwards enter'd upon a fine great Plain, where we encamp'd near a Village called *Curtanos*. The 24th we fet out at Four in the Morning from the Plain of *Curtanos*, and pafs'd over a Mountain, and through Valleys, which were very rugged; through which runs, on the right of the Road, a River, which is very red with the great quantity of Bole it wafhes off, and carries with it. It winds thro very dangerous Paffages, where Beafts of Burden can hardly pafs one after another. Thefe Paffages brought us at length to

* the

the foot of other Mountains, very rugged and pointed; on the higheſt of which, is built the Town of *Chonac,* or *Couleiſar,* a ſmall Place, in form of an Amphitheater, and terminated by an old Caſtle. The River, which appears all bloody, runs along at the bottom of the Mountain, and renders the Paſſage much more frightful. The Neighbourhood is horribly ſteep, but on a ſudden the Situation is chang'd; for as ſoon as we are paſt *Chonac,* we come into one of the moſt pleaſant Valleys in *Aſia,* full of Vineyards and Orchards. This Alteration, which we did not expect, made a very agreeable Contraſt, which continued even to *Agimbrat,* or *Agimourat,* a ſmall Town, an Hour and a half from *Chonac. Agimbrat* is upon a Mountain like a Pye ſqueez'd flat, at the foot of which runs the ſame River. A Rock riſes on the ſide of this Town, on which there ſtands an old ruin'd Caſtle, which antiently guarded the Paſſage of the Valley. We ſaw nothing but fine Plants all this Journey: The Vineyards are furniſh'd with *Peaches, Apricocks,* and *Plumbs.* Our Inn was very agreeable: 'Tis a fine Caravanſera at the ſide of a River, with a double Nave, like the great Hall in the Palace at *Paris*; the Vault is of Free-ſtone, and the Archings are well moulded. But this Building, tho it be ſurprizingly beautiful for the Place, receives Light only by a Sky-light; and we lodg'd there on a Bench which runs all round both Naves. We that lov'd to be cool, went and lay in the Court, where we yet continued ſenſible of the great Heat of the Day: But we were oblig'd to leave our Lodging an Hour before Day, and to come and breathe an Air infected with the Breath of all the Horſes and Mules of the Caravan; for the Cold had benumb'd us, and unhappily we had nothing to drink but Water cooled with Ice. As this Country is only inhabited by *Turks,* they ſell their Wine by Wholeſale to the *Armenians*; and after the Sale is made, one could not get a quarter of a Pint to ſave one's Life: We ſatisfied our ſelves with eating Raiſins, tho they were ſoft, and too ſweet. They told us the Vines were of little conſequence, and not very profitable.

THE 25th of *September* we kept the ſame Vale from Five in the Morning till Eight. The red River run on the right; but we left it at a Village which takes up almoſt all the Bottom of the Valley. This River runs towards the North, and throws it ſelf, as they told us, into one of thoſe which empty themſelves into the *Black-Sea.* We did not

*   trouble

Chonac or Couleisar.

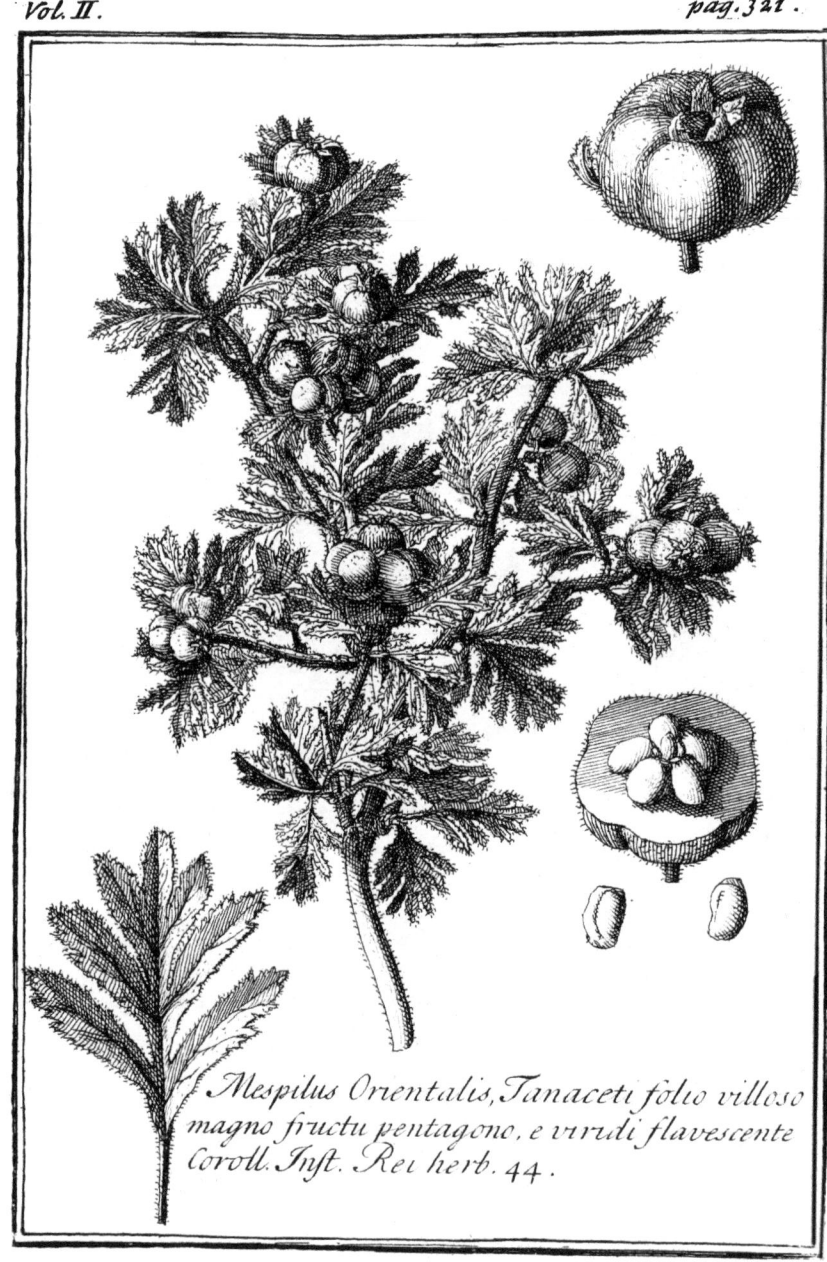

Mespilus Orientalis, Tanaceti folio villoso
magno fructu pentagono, e viridi flavescente
Coroll. Inst. Rei herb. 44.

trouble our felves much about this, becaufe the Merchants of the Caravan Lett. IX. are not able to give much Light into fuch kind of matters : But we were very uneafy to know what Road we took; becaufe which way foever we turn'd our Eyes, we could fee nothing but the Opening where the River emptied it felf. Our *Armenians* quickly fhewed us the Road; and the Head of the Caravan began to afcend up one of the higheft Mountains we had yet pafs'd fince we came from *Erzeron.* We faw there a great many *Oaks* and *Pines.* But the Defcent was very frightful; and we en-camp'd in a kind of Abyfs at the foot of certain Mountains, not quite fo high as this.

T H E S E Mountains produce a fine Sort of *Azarolier,* or *Medlar-Tree :* There are fome as big as Oaks. Their Trunk is cover'd with a cleft greyifh Bark; the Branches are bufhy, and fpreading out on the fides. The Leaves are in Bunches, two inches and a half long, fifteen lines broad, pale-green, fhining, a little hairy on both fides, commonly di-vided into three Parts, even to the Rib; and thefe Parts indented very neatly on the Edges, pretty much like the Leaves of *Tanfy;* the Part at the end of the Leaf is again divided into three Parts. The Fruit grows two or three together at the Ends of young Shoots, and refemble fmall Apples, of an inch diameter, rounding with five Coins, like the Ribs of a Melon, a little hairy, pale-green, inclining to a yellow, with a Na-vel rais'd of five Leaves, four lines long, one line and a half broad, and indented like the Leaves of the Tree. We fometimes find one or two of thefe Leaves grow out of the Flefh of the Fruit, or its Stalk. This Fruit, tho agreeable, is not fo pleafant as our *Medlar;* but I be-lieve it would be excellent if it were cultivated. The *Armenians* do not only eat as much of this as they can, but do likewife fill their Bags. The Middle of this Fruit is fill'd with five fmall Stones, four lines long, rounding on the Back, a little flat on the Sides, fharp on that part which lies toward the Middle of the Fruit, very hard, and fill'd with a white Marrow, or Pith. This Tree has no Prickles; its Leaves are unfavoury, and of a mucilaginons Tafte.

T H E other Kinds of *Medlar-Tree* have a red Fruit; and differ from one another only in the Bignefs of their Fruit, whereof fome are an inch in diameter, and others not above feven or eight lines thick.

Thefe fort of Trees, which are not higher than Plumb-Trees, have a Trunk as big as a Thigh, cover'd with a greyifh cleft Bark. The Branches are bufhy, ending in hard Prickles, blackifh, and fhining. The Leaves grow in Bunches, like thofe of the *Azarolier*, or *Medlar-Tree*, one inch and a half long, pale-green, hairy, and downy on both fides, cut into three Parts, the Middle whereof is again cut into three Parts, and thofe on the fides cut into two. The Fruit grows four or five together, raifed into five Coins or Wedges, rounding, red, hairy, with a Navel furnifh'd with five pointed Leaves: They are a little fharp, more agreeable than thofe of the preceding Species. Their Flefh is yellowifh, and inclofes five fmall Stones, very hard, fill'd with a white Pith.

THE 26th of *September* we fet out about Five of the clock, and did not make any ftop till Noon, which tir'd us much; for we travell'd all the while in the fame Vale, which is, as I may fay, water'd, and which we expected to leave every moment; tho it made fo many Turnings and Windings, that we were forc'd to encamp there this Day too upon the Banks of a River. In this Road we faw Tombs of Stone, built after the *Turkifh* Manner, without Mortar. They told us that poor murder'd Merchants were buried there; for this Route was formerly one of the moft dangerous in *Anatolia*. At prefent the People of the Country, who from time to time rob feveral little Caravans, fire upon ftrange Robbers, and have almoft deftroy'd them. 'Tis a Maxim among them, That every one fhould rob in his own Country: So that one would run a great hazard to pafs this way without a good Guard. Otherwife the Country is very pleafant. And I had forgot to mention the vaft Quantity of Partridges we faw all along the Road, fince we left *Erzeron*.

BESIDE the common *Oaks*, and that which bears the *Velanede*, we faw feveral other Kinds in this Valley, efpecially thofe with Leaves of three or four inches long, and two broad, cut almoft to the Rib, in a manner much like the Slafhes of the *Acanthus*. The Rib is pale-green, and begins by a Stalk feven or eight lines long; but the Leaves are fmooth, and dark-green above, but whitifh beneath; their Slafhes are fometimes cut into three Parts at the Point. The Acorns grow commonly by two and two, in a great many Pairs, heap'd one upon another, and

faften'd

faften'd to the Branches without a Foot-ftalk.　Each Acorn is fifteen Lett. IX.
lines long, eight or nine in diameter, and half way out of the Cup,
rounding, and terminated by a fmall Nib.　The Cup is fifteen or fixteen
lines in diameter, about an inch deep, adorn'd with Threds after the manner of a Perriwig, half an inch long, efpecially towards the Edges, curled
fome upward, fome downward, and as it were frizled up, half a line
thick at their Bafe, but taper quite to the end.　On the fame Stalk are
fometimes found Acorns which are fhorter and rounder.　The Leaves of
this Tree are of an infipid mucilaginous Tafte.

THE 28th of *September* our Route was of eight or nine Hours,
almoft all the while in the fame Valley; which after having widen'd and
narrow'd it felf in many places, opens at length into a fort of unculti-
vated Plain, where we took notice of the fame Species of Oaks.　The Ri-
ver hitherto run all the way on our Left; we forded it an Hour from our
Inn, and left it on the Right in this Plain.　Part of the Caravan went this
Day to lodge at *Tocat*.　They caus'd us to encamp near a Village call'd
Almous, in the midft of Oaks with the great and with the fmall Leaves.
Among many other rare Plants, we obferv'd *Sage with large frizled Sickles*,
Juniper with red Berries, the *Spindle-Tree*, *Alder-Tree*, *Cornel-Tree*, the
Common Turpentine-Tree, *Melilot*, *Burnet*, *Wild Succory*, *Savory*, *Jerufa-
lem Oak*, the *Female Fern*, and I know not how many very common
Plants.　But nothing pleas'd us better than that Kind of *Thapfia*, of
which *Rauvolf* gives the Figure, under the Name of *Gingidium Diofco-
ridis*.　The Defcription whereof is as follows:

ITS Root is but one line thick, whitifh, three or four inches long,
furnifh'd with fome Fibres.　The Stalk, of the moft part of what we
found, was not above a fpan high, twifted, one line thick, accompanied
with Leaves like thofe of the *Scandix Cretica minor* C. B. two or three
inches long, which enwrap the Stalk in a fort of Sheath of half an inch
long.　The *Umbellæ* are an inch and a half in Bignefs, furrounded at the
Bafe with five Leaves, cut like the others, but feven or eight lines long,
folded in Gutters from their Beginning.　Each Furrow is terminated by
two Leaves like thofe which accompany the Flowers.　They were gone
off, as well as the Seed, which we gather'd up from the Ground in
great quantity.　Thefe Seeds are oval and flat.

　　　　　　　　　　THE

THE 28th of *September* we took Horſe at One in the Morning, and reach'd *Tocat* about Ten. After having paſs'd very narrow Valleys, co-ver'd with Oaks, we again found our River, which we forded twice. It is called *Toſanlu,* and runs into the *Iris* of the Antients, which the *Turks* call *Caſalmac.* At length we enter'd a larger and more beautiful Valley than the reſt had been, which led to *Tocat.* But this City did not ap-pear till we came to the Gates of it, for it is ſituate in a Nook among great Mountains of Marble. This Nook is well cultivated, and fill'd with Vineyards and Gardens, which produce excellent Fruit. The Wine would be admirable, if it were not ſo ſtrong.

THE City of *Tocat* is much bigger and pleaſanter than *Erzeron.* The Houſes are handſomely built, and for the moſt part two Stories high; they take up not only the Land which lies between theſe rugged Hills, but likewiſe ſtretch themſelves along the tops of the Hills, in form of an Amphitheatre, in ſuch manner, that there is not a City in the World of a Situation ſo ſingular. Not to loſe any Ground, they have even built upon two very frightful, rugged, and perpendicular Rocks of Marble, for one ſees an old Caſtle on each of them. The Streets of *Tocat* are well enough pav'd, which is very rare in the *Levant.* I believe the Inhabitants have been oblig'd out of neceſſity to have them pav'd, that the Rains in tempeſtuous times might not lay open the Foundations of their Houſes, and overflow their Streets. The Hills on which the City is built, have ſo many Springs, that each Houſe has its Fountain. Not-withſtanding this great Quantity of Water, they could not put out a Fire, which a little before our Arrival there conſumed the fineſt part of the City and Suburbs. Several Merchants were ruin'd by it, their Ware-houſes being at that time full of Goods; but they began to rebuild it, and they hop'd that quickly there would be no Sign of the Fire left. They find Timber and other Materials enough about the City.

THERE is at *Tocat* a Cadi, a Vaivode, an Aga of the Janizaries, with about a thouſand Janizaries, and ſome Spahi's. They reckon there are twenty thouſand *Turkiſh* Families, four thouſand *Armenian* Families, three or four hundred Families of *Greeks,* twelve Minaret Moſques, and an infinite Number of *Turkiſh* Chappels. The *Armenians* have ſeven Churches there, the *Greeks* only one ſorry Chappel, which they boaſt

to

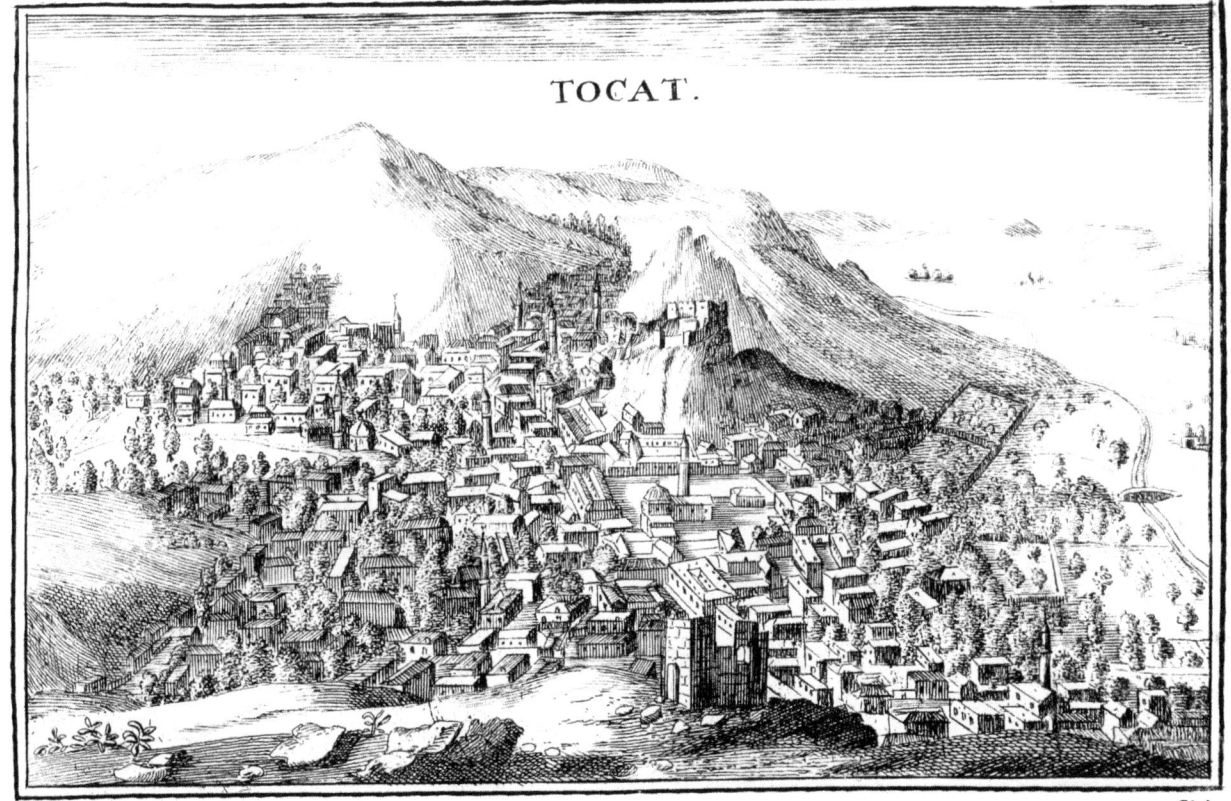

TOCAT.

Turkish Women of TOCAT

to have been built by the Emperor *Juftinian*. It is governed by a Me- Lett. IX.
tropolitan dependent on the Archbifhop of *Nicfara*, or to fpeak more
properly, of *Neocæfarea*, an antient City almoft ruined, two days Journey from *Tocat*.

NICSARA is ftill the Metropolis of *Cappadocia*, and it will never be
forgot, that in the third Century it had St. *Gregory Thaumaturgus*, or *the
Worker of Miracles*, for its Paftor. *Niger*, and fome other Geographers,
had no manner of reafon to confound this City with *Tocat*. The Archbifhop of *Nicfara* has the fifth Place among the Prelates, who are under
the Patriarch of *Conftantinople*.

BESIDES the Silks of the Country, which are very confiderable
they ufe at *Tocat* every Year eight or ten Loads of that of *Perfia*. All
this Silk is made up in flight Pieces into fewing Silk, or Silk to make
Buttons. This Trade is very good; but the chief Trade of *Tocat* is in
Copper Veffels, as Kettles, drinking Veffels, Lanthorns, Candlefticks, which
are made here very handfome, and fent to *Conftantinople*, and into E-
gypt. The Workmen of *Tocat* have their Copper from the Mines of *Gu-
mifcana*, which are three days Journey from *Trebifond*, and from thofe
of *Caftamboul*, which are much richer than the other, ten days Journey from *Tocat*, on the fide towards *Angora*. They likewife at *Tocat* prepare much of the yellow *Turky* Leather, which is carried by Land to *Sam-
fon* upon the *Black Sea*, and from thence to *Calas*, a Port in *Wallachia*.
They alfo bring thence a great deal of the red fort, which the Merchants of *Tocat* convey from *Diarbec* and *Caramania*. They inform'd us
that they dy'd the Leather yellow with *Fuftet*, and red with *Madder*.
The painted Cloth of *Tocat* is not fo beautiful as that of *Perfia*, but it
ferves the *Mufcovites* and *Crim-Tartars*. They are likewife carried into
France, and are thofe which they call there *Toiles de Levant*. *Tocat* and
Amafia furnifh more of them than all the reft of the Country.

TOCAT ought to be look'd on as the Center of the Trade of the
Leffer Afia. The Caravans of *Diarbekir* come thither in eighteen Days;
a Horfeman will go it in twelve. They are fix Days going from *Tocat*
to *Sinope*; Footmen go it in four Days. The Caravans go from *Tocat*
to *Prufa* in twenty Days; Horfemen in fifteen. They who travel directly from *Tocat* to *Smyrna*, without going to *Angora* or *Prufa*, are feven and

twenty Days upon the Road with Mules, and forty with Camels; but they run great hazards of the Robbers. Our Caravan was bound for *Smyrna*, but part went to *Prufa*, and part to *Angora*, to avoid the Robbers. Our *Armenians* aſſur'd us they got a great deal more by carrying their Silk to *Smyrna*; for they bought it at *Gangel*, on the Frontiers of *Perſia*, at the rate of twenty Crowns the Batman; ſo that ſelling the ſame Weight at *Smyrna*, at the rate of thirty Crowns, they gain'd three Crowns clear of all Charges they were at in their Journey. This Profit is very conſiderable; for a Batman weighs but ſix Oques, that is to ſay, eighteen Pounds twelve Ounces; and a Horſe carrying ſix hundred Pounds weight, and a Camel a thouſand, there will be an hundred Crowns gain'd by every Horſe-Load, and five hundred Livres by every Camel's Load. The Merchants who carry ten Loads of Silk, gain at this rate a thouſand Crowns, if they uſe Horſes; and five thouſand Livres, if they uſe Camels; without reckoning the Advantages made by ſuch Goods as they carry back.

TOCAT belongs to the Government of *Sivas*, where there is a Baſſa, and an Aga of the Janizaries. The *Greeks* of this Province pay Capitation for four thouſand. *Sivas*, according to their Tradition, is the antient City of *Sebaſtia*, which *Pliny* and *Ptolemy* place in *Cappadocia*. This City is but two Days Journey from *Tocat*, toward the South; and *Amaſia*, another antient City, is three Days Journey from *Tocat*, toward the North-Weſt: but theſe two Cities, tho antient, are much leſs than *Tocat*. *Sivas* is very ſmall at preſent, and would hardly be known, if the Baſſa did not keep his Reſidence there. *Ducas*, who wrote the *Byzantine* Hiſtory from *John Paleologus* to *Mahomet* II. affirms that *Bajazet* took *Sivas* in 1394. *Tamerlane* beſieg'd it a little after in ſo ſingular a manner, that our Engineers will not be diſpleas'd to have an account of it.

TAMERLANE caus'd the Walls of the Place to be undermin'd, and ſupported them with Pieces of Wood, as they took out the Stone. The Workmen approach'd it under-ground, by Paſſages which open'd at a Mile diſtance from the City, without being ſuſpected by the Inhabitants. When the Work was finiſh'd, the Place was ſummon'd to ſurrender. The Beſieg'd knowing nothing of their Danger, and not ſeeing their Walls any way damag'd, believ'd they could defend themſelves ſome time; but

but were ſtrangely ſurpriz'd to ſee their Walls fall on a ſudden, after the Lett. IX.
Beſiegers had ſet fire to the Wood which ſupported them.	They entred
the Town, and made a dreadful Slaughter ; and they who eſcap'd it were
however deſtroyed in a manner unheard of before that time.	They tied
them faſt with Cords in ſuch manner, that their Head was brought between
their Thighs, and their Noſe to their Fundament ; and in this Poſture
they were thrown by dozens into Ditches, which they cover'd with
Planks, and then with Earth, and ſo left them to die gradually.	The
City was raz'd, and has not been rebuilt ſince, tho it preſerves its Rank
and Dignity.

THERE might be many very remarkable things ſaid of *Amaſia*, but
this is not the Place : I only add that *Strabo*, the moſt famous of the
antient Geographers, tho originally of *Crete*, was a Native of this Place.
I don't know whether he has made any mention of *Tocat* ; all the *Greeks*
of the Place, of whom we enquir'd, told us it was formerly call'd *Eu-
doxia* or *Eutochia* : Is not this the City of *Eudoxiana*, which *Ptolemy* mentions in *Galatia Pontica?* *Paulus Jovius* calls *Tocat Tabenda*, 'tis like be-
cauſe he thought this was the City this Geographer calls *Tebenda*.	One
ſhould probably find the true Name of *Tocat* upon ſome of the Inſcriptions, which, as they told us, are to be ſeen in the Caſtle ; but the
Turks would not give us entrance.	They had juſt been taxing the *Ar-
menian* Catholicks of this City, after a great Perſecution, which had been
rais'd againſt them at *Conſtantinople* ; and therefore all over *Aſia* the *Franks*
were not ſo civilly us'd as they were wont to be.

AFTER the bloody Battel of *Angora*, where *Bajazet* was made Priſoner to *Tamerlane*, Sultan *Mahomet*, who after the *Interregnum*, and the
Death of all his Brethren, reign'd peaceably under the Name of *Mahomet* I.
this Sultan, I ſay, who was one of *Bajazet*'s Sons, at the Age of fifteen
Years, with the few Troops he could get together, march'd Sword in hand
among the *Tartars*, who then poſſeſs'd the Country, and came to *Tocat*, of
which he was Governour till his Father's Misfortune, who had obtain'd
it ſome time before ; ſo that this City was the Capital of the *Turkiſh*
Empire : and *Mahomet* I. having defeated his Brother *Muſa* or *Moſes*,
caus'd *Mahemet Bey* and *Jacob Bey*, who had been in his Brother's Inte-
reſt, to be put into the Priſon of *Tocat*, call'd the *Great Cord*.	It appears
by

by this, that the City did not at that time fall into the hands of *Ta-merlane*, but that it was under *Mahomet* II. *Jufufzes Begue*, General of the Forces of *Ufum-Caffan*, King of the *Parthians*, ravifh'd this City, fays *Leunclavius*, and pour'd into *Caramania*. Sultan *Muftapha*, Son of *Mahomet*, defeated him in 1473, and fent him Prifoner to his Father, who was at *Conftantinople*.

WE in vain fought for Company to go to *Cæfarea* of *Cappadocia*. This City is but fix Days Journey from *Tocat*, and has not chang'd its Name; for the *Greeks* call it *Kefaria* ever fince the time of *Tiberius*, who chang'd the antient Names of *Euzebia* and *Mazaca*. *Cæfarea* had the Happinefs to have the great St. *Bafil* for its Paftor; and its Arch-bifhop to this Day holds the firft Rank among the Prelates who are under the Patriarch of *Conftantinople*. They affur'd us there were Infcriptions at *Cæfarea*, which made mention of St. *Bafil*; but we could not go out of the Country of *Tocat*. This Country produces a great many fine Plants, and efpecially Vegetations of Stone, of a furprizing Beauty. We found ftrange things in breaking of Pebbles and Pieces of Rocks, cover'd over with Chryftallizations, which were very charming. I have fome of them in my Cabinet which are like the candied Citron-Peel; fome are fo like Mother of Pearl, that one may eafily take them for thofe Shells petrified. Some are of a Gold-colour, which differ only in their Hardnefs from candy'd Orange-Chips.

THE River which paffes by *Tocat* is not the *Iris* or *Cafalmac*, as Geo-graphers fuppofe; but the *Tofanlu*, which paffes alfo by *Neocæfarea*, and without doubt is the *Loup* which *Pliny* mentions, and which throws itfelf into the *Iris*. This River does much mifchief in time of great Rains, and when the Snows melt. They told us there are three Rivers which unite towards *Amafia*, the *Couleifar-fou*, or *the River of* Cho-nac; the *Tofanlou*, or that of *Tocat*; and the *Cafalmac*: this laft keeps its Name even to the Sea.

WE fet out from *Tocat* to *Angora* the 10th of *October* 1701, with a Caravan made up of new Comers, and thofe we had follow'd to *Tocat*. Thefe new Comers had been four and twenty Days coming from *Gangel* to *Erzeron*, and confequently had made their Journey fix Days longer than otherwife they had need, to avoid the Taxes at *Teflis*, where they

pay

pay very confiderable Duties. They had with them feventy five Hor- Lett. IX.
fes or Mules laden with 150 Bales of Silk, which weigh'd each fix and
twenty Batmans. At going out of *Tocat*, we entred upon a fine Plain,
in which the River winds: This perhaps is the Plain which *Paulus Jovius* calls the *Fields of the Geefe*, wherein the Battel was fought between the Troops of *Mahomet* II. and thofe of *Uzum-Caffan*, King of
Perfia.

AFTER travelling four Hours, we encamp'd near the Village of *A-
gara*, in whofe Churchyard are feen fome Pieces of antient Columns and
Cornifhes of white Marble, and of a fine Profil, but without Infcriptions.
All the Mountains round about are of Marble, as at *Tocat*. The Bole, I
doubt not, is plentiful, for there are Places very fteep and perpendicular, which are of a bright red, like the Rocks of which *Paulus Jovius* fpeaks, in the Caverns whereof *Techellis*, the famous *Mahometan*, Difciple of *Hardual* the great Interpreter of the Law, retir'd, to give himfelf up to Meditation and Prayer, and to efcape the Perfecutions of thofe
who oppos'd the Doctrines of his Mafter.

THE 11th of *October* we continued our Route in the Plain of *Tocat*,
which grows narrower within fix Miles on this fide of *Turcal*, and widens
again as we come nearer to it. *Turcal* is a fine Borough, fifteen Miles
from *Agara*, fituate round and on the top of a fteep Rock, feparate from
others about it, terminated by an old Caftle, and water'd at bottom by
the River of *Tocat*. All this Part is full of good Vineyards, the Fields
are well cultivated, the Villages numerous, and Pieces of antique Columns are common in their Churchyards, which is a fign the Country was
formerly inhabited by rich People. When we are pafs'd *Tocat*, we hear no
more of the *Curdes*, but enough of the *Turcmans*, that is to fay, of another
kind of Robbers more dangerous than the former, becaufe the *Curdes* fleep
in the Night, but the *Turcmans* rob both Night and Day. However we
encamp'd without any fear in the Plain half a League below *Turcal*. The
next Day we enter'd upon a very narrow Valley, bounded by a confiderable Mountain, from whence we defcended into another winding
Valley, where our Caravan ftopt. The whole Country is very pleafant,
and cover'd with Woods, but the Pines and Oaks are fmaller than in
other Places. The River of *Tocat* runs towards the North at *Turcal*, and

throws itfelf into the *Cafalmac* towards *Amafia*. We left it to the Right, to follow the Road to *Angora*, and met with nothing remarkable all the reft of our way to the City. We heard the Partridge, and Game of all forts is there Plenty enough, as likewife in all parts of *Natolia*.

THE next day we faw nothing but Oaks and Pines for nine Hours Journey, fometimes in fmall Valleys, and fometimes on Mountains of a confiderable Height. We faw but one pretty large Plain, wherein is the Village *Geder*, upon a fmall River of the fame Name. When we were paft this Village, there was nothing but fteep Rocks to the Right and Left, adorn'd with fome Thickets.

THE 14th of *October* the Landskip was the fame as the Day before, but our Journey was but of about five Hours. We encamp'd in a pleafant Plain near the Village of *Emar-Pacha*. All the Thiftles were cover'd with a very pretty fort of fmall *Buccinum*, only one inch long, and three or four lines in diameter, almoft cylindrical, greyifh, turn'd like a Skrew in nine narrow Windings, and ending in an obtufe Point. The Mouth of this Shell is more remarkable than all the reft; it is turn'd to the right, two lines and a half long, pointed at bottom, rounding towards the top, and adorned with two or three Teeth. This Shell is common in the Ifles of the *Archipelago*; and *Columna* has caus'd one to be engrav'd, which is very much like this we are fpeaking of. Tho it does not feem to be any thing extraordinary, that thefe Shells fhould have their Mouths turn'd to the right or left, yet it is very certain that the Author of Nature has made very few of thefe Shells with their Mouths and Windings turn'd to the right; and the Curious are very defirous of fuch. Among a great number of forts of *Buccinum*, which I have in my Cabinet, there are not above three or four which have the Mouth and Winding turn'd in this manner; namely, the fmall one we have been fpeaking of, another kind of about two inches long, and one thick, of a fhining yellow, or marbled with oblique tawny and yellowifh Bands or Stripes, white round the Mouth. The moft confiderable is all tawny, five inches high, and two thick, with a Mouth which has no Border or Ledge; whereas the others have the Mouth rais'd with a fort of Border, and the Winding is eight or nine times round.

†

THE 15th of *October* we travell'd thro horrid Defiles which run in-to a fine Plain. After eight Hours Journey, we encamp'd below *Sike*. The next Day we pitch'd our Tents near *Tekia*, another Village, four Hours from the former, and in the same Plain. All the Country is plea-sant, and well cultivated. The wild Pear-trees are cover'd over with Misletoe; and I obferv'd upon their Trunks, tho the Bark was hard, the first shootings of the Seed, which I had long fought, but could never find in *France*, where this Plant is so common. These Seeds, which are of the shape of a Heart, were out of their Cafes, and stuck by their Clamminefs to the Trunks and Branches of thefe Trees, when the Wind, or any other Caufe shook them out. Each Seed was laid in such manner, that the Point of the Root began to pierce into the Bark, whilft the Eye of the Seed shot out and unfolded itfelf. All this confirm'd me in my Opinion, which I had mentioned concerning the Multiplication of Misletoe, in my *History of Plants which grow about Paris*.

OUR Journey of the 17th of *October* was about twelve Hours. We pafs'd this Day thro nothing but small Vales cover'd with Oaks and Pines. The next Day the Profpect was very different, for we travell'd nine Hours in a flat Country, meanly cultivated, without Trees or Bushes, with some small Rifings full of foffile Salt. This Salt, which is chryftal-liz'd in Bottoms where the Rain-water ftagnates, mixes with the Moi-fture of the Earth, and caufes it to produce fuch Plants as love the Sea-fide, fuch as the *Salt-wort* and *Limonium*. I obferv'd the fame thing upon the Mountain of *Cardonna*, fituate on the Frontiers of *Catalonia* and *Ar-ragon*, which is nothing but a prodigious Mafs of Salt.

THE 19th of *October* we quitted this Salt Country, to enter again into Valleys and Plains, cover'd with divers forts of Oaks. We encam-ped near the Village of *Beglaife* after feven Hours Journey. The Route of the next Day was of twelve Hours, in Plains divided by fmall Hills, adorn'd with Woods of Oaks with Leaves like to ours, tho they don't grow much higher than our Underwoods. We this Day forded the River *Halys*, or the *Cafilrimac* of the *Turks*, which turns its Courfe towards the North, by reafon of a Mountain directly oppofite to the great Road. The *Cafilrimac* is not deep, but it feem'd as wide as the *Seine* at *Paris*,

and

and they told us that it runs but one Day's Journey from *Cefarea*. From the top of this Mountain, we fell, as I may fay, into a horrible Bottom, and ftopp'd at the Village *Courbaga*. Hence the Country is very rugged and unpleafant, till within two Leagues of *Angora*. We arriv'd at this famous City the 22d of *October*, after four Hours Journey, thro a Valley very well cultivated in many Places.

A N G O R A, or *Angori*, as fome pronounce it, which the *Turks* call *Engour*, delighted us more than any other City in the *Levant*. We imagin'd the Blood of thofe brave *Gauls*, who formerly poffefs'd the Country about *Touloufe*, and between the *Cevennes* and the *Pyrenees*, ftill ran in the Veins of the Inhabitants of this Place. Thofe generous *Gauls*, confin'd in their own Country too much for their Courage, fet out to the number of thirty thoufand Men, to go and make Conquefts in the *Levant*, under the Conduct of many Commanders, of whom *Brennus* was Chief. Whilft this General ravag'd *Greece*, and plunder'd the Temple of *Delphos* of its immenfe Riches, twenty thoufand Men of this Army march'd into *Thrace* with *Leonorius*, who, as a *Gaul*, doubtlefs call'd himfelf *Leonorix*; and I would willingly, to accommodate the Name to our Language, call *Leonor*. One might fay the fame of the other Chief who followed him: the *Latin* Authors call him *Lutarius*, from the Word *Lutarix*, which anfwers much better to our old *French* Terminations.

THESE two Chiefs fubdued the whole Country to *Byzantium*, and went down to the *Hellefpont*. Glad to find that *Afia* was not feparatet from *Europe* but by an Arm of the Sea, they fent to *Antipater*, who commanded on the Coaft of *Afia*, and who might oppofe their Paffage. This Affair went on but flowly, and probably *Antipater* thought he could not well agree with fuch fort of Guefts: the two Kings feparated themfelves. *Leonorius* return'd to *Byzantium*. *Lutarius* fome time after receiv'd an Embaffy from the *Macedonians*, fent by *Antipater* in two Ships and three Shallops. Whilft they obferv'd the Troops of the *Gauls*, *Lutarius* loft no time, but pafs'd them over into *Afia* Night and Day in thofe Veffels. *Leonorius* haftned into *Bithynia*, with his Forces, being invited thither by King *Nicomedes*, who made confiderable Ufe of thefe two Bodies of *Gauls* againft *Zipoetes*, who then poffefs'd Part of his Country.

*

THE

ANGORA

THE *Gauls* fpread Terror all over *Afia*, even to Mount *Taurus*, as we learn from *Titus Livy*, whom I follow clofe in this Expedition. Of the twenty thoufand *Gauls* who went from *Greece*, there remain'd hardly more than half the Number; but all things gave way to their Valour, and they put the whole Country under Contribution. In fine, there being three forts of *Gauls* among them, they divided their Conquefts in fuch manner, that one fort fix'd upon the Coaft of the *Hellefpont*; another inhabited *Æolia* and *Ionia*; and the moft famous, who were called *Tectofages*, penetrating further, extended themfelves to the River *Halys*, one Day's Journey from *Angora*, which is the antient *Ancyra*. This River is reprefented upon a Medal of *Geta*, under the form of an old Man lying half along, holding a Reed in his right Hand. Thus our *Touloufians* poffefs'd *Phrygia major* to *Cappadocia* and *Paphlagonia*; and all the Country thro which they had fpread themfelves, was call'd *Galatia* or *Gallo-Grecia*, as much as to fay, *Greece of the Gauls*. *Strabo* affirms, that they divided their Conquefts into four Parts, that every one had its King and Officers Civil and Military; and above all, that they continued to do Juftice in the midft of a Wood of Oaks, according to the Cuftom of their Anceftors: there was no want of this fort of Trees about *Ancyra*. *Pliny* makes mention of feveral People among the *Gauls*, who perhaps bore the Name of their Chiefs: it is probable they were only larger Divifions of the fame People.

MEMNON reports, that the *Trocmian Gauls* built the City of *Ancyra*, but I believe this Paffage of that Author is corrupted in the Extract *Photius* has given us of it; for befides that they fix'd themfelves upon the Coafts of *Phrygia*, *Pliny* fays exprefly, that *Ancyra* was the Work of the *Tectofages*. The following Infcription, which is upon a Column, fet in the Wall of this City, between the *Smyrna* Gate and that of *Conftantinople*, mentions only the *Tectofages*, and does them a great deal of Honour.

Η ΒΟΥΛΗ ΚΑΙ Ο ΔΗ-	*Senatus Populufque*
ΜΟΣ ΣΕΒΑΣΤΗ	*Sebaftenorum*
ΝΩΝ ΤΕΚΤΟΣΑ-	*Tectofagum*
ΓΩΝ ΕΤΙΜΗΣΕΝ	*honoravit*

M. KOK

Lett. IX.

M. KOKKHION	*M. Cocceium*
ΑΛΕΞΑΝΔΡΟΝ ΤΟΝ	*Alexandrum*
ΕΑΥΤΩΝ ΠΟΛΙΤΗΝ	*Civem fuum*
ΑΝΔΡΑ ΣΕΜΝΟΝ ΚΑΙ	*virum honorabilem*
ΤΩΝ ΗΘΩΝ ΚΟΣΜΙΟ·	*Et morum elegantia*
ΤΗΤΙ ΔΟΚΙΜΩΤΑΤΟΝ.	*Spectabiliſſimum.*

MOREOVER, when *Manlius*, the *Roman* Conful, had defeated a Party of the *Gauls* at Mount *Olympus*, he came to attack the *Tectofages* at *Ancyra*. It is probable the *Tectofages* did only rebuild this City; for long before their coming into *Aſia*, *Alexander* the Great gave Audience here to the Deputies from *Paphlagonia*. 'Tis furprizing that *Strabo*, who was of *Amaſia*, has made no mention of *Ancyra* but only as a Caftle of the *Gauls*, tho he liv'd under *Auguſtus*, to whom they confecrated in the middle of *Ancyra* that fine Building of Marble, which I fhall fpeak of prefently. Perhaps *Strabo* was not pleas'd with the *Gauls*, who, it may be, had us'd the Inhabitants of *Amaſia* but ill. *Titus Livy* is more juft to *Ancyra*, and calls it an *Illuſtrious City.*

OF all the Kings of *Aſia*, *Attalus* was the only one who vigoroufly oppos'd the *Gauls* in their Enterprizes, and had the good Luck to beat them; but they fupported themfelves powerfully till the Defeat of *Antiochus* by *Scipio*. The *Gauls* made the beft part of the Troops of this Prince, and flatter'd themfelves that the *Romans* would not penetrate fo far as into their Country: But the Conful *Manlius*, under pretence that they had affifted *Antiochus*, declared War againft them, and defeated them at Mount *Olympus*. He penetrated even to *Ancyra*, which he took, according to *Zonaras*, and oblig'd them to accept of Peace upon his own Terms. The four Provinces of *Galatia* were reduc'd to three, fays *Strabo*; afterwards to two; and then to one Kingdom, over which the *Romans* put *Deiotarus*: His Son *Amyntas* fucceeded him. At length *Lelius Marcus* fubdu'd *Galatia* under *Auguſtus*. It was reduced to a Province, and taken from *Pylemenes*, Son of *Amyntas*. The Name *Pylemenes* was fo common to the Kings of *Paphlagonia*, that this Province was called *Pylemenia*. Thus ended the Empire of the *Galatians*, who had made even the Kings of *Syria* their Tributaries; without whom the Kings of

Aſia

Afia could not make War, and who fupported the Majefty of Kings, as Lett. IX. *Juftin* expreffes himfelf.

THE Emperor *Auguftus* did, no doubt, beautify *Ancyra*, feeing *Tzetzes* calls him the Founder of it; and it was probably in acknowledgment that the Inhabitants confecrated to him the greateft Monument ever yet in *Afia*. You fhall judge, my Lord, of this Beauty of the Building by the Defign of it, which you commanded me to take. It was all of white Marble, in large Pieces; and the Corners of the *Veftibulum*, which yet remain, are alternately of one Piece, returning with a Corner, in manner of a Square; the Sides or Legs of which are three or four feet long. Thefe Stones are moreover cramp'd together with Pieces of Copper, as appears by the Hollows in which they lay. The chief Walls are ftill thirty or five and thirty feet high. The Front is entirely deftroy'd; there remains only the Door by which they went out of the *Veftibulum* into the Houfe. This Door, which is fquare, is twenty four feet high, and nine feet two inches wide; and its Pofts, which are each of one Piece, are two feet three inches thick. On the fide of this Door, which is full of Ornaments, was cut above feventeen hundred Years ago the Life of *Auguftus* in fine *Latin*, and handfome Characters. The Infcription is in three Columns on the Right and Left: But befides the defac'd Letters, 'tis full of great Hollows, like thofe wherein they caft Bullets for Cannon. Thefe Hollows, which have been made by the Peafants, to get out the Pieces of Copper with which the Stones were cramp'd together, have deftroy'd half the Letters. The Facings of Stone are of an oblong Square, very neat, jetting out one inch. Without reckoning the *Veftibulum*, this Building is within-fide fifty two feet long, and thirty fix and a half wide. There remain ftill three grated Windows of Marble, with great Squares, like thofe of our Windows. I don't know how thefe were furnifh'd, whether with a tranfparent Stone, or with Glafs.

ONE fees within the Circumference of this Building the Ruins of a poor Chriftian Church, near two or three forry Houfes, and fome Cowhoufes. This is whar the Monument of *Ancyra* is come to; which was not a Temple of *Auguftus*, but a Publick Houfe, or *Prytaneum*, wherein they ate on the great Feafts of the publick Games, which were frequently celebrated in this Place, as appears by the Medals of *Nero*, *Caracalla*,

Decius,

Decius, *Valerianus* the elder, *Gallienus*, and *Saloninus*. The Legends fhew the Games wherein they exercis'd themfelves.

WE might perhaps difcover fomething more particular concerning this Edifice, if we could find out the meaning of divers *Greek* Infcriptions which are cut on the out-fide of the Walls; for this Building undoubtedly ftood alone. At prefent we find thefe Infcriptions in the Chimneys of feveral particular Houfes, where they are cover'd with Soot. Thefe Houfes ftand againft the chief Wall on the Right.

THE Infcription we mention'd above, which contains the Life of *Auguftus*, is to be found in the *Monumentum Ancyranum Gronovii*, and in *Gruter*. *Leunclave* had it of ² *Clufius*, who, befide the great Knowledge he had in Plants, was well acquainted with Antiquity: And *Fauftus Verantius*, who communicated this valuable Piece to *Clufius*, had it from his Uncle *Antonius Verantius*, Bifhop of *Agria*, and Ambaffador of *Ferdinand* II. to the Porte. This Prelate caus'd it to be tranfcrib'd as he paffed by *Angora*. *Busbequius* took a Copy of it; and fancies the Houfe we fpeak of was rather a *Prætorium*, than a Houfe defign'd for the Feafts of the publick Games.

² Charles de l'Eclufe.

WHAT we have been faying, fufficiently fhews that *Ancyra* was one of the moft illuftrious Cities of the *Levant*. Its Inhabitants were the principal *Galatians*, whom St. *Paul* honour'd with an Epiftle; and the Councils which have been there held, make it as confiderable among Chriftians, as any other Things which have been there tranfacted. It appears by the Medals of *Ancyra*, that it fupported its Honour under the *Roman* Emperors. There are fome with the Heads of *Nero*, *Lucius Verus*, *Commodus*, ³ *Caracalla*, *Geta*, *Decius*, *Valerianus*, *Gallienus*, *Saloninus*. *Ancyra* took the Name of *Antoniniana* in acknowledgement of the many Favours heap'd upon it by *Antoninus Caracalla*. It was declared the Metropolis, that is, the Capital of *Galatia*, under *Nero*, and has always preferv'd that Title. There is mention made of it on a Medal of *Antinous*, and of *Julius Saturninus* one of its Governors. He is nam'd in the following Infcription, which is upon Marble fet in the Walls of the City. *Gruter* gives it thus:

Monumentum Ancyranum.

188.

ΑΓΑΘΗΙ ΤΥΧΗΙ	*Bonæ fortunæ*
Η ΜΗΤΡΟΠΟΛΙΣ	*Metropolis*
ΙΟΥΛΙΟΝ	*Julium*
ΣΑΤΟΡΝΕΙΝΟΝ	*Saturninum*
ΤΟΝ ΗΓΕΜΟΝΑ.	*Ducem.*

THE Name of Metropolis is also to be found upon a Tomb-stone in the Church-yard belonging to the Christians without the City.

Λ. ΦΟΥΛΟΥΙΟΝ ΡΟΥ	*Lucium Fulvium*
ΣΤΙΚΟΝ ΑΙΜΙΛΙΑ-	*Rusticum Æmilianum*
ΝΟΝ ΠΡΕΣΒ. ΣΕΒΑ..	*Legatione functum*
ΤΗΣ ΤΡΑΥΠΑΤΟΝ Η ΒΟΥ	*ter Proconsulem*
ΛΗ ΚΑΙ ΔΗΜΟΣ ΤΗΣ ΜΗ	*Senatus Populusque*
ΤΡΟΠΟΛΕΩΣ ΑΓΚΥ-	*metropoleos Ancyræ*
ΡΑΣ ΤΟΝ ΕΑΥΤΩΝ	*Benefactorem suum;*
ΕΥΕΡΓΕΤΗΝ ΕΠΙΜΕ-	*Curante Trebio*
ΛΟΥΜΕΝΟΥ	*Alexandro.*
ΤΡΕΒΙΟΥ ΑΛΕΞΑΝΔΡΟΥ.	

° For τςὶς Ἀν. θὐπαιον.

THE following is cut on a Pedestal, which serves for a Trough in the Caravansera where we lodg'd.

ΔΙΙ ΗΛΙΩ ΜΕΓΑΛΩ ΣΑΡΑΠΙΔΙ ΚΑΙ ΤΟΙΣ ΣΥΝ-
ΝΑΙΟΙΣ ΘΕΟΙΣ ΤΟΥΣ ΣΩΤΗΡΑΣ ΔΙΟΣΚΟΥΡ-
ΟΥΣ ΥΠΕΡ ΤΗΣ ΤΩΝ ΑΥΤΟΚΡΑΤΟΡΩΝ ΣΩΤΗ-
ΡΙΑΣ ΚΑΙ ΝΕΙΚΗΣ ΚΑΙ ΑΙΩΝΙΟΥ ΔΙΑΜΟΝΗΣ Μ
ΑΥΡΗΛΙΟΥ ΑΝΤΩΝΕΙΝΟΥ ΚΑΙ Μ. ΑΥΡΗ-
ΛΙΟΥ ΚΟΜΜΟΔΟΥ ΚΑΙ ΤΟΥ ΣΥΜΠΑΝΤΟΣ
ΑΥΤΩΝ ΟΙΚΟΥ ΚΑΙ ΥΠΕΡ ΒΟΥΛΗΣ ΚΑΙ
ΔΗΜΟΥ ΤΗΣ ΜΗΤΡΟΠΟΛΕΩΣ ΑΓΚΥΡΑΣ.
ΑΠΟΛΛΩΝΙΟΣ ΑΠΟΛΛΩΝΙΟΥ.

Jovi Soli magno Sarapidi & ejusdem
Templi Diis; servatores Dioscuros

Pro salute Imperatorum
Et victoria & perennitate
M. Aurelii Antonini & M. Aure-
lii Commodi & pro universa
ipsorum domo & pro Senatu
Populoque metropoleos Ancyræ,
Apollonius Apollonii F.

THIS is found on the Walls of a square Tower, between the Gate of the Gardens, and the Gate of *Esset.*

Caracylæam,	ΚΑΡΑΚΥΛΑΙΑΝ
Sacerdotum principem,	ΑΡΧΙΕΡΕΙΑΝ
ex regibus ortam,	ΑΠΟΓΟΝΟΝ ΒΑ
filiam Metropoleos,	ΣΙΛΕΩΝ ΘΥΓΑ-
Uxorem Julii	ΤΕΡΑ ΤΗΣ ΜΗΤΡΟ-
Severi	ΠΟΛΕΩΣ ΓΥΝΑΙ-
Græcorum primi.	ΚΑ ΙΟΥΛΙΟΥ ΣΕ
	ΟΥΗΡΟΥ ΤΟΥ ΠΡΩ-
	ΤΟΥ ΤΩΝ ΕΛΛΗ-
	ΝΩΝ *ΥΠΕΡΡΑ.

ΑΝΚΥΡΑΣ
ΜΗΤ. Β. Ν.
Ancyra Me-
tropolis bis
Neocoræ.

THE Legend of a Medal of the elder *Valerianus* notes that *Ancyra* was twice *Neocore.* It received this Honour the first time under *Caracalla,* and the second time under *Valerianus* the elder. The Reverse of this Medal represents three Urns, out of each of which spring two Palms.

THE *Greeks* call those *Neocores,* who have the Care of the Temples, common to a whole Province, and wherein they assembled on occasion of the publick Games. This Charge of *Neocore* answer'd almost to that of Churchwarden: But when afterwards they took to deifying of the Emperors, those Cities which asked Permission to prepare Temples in their Honour, were likewise called *Neocores.*

THE Situation of *Ancyra* in the middle of *Asia minor,* has frequently expos'd it to great Ravages. It was taken by the *Persians* in 611, in the time of *Heraclius,* and ruin'd in 1101, by that dreadful Army of *Normans* or *Lombards,* as M. *du Cange* will have it, commmanded by

' Alexiad.
lib. xi.
' Notæ in
Alexid.

*

Tzitas

Tzitat and the Count *de S. Gilles*, who was afterwards known by the Name of *Raimond*, Count of *Touloufe* and *Provence*, at the time when Baldwin, Brother of *Godfrey* of *Bologne*, was chofen King of *Jerufalem*. This Army, which confifted of an hundred thoufand Foot, and fifty thoufand Horfe, after the Expedition of *Angora*, paffed the River *Halys*; but was fo beaten by the *Mahometans*, that the Generals found a great deal of difficulty to retire to *Conftantinople* near *Alexis Comnenus*.

THE *Tartars* made themfelves Mafters of *Ancyra* in 1239. It was afterwards the chief Seat of the *Ottomans*; for *Orthogul*, Father of the famous *Ottomans*, fettled himfelf here; and his Succeffor feized not only *Galatia*, but likewife *Cappadocia* and *Pamphylia*. *Angora* was fatal to the *Ottomans*, and the Battel which *Tamerlane* obtain'd there over *Bajazet*, had well nigh deftroy'd their Empire. *Bajazet*, the haughtieft Man in the World, too confident in himfelf, left his Camp to go a hunting. *Tamerlane*, whofe Troops began to want Water, laid hold on this Opportunity, and rendring himfelf Mafter of the fmall River which run between the two Armies, three Days after forc'd *Bajazet* to give him Battel, to prevent his Army from dying of Thirft. His Army was cut to pieces, and the Sultan taken Prifoner, the 7th of *Auguft*, 1401. After the Retreat of *Tamerlane*, the Children of *Bajazet* retir'd whither they could. *Mahomet* fecured to himfelf *Galatia*, which his Brother *Efes* had difputed with him: He made ufe of *Temirte*, an old Captain, who had ferv'd under *Bajazet*; and *Temirte* overcame *Efes* at *Angora*, and caufed his Head to be cut off.

ANGORA, at prefent, is one of the beft Cities in *Anatolia*, and every where fhews Marks of its antient Magnificence. One fees nothing in the Streets but Pillars and old Marbles; among which there is a Species of reddifh Porphyry, mark'd with White, like that at *Pennes*, near *Marfeilles*. One finds likewife at *Angora* fome Pieces of red and white *Jafper*, with large Spots, like that of *Languedoc*, The greateft Part of the Pillars are fmooth and cylindrical; fome are channelled fpirally; the moft fingular are oval, adorn'd with a *Plate-band* before and behind, which alfo runs all along the Pedeftal and the Capital. They feem'd to me beautiful enough to be engrav'd: I think no Architect has fpoken

X x 2

of

of this Order. There is nothing fo furprizing as the Steps of the Door of a Mofque: They are fourteen in Number, and confift only of Bafes of Marble-Pillars, plac'd one upon another. Tho at prefent the Houfes are made of Clay, yet one fees in them oftentimes very fine Pieces of Marble.

THE Walls of the City are low, and furnifh'd with very forry Battlements. They have indifferently made ufe of Pillars, Architraves, Capitals, Bafes, and other antient Pieces, intermingled with Mafonry, to build the Wall, efpecially in the Towers and Gates, which neverthelefs are not at all the more beautiful; for the Towers are fquare, and the Gates plain. Tho they have put many Pieces of Marble into this Wall with the Infcriptions inwards, there are however many whofe Infcriptions may be read: They are moftly *Greek*, and fome *Latin*, *Arabick*, or *Turkifh*. The following Infcription is very near certain Lions of Marble, very much disfigur'd at the Port of *Kefaria*.

ΚΑΙΡΕ ΠΑΡΟΔΕΙΤΑ.   *Salve Viator.*

UNDERNEATH thefe Words is a Head in *Bas-relief*, of which we know nothing; but underneath are the following Words:

ΜΑΡΚΕΛΛΟC	*Marcellus*
CΤΡΑΤΟΝΕΙΚΗ	*Stratonice*
ΓΛΥΚΥΤΑΤΗ Γ	*Dulciſſimæ*
ΥΝ . . . . . ΜΝΗΜΗC	*Conjugi Memoriæ*
ΧΑΡΙΝ	*Cauſa*

AT the Port of the Gardens one reads the following Infcription

ΑΓΑΘΗΙ ΤΥΧΗΙ
ΤΟΡΝΕΙΤΟΡΙΑΝΟΝ, ΕΠΙΤΡΟΠΟΝ ΤΩΝ ΚΥΡΙ
ΩΝ ΗΜΩΝ ΕΠΙ ΑΘΥΛΩΝ
ΤΟΝ ΔΙΚΑΙΟΝ ΚΑΙ ΣΕΜΝΟΝ Κ ΑΙΛΙΟΣ
ΑΓΗΣΙΛΑΟΣ ΤΟΝ ΕΑΥΤΟΥ ΦΙΛΟΝ ΚΑΙ
ΕΥΕ . . . .

Bonæ

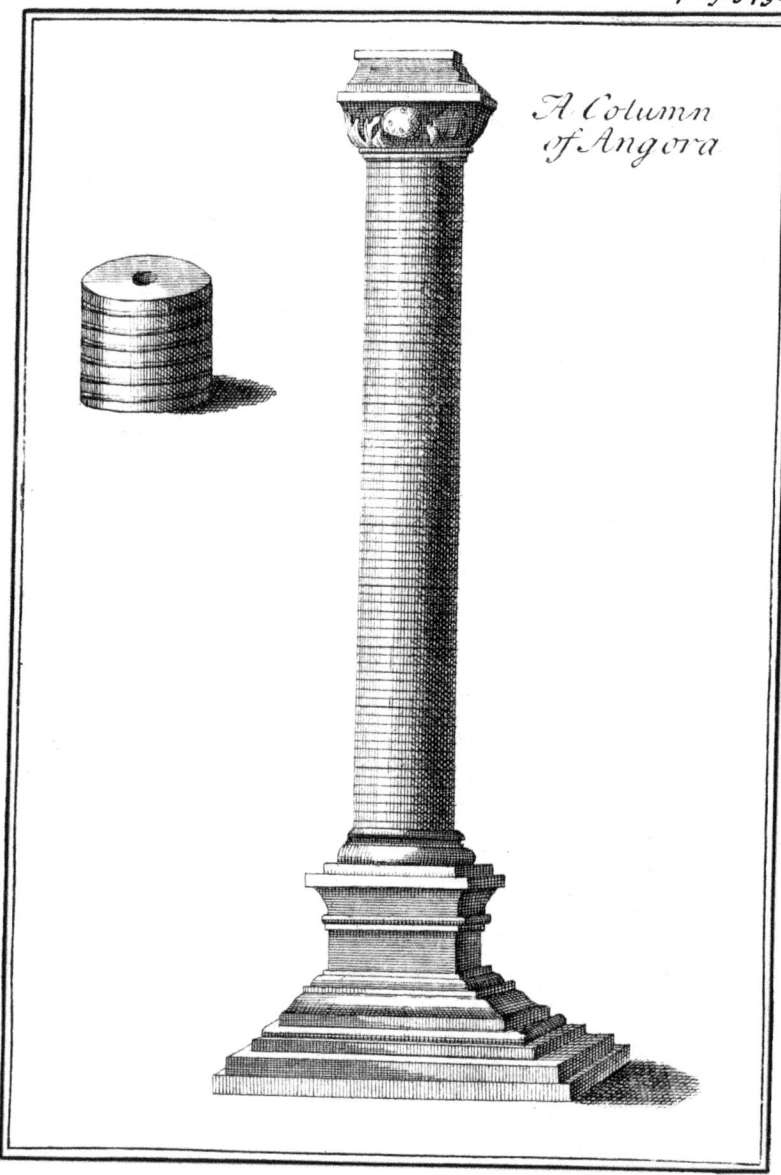

A Column
of Angora

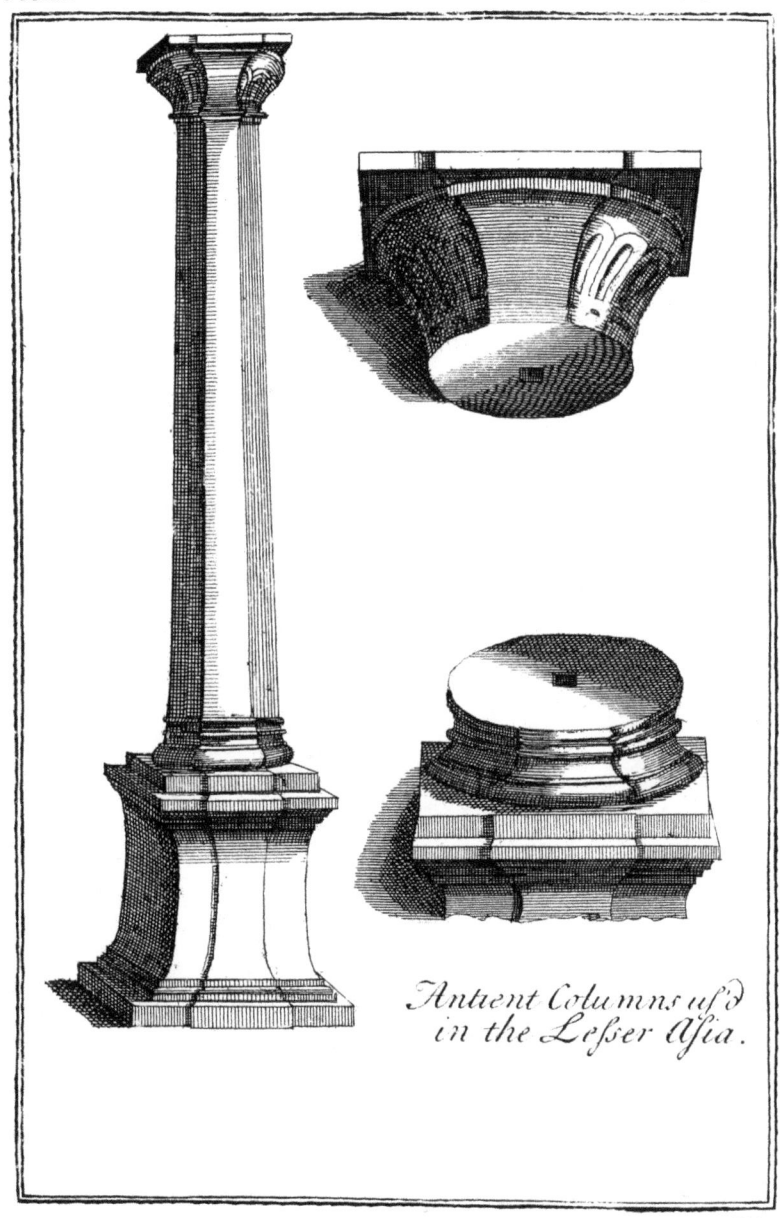

Antient Columns us'd
in the Leſſer Aſia.

Bonæ fortunæ
Tornitorianum curatorem Domi-
norum noſtrorum . . . .
juſtum & illuſtrem, C. Ælius
Ageſilaus amicum ſuum &
beneficum..

WE read below the Tower, as we paſt to the Port of *Eſſet,* upon a Pillar work'd into the Wall, theſe following Words;

I M P.   C Æ S.
. . . . . .
E T   I M P R O . . . .
G A L L I E N O

THE reſt is on that Part of the Pillar which is in the Wall..

THERE remain three Medals ſtamp'd with the Head of this Empe-ror, and with the Legend of *Ancyra,* wherein this City is called a Metro-polis. The Reverſe of the firſt repreſents three Urns with Palms; that of the ſecond a Wolf, with *Romulus* and *Remus* ſucking: On the third is the Figure of *Apollo* ſtanding, and naked, holding in his Right-hand a Crown, and leaning his Left-Elbow on a Pillar, which has on it a Lyre. There is a fourth, in the King's Cabinet, with the ſame Reverſe as the firſt;, but the Legend expreſſes that the City is *Neocore* for the ſecond time.

THE three Lions which are at the *Smyrna* Port are handſome enough. There is upon a Piece of a broken Architrave, which ſerves for the Lintel of a Door, the following imperfect Line, written in great Characters;

. . . Β Α Σ Τ Ω   Ε Υ Σ Ε Β Ε Ι   Ε Υ Υ . . . .

I WILL ſet down ſome other Inſcriptions which are to be met with on the ſame Walls between the *Smyrna* Port, and that of *Conſtantinople.*

UPON.

UPON a Pedeſtal :

ΘΕΟΙΣ ΚΑΤΑΧΘΟΝΙ-	*Dis manibus*
ΟΙΣ ΚΑΙ ΚΑΠΙΤΟΝΙ	*Et Capitoni*
ΠΑΣΙΚΡΑΤΟΥΣ	*Paſicratis F.*
ΑΝΔΡΙ ΓΕΝΝΑΙΩ	*Viro generoſo*
ΚΑΙ ΑΓΑΘΩ ΠΟΥ	*& probo Pu-*
ΒΛΙΟΣ ΑΔΕΛΦΟΣ	*blius frater*
ΑΥΤΟΥ ΚΑΙ ΠΑΣΙ	*ejus & Paſi-*
ΚΡΑΤΗΣ ΚΑΙ ΜΗ-	*crates & Me-*
ΝΟΔΩΡΟΣ ΥΙΟΙ	*nodorus filii*
ΑΥΤΟΥ ΠΕΡΤΙΝΗ	*ejus ....*
¦ΜΝΗΜΗ ΕΙΧΑ	*Memoriæ gratia.*

¹ For μνήμης ἕνεκα.

UPON another Pedeſtal adorn'd with a Feſton ;

D. M.
VENTIDIA CAR
PILLA
VIXIT ANNIS
XXXIII M VIII
D VI
T. LIVIUS CARPUS
PATER EJ....
DIONYSIUS UXORI CARISSIMÆ.

ON the Inſide of the ſame Walls,

ΔΙΟΤΕΙΜΟϹ ΔΙ	*Diotimus Dio*
ΟΤΕΙΜΟ ΚΑΙ ΛΟ	*timo & Lotatio*
ΤΑΤΙΟ ΙΔΙΟΙϹ	*propriis*
ΓΟΝΕΥΣΙ ΜΝΗ	*parentibus*
ΜΗϹ ΧΑΡΙΝ	*memoriæ gratia.*

IN the fame Place upon a Stone fet in the Wall:

EUTYCHUS
NEREI
CAESARIS
AUG.
SER. VIC.
FILIO.

THE Caftle of *Angora* has a triple Enclofure, and the Walls arc of large Pieces of white Marble, and a Stone much like to Porphyry. They fuffer'd us to go all over it; and they carried us in the firft Enclofure to an *Armenian* Church, built, as they pretend, under the Name of the Crofs, twelve hundred Years ago. It is very fmall and dark, enlighten'd partly by a Window, which receives the Light only thro a great fquare piece of Marble like to Alabafter polifh'd, and fhining like Ifinglafs; but it is dull within, and the Light which paffes thro is fenfibly reddifh, and inclining to a Cornaline. The Sun did not fhine on it when we obferv'd it; it is perhaps of that fort of Marble, which *Pliny* calls *Sphingites.* This whole firft Enclofure is full of Pedeftals and Infcriptions; but what part of *Angora* is without them? A good Antiquary would find what would employ a whole Year to tranfcribe. We copied out the following.

THE Infcription, which mentions *Julian* the Apoftate, is upon a Stone fix'd in the Wall and plaifter'd, the Letters are very ill made.

DOMINO TOTIUS ORBIS
JULIANO AUGUSTO
EX OCEANO BRI
TANNICO VIS PER
BARBARAS GENTES
STRAGE RESISTENTI

¹ *For* VIIS.

U M

UM PATEFACTIS-----

```
· ·  ·  ·  ·  ·  ·  ·
·  ·  ·  ·  ·  ·  ·  ·
·  ·  ·  ·  ·  ·  ·  ·
·  ·  ·  ·  ·  ·  ·  ·
·  ·  ·  ·  ·  ·  ·  ·
```

PROBABLY it was made at the time when this Emperor was at *Ancyra.*

UPON a Pedeſtal in the Encloſure of a Moſque of the ſame Caſtle:

ΤΑΦΟΝ ΤΟΝ	
ΕΝΘΑ ΠΛΗΣΙ-	*Sepulchrum hoc*
ΟΝ ΒΩΜΟΝ ΑΘ	*& aram ſimul*
ΜΑ ΕΤΕΥΞ ΚΑ-	*excitavit in terra*
ΤΑ ΤΗΣ ΚΛΑΥΔΙΑ Η	*Claudia, Dexas*
ΚΑΙ ΔΕΞΑΣ ΑΘΗ	*item vocata,*
ΝΙΩΝ ΓΛΥΚΥΤΑΤΩ	*Athenioni dulciſſimo*
ΚΑΙ ΦΙΛΤΑΤΩ ΑΓΝΩ	*& amabiliſſimo*
ΤΕΝΟΜΕΝΩ ΣΥΜ-	*Caſtoque Conjugi,*
ΒΙΩ ΜΝΗΜΗΣ	*Memoriæ cauſa.*
ΧΑΡΙΝ	

UPON a Pedeſtal in the Encloſure of the Caſtle:

ΑΠΟΛΛΩΝΙΟC ΕΥΤΥ-	*Apollonius Euty-*
ΧΟΥ ΚΛΑΥΔΙΑ ΙΟΥ-	*chis F. Claudiæ Ju-*
ΛΙΤΤΗ ΣΥΜΒΙΩ Α-	*littæ conjugi opti-*
ΓΑΘΗ ΤΟΝ ΒΩΜΟΝ	*mæ hanc aram*
ΚΑΙ ΤΗΝ ΟCΤΟΘΗ-	*& hoc monumen-*
ΚΗΝ ΜΝΗΜΗC ΧΑ-	*tum memoriæ cauſa*
ΡΙΝ ΛΝΕCΤΗ-	*poſuit.*
CΕΝ.	

UPON another Pedeftal in the fame Caftle:

ΑΡΧΗΣΑΝΙΑ
ΚΑΙ ΑΣΤΥΝΟ-
ΜΗΣ ΑΝΤΑΚΑΙ
ΙΕΡΑΣΑΜΕΝΟΝ
ΔΙΣ ΘΕΑΣ ΔΗΜΗ-
ΤΡΟΣ ΤΙΜΗΘΕΝ
ΤΑ ΕΝ ΕΚΚΛΗΣΙ-
ΑΙΣ ΠΟΛΛΑΚ
ΦΥΛΗ ΕΝΑΤΗ
ΙΕΡΑ ΒΟΥΛΑΙΑ
ΤΟΝ ΕΑΥΤΗΣ
ΕΥΕΡΓΕΤΗΝ.

UPON a Stone of an antient Building, which the *Turks* call *Me-ferefail*:

D. M.
Q. AQUILIO LUCIO
LEG. II AUG.
SEVERIA MAPTINV
LA CONIUNX. ET
AQUILIA SEVERINA
FILIA ET HERES
F. C.

IN the Chamber of a private Perfon who lodges in this Houfe, on a Stone behind the Door:

G. Longino Pau-	Γ. ΛΟΝΓΕΙΝΩ ΠΑΥ-
lino G. Longi-	ΛΕΙΝΩ Γ. ΛΟΝΓΕΙ-
nus Sagaris, &	ΝΟΣ ΣΑΓΑΡΙΣ, ΚΑΙ
G. Longinus	Γ. ΛΟΝΓΕΙΝΟΣ
Claudianus	ΚΛΑΥΔΙΑΝΟΣ

Patri, me-	ΠΑΤΡΙ ΜΝΗ.
moriæ cauſa.	ΜΗΣ ΧΑΡΙΝ.

IN the ſame Building, upon a Stone in the Wall:

Flavio Sabi-	ΦΛΑΟΥΙΩ ΣΑΒΕΙ.
no genere Nico-	ΝΩ ΓΕΝΕΙ ΝΕΙΚΟ
medienſi, Filia	ΜΗΔΕΙΗ ΘΥΓΑΤΗΡ
Cippum (ſupple poſuit)	ΤΗΝ ΣΤΗΛΗΝ
memoriæ cauſa.	ΜΝΕΙΑΣ ΧΑΡΙΝ.
Qui expilaverit	ΟΣΑΝ Δ ΕΣΚΥΛΗΤΟ
Sepulchrum dabit	ΜΝΗΜΑ ΔΩΣΕΙ ΕΙΣ
ad fiſcum denaria bis	Τ´ΟΝ ΦΙΣΚΟΝ. B̅. Φ.
mille quingenta.	

UPON three different Stones of the ſame Building:

D. M.
C. JUL. CANDIDO
P. P. LEG. XVII. GEM.
HEREDES EX TES
TAMENTO FECE
RUNT.

ΛΟΥΚΙΟΣ	*Lucius*
ΣΕΡΗΝΙΑ ΣΥΝΒΙΩ	*Sereniæ Conjugi*
ΑΝΕΣΤΗΣΑ ΜΝΗ	*erexi, memoriæ*
ΜΗΣ ΧΑΡΙΝ	*gratia, proſpere*
ΔΙ ΕΥΤΥΧΙΤΕ	*agite.*

D. M.
C. SECUNDI
NIO JULIANO
EQUITI LEG
XXII. PR. P. P. AN
N XXXV. STIP. XV.

C. SERA.

C. SERANIVS VE
CTIVS SECVNDVS
HERES ET CONLEGA
F. C.

THE Churchyard belonging to the Chriſtians is ſo full of *Greek* and *Latin* Inſcriptions, it can never be exhauſted; but the greateſt Part are Epitaphs of Perſons, about whom we are not concern'd.

UPON a Tomb-Stone:

D. M.
ASTIO AVG
LIB. TAR.
VENNONIA AETETE
CONJVGI
PIENTISSIMO FECIT.

UPON another Tomb-ſtone:

Valens & San-
batus propriæ ma-
tri hanc aram
erexerunt memoriæ
cauſa.

ΟΥΑΛΗϹ ΚΑΙ ΣΑΝ·
ΒΑΤΟΣ ΤΗΕ ΔΙΑ ΜΗ·
ΤΡΙ ΑΕϹΤΗϹΑΝ ΤΟΝ
ΒΩΜΟΝ ΜΝΗΜΗϹ
ΧΑΡΙΝ.

[1] For τῇ ἰδίᾳ.
[2] For ἀνέστησαν

UPON another Tomb-Stone:

C. IVI' SENECIO
NEM: VE
PROC PROV: GA
LAT. ITEM VICEPRAE
SIDIS EJUSD. PROV.
ET PONTI
ZENO AUC CUB
TABULAR

PROV.

PROV: EJUSD: PRÆPO
SITO INCOMPARABILI.

WITHOUT the City, about the Convent of St. *Mary* of the *Armenians*, among very fine antique Marble, Pillars, Architraves, Bafes, Capitals, which are near the little River of *Chibouboujou*, are to be feen many Infcriptions; the moft remarkable of which is this of *M. Aurelius.*

IMP. CAESARI
M. AURELIO
ANTONINO. IN
VICTO. AVGVSTO
PIO FELICI
AEL. LYCINVS. V.I.
DEVOTISSIMVS
NVMINI EJVS.

PERHAPS the Buft which is near, is that of this Emperor. It is a Buft with a full Face, of two feet high, and twenty inches wide; but it has been very much abus'd. The Marble is grey, vein'd with white, as is likewife the Pedeftal on which it ftood.

HERE is an Infcription which we found upon another Pedeftal, lying on a Tomb-ftone near the Convent.

Γ. ΑΙΛ. ΦΛΑΟΥΙΑΝΟΝ	
ΣΟΥΛΠΙΚΙΟΝ ΔΙΣ Γ.	*Gaium Ælium Flavianum*
ΛΑΤΑΡΧΗΝ ΤΟΝ Α	*Sulpicius bis Galat-*
ΓΝΟΤΑΤΟΝ ΚΑΙ ΔΙ	*archen caftiſſimum*
ΚΑΙΟΤΑΤΟΝ	*& juſtiſſimum*
ΦΛΑΟΥΙΑΝΟΣ	*Flavianus*
ΕΥΤΥΧΗΣ	*Eutyches*
ΤΟΝ ΓΛΥΚΥΤΑΤΟΝ	*Dulciſſimum*
ΠΑΤΡΟΝΑ	*Patronum.*
ΔΙΕΥΤΥΧΙ.	

THESE

THESE two modern Epitaphs are in the fame Churchyard:

HIC IACET INTERRATVS
D. IOANNES ROOS
SCOTVS QVI OBIIT IN AN
GORA DIE 22. JVNII ANNO
DOMINI M. DC. LXVIII.
ÆTATIS SVÆ XXXV.
ANNORVM
HODIE MIHI: CRAS TIBI.

HIC IACET
SAMVEL FARRINGTON
ANGLVS, ACIDWALLI
FARRINGTON MERCA
TORIS LONDINENSIS
FILIVS: OBDORMIVIT
IN CHRISTO, ANNO
ÆTATIS XXIII.
SALUTIS MDCLX.

YOU will find here, my Lord, the Defign of a Pillar which is handfome enough, and is fet up hard by the Monument of *Auguftus*, with which I have had the honour to entertain you. This Pillar is made of fifteen or fixteen Pieces of white Marble, about twenty inches each in height; the Bafe and the Capital are of the fame Stone. The Capital which is fquare, is adorn'd at each Corner with a Leaf of the *Acanthus*, and a kind of indifferent Efcutcheon, whofe Ornaments are effac'd: There is no Infcription on it. The *Turks* call this Pillar *The Maidens Minaret*, becaufe they imagine it once fupported a Maid's Tomb-ftone.

THE Baffa of *Angora* has thirty or five and thirty Purfes Income. The Janizaries are there under the Command of a *Sardar*, but there are but about three hundred. They reckon there are in this City forty thoufand Souls among the *Turks*, four or five thoufand *Armenians*, and fix

hundred

hundred *Greeks*. The *Armenians* there have feven Churches, without reckoning the Monaftery of *St. Mary*. The *Greeks* have but one Church in this City, and one in the Caftle.

A N'G O R A is four great days Journey from the *Black Sea* the fhorteft way. The Caravan from *Angora* to *Smyrna* is twenty days paffing ; and the antient City of *Cotyæum*, which the *Turks* ftill call *Cataya*, is half way. The Caravans go from *Angora* to *Prufa* in ten days, from *Angora* to *Kefaria* in eight, from *Angora* to *Sinope* in ten, from *Angora* to *Ifmith*, or the antient *Nicomedia*, in nine days; and from *Angora* to *Affamboul* in twelve or thirteen Days.

T H E Y breed the fineft Goats in the World in the Champaign of *Angora*. They are of a dazzling white ; and their Hair, which is fine as Silk, naturally curl'd in Locks of eight or nine inches long, is work'd up into the fineft Stuffs, efpecially Camlet : but they don't fuffer thefe Fleeces to be exported unfpun, becaufe the Country People gain their Livelihood thereby. *Strabo* feems to have fpoken of thefe fine Goats : *In the Neighbourhood of the River* Halys, *fays he*, *they breed Sheep, whofe Wool is very thick and foft ; and befides, there are Goats, not to be met with any where elfe*. However it be, thefe fine Goats are not to be feen only within four or five days Journey of *Angora* and *Beibazar* ; their Young degenerate if they are carried farther. The Thred made of this Goat's Hair is fold from four Livres to twelve or fifteen Livres the Oque; there is fome fold even for twenty or five and twenty Crowns the Oque, but this is only made up into Camlet for the Ufe of the Grand Signior's Seraglio. The Workmen of *Angora* ufe this Thred of Goat's Hair without mixture, whereas at *Bruffels* they are oblig'd to mix Thred made of Wool, for what reafon I know not. In *England* they mix up this Hair in their Perriwigs, but it muft not be fpun. In this confifts the Riches of *Angora* ; all the Inhabitants are employ'd in this Trade. 'Tis with reafon that they prefer the Goat's Hair of *Angora* to that of *Cougna*, which is the antient City of *Iconium*, where *Cicero* affembled the *Roman* Army ; for the Goats of *Cougna* are all either brown or black.

T H E 2d of *November* we fet out from *Angora* for *Prufa* or *Brouffe*, as the *Franks* call it, accompanied only with one *Turkifh* Carrier, and one *Greek* Servant who did not underftand *French*, fo that we were oblig'd

A Goat of Angora.

'lig'd to wait on our felves. We travell'd this Day but about four Hours, Lett. IX. in a fine flat Country well cultivated. We lay at *Soufons*, a forry Village, where we join'd fome Perfons of *Kefaria*, who were going to *Prufa*. The 3d of *November* we travelled feven Hours on beautiful Plains, with only one fmall Hill, on this fide of *Aaias*, a pretty handfome City in a Bottom, whofe Gardens are pleafant, and where there are a great many old Marbles. The next Day we arrived at *Beibazar* after nine Hours, Journey.

B E I B A Z A R is a fmall City built on three fmall Hills, pretty near equal to one another, in a clofe Valley. The Houfes are of two Stories, neatly cover'd with Planks, but you are always going up and down. The River of *Beibazar* runs into the *Aiala*, after it has turn'd feveral Mills, and made fruitful many Parcels of Land, which are divided into Orchards and Kitchen-Gardens. Hence come thofe excellent Péars fold at *Conftantinople*, by the name of *Angora Pears* : but they are very backward, and we had not the good Fortune to tafte them. All this Country is dry and bare, except the Orchards. The Goats eat nothing but the young Shoots of Herbs ; and perhaps 'tis this which, as *Busbequius* obferves, contributes to the Confervation of the Beauty of their Hair, which is loft when they change their Climate and Pafture. The Goatherds of *Beibazar* and *Angora* often comb them and wafh them in the Brooks. This Country puts me in mind of the *Land without Wood* which *Titus Livy* fpeaks of, which can't be far from *Beibazar*, becaufe the River *Sangaris* roll'd its Waters thither : they burn nothing but Cow-dung here, as well as in many other Parts of *Afia*.

W E left *Beibazar* the 6th of *November* about nine in the Morning, and about four in the Evening lodg'd in an old Building which was forfaken, and without a Covering : however, the Country is fine and well cultivated, but rais'd into feveral fteep Hillocks. There we pafs'd the River of *Aiala*, thro a deep Ford; its Waters overflow the Land when one pleafes, but it is to raife excellent good Rice. It runs into the *Black Sea*, and we had encamp'd at the Mouth of it in our way to *Trebifond*.

W E took horfe about Six in the Morning, and arriv'd the 7th of *November*, at half an hour after One, near the Town of *Kahe*, in a Kan

with.

without Benches, or rather in a great Stable. The Country began to be rais'd into Mountains, cover'd with Pines and Oaks, which are never cut, and which are yet hardly higher than our Underwoods, the Land is so poor and unfruitful. The 8th we lay at *Caragamous*, after a Journey of ten Hours cross one of the finest Plains in *Asia*; but uncultivated, without Trees, very dry, tho marshy in some Places, and interspers'd with low small Hills. The old Marbles, which are in the Church-yards, plainly shew that there has been formerly some famous City: But how should we come at the Name of it, supposing it might be found upon some Inscription? For we did not stop there at all, and the Carriers thought of nothing but how to escape the Robbers.

THE 9th of *November* we pursu'd our Route for seven Hours on the same Plain. We discover'd there several Villages, whose Fields are water'd by a little River, which winds very agreeably. We stopp'd at *Mounptalat*, in a sorry Kan, instead of proceeding, as we hop'd to have done, to *Eskissar*, which is a League farther. All the Places which the Turks call *Eskissar* are remarkable for their Antiquity, as are likewise those the *Greeks* call *Paleocastron*, for both these Words signify *Old Castle*. They told us *Eskissar* was a tolerable good City, full of antient Marbles: It is to the Left of the great Road to *Prusa*: Is it not the famous *Pessinunte*? Our Journey of the 10th of *November* was twelve Hours, among beautiful Plains, border'd with small Woods. We were pleasantly lodg'd at *Boutdouc* in a Caravansera cover'd with Lead, as was the Dome of the Mosque. The Church-yards abound with Pillars; and one sees nothing but old Marbles about the Town, but without Inscriptions. Our Journey the 11th of *November* was equal to that of the Day before. We retir'd at *Kpursounou* into a tolerable good Caravansera, on the other side of a small River. 'Tis a Country full of Woods, especially of Oaks. The 12th of *November* we arriv'd at *Acsou*, which signifies a *white Water*. 'Tis a Village five Hours from *Prusa*, in a well-cultivated Plain, and well peopled: After which we met with nothing but Woods of great and small Oaks, of different Kinds. We had all this day Mount *Olympus* on the Left: It is a vast Range of Mountains, on the top of which is nothing to be seen but old Snow, in a very great quantity.

IT

A View of Prusa from y.^e Road to Angora.

IT is a great while, my Lord, fince I talk'd as a Botanift; though Lett. IX. we faw fome very fine Plants after we left *Tocat*, intermix'd with moft of thofe we had met with in *Armenia*, and many others not rare in *Europe*. As we drew near to Mount *Olympus*, we faw nothing but Oaks, Pines, Thyme of *Crete*, *Laudanum Ciftus*, another fine Species of *Ciftus*, which *J. Bauhin* calls the *Ciftus of Crete with large Leaves*, which grows not only *Ciftus ledon* about *Montpellier*, but alfo the Abby of *Fontfrede*, and throughout *Roufillon*. *Creticum lati folium,* J. B. C. *Bauhin* juftly obferves, that *Belonius* had found it upon Mount *Olym-pus*; but *Bauhin* confounds it with the *Laudanum Ciftus*, which *Belonius* and *Profper Alpinus* have mentioned. The *Alder-Tree*, *Dwarf-Elder*, the *Male* and *Female Cornel-Tree*, *Fox-gloves*, with a Flower of a rufty Co-lour, *Pifs-a-beds*, *Succory*, *Knee-holm*, Brambles, are common in the Neighbourhood of Mount *Olympus*. But what a Number of rare Things are there befides thefe? I muft referve them for the *Hiftory of the Plants in the* Levant, which I hope to write.

AT length we arriv'd at *Prufa*, after a Journey of five Hours thro Defiles cover'd with Woods, which abut upon this fine Plain to the North of Mount *Olympus*. We began to fee there Plants and Chefnut-Trees as tall as the Fir-Trees upon the Mountain. It's true, the Lands are in fome meafure incommoded by the Stones which the Waters car-ry down; but in proportion as we approach to *Prufa*, the Fields are co-ver'd with Mulberry-Trees and Vineyards. Moft of the Mulberry-Trees are low, and, as it were, planted in Nurferies. The largeft are fet one near another, and form fmall Forefts, divided by large Thorn-bufhes; among which grows a Species of *Apocin*, which not only twines along the Hedges, but alfo creeps up the higheft Trees. In our Approach to *Prufa*, on the fide of *Angora*, we could fee but a part of the City thro the Woods of high Trees. The fineft part of it, which is the Seraglio-Quarter, does not appear; which is the reafon that I have the Honour to fend you two different Plans: The firft defign'd to the North-Eaft, on the Way from *Angora*; the other on the fide of the Baths, to the North-North-Weft.

PRUSA, the Capital of antient *Bithynia*, is the biggeft and moft magnificent City in *Afia*. This extends it felf Weft to Eaft, at the foot of the firft fmall Hills of Mount *Olympus*, of an admirable Verdure.

Thefe

These Hills are, as we may say, so many Steps up to that famous Mountain. On the North-side, the City stands upon the Edge of a large fine Plain, full of Mulberry and Fruit-Trees. It seems as if *Prusa* was made purposely for *Turks*; for Mount *Olympus* sends out so many Springs, that every House has its own Fountains: I never saw a City which had so many, except *Granada* in *Spain*. The most considerable Spring of *Prusa* is to the South-West, near a small Mosque. This Spring, which sends out a Stream as big as a Man's Body, runs in a Channel of Marble, and so spreads it self over the City. They say there are above three hundred *Minarets* there. The Mosques are very fine; for the most part cover'd with Lead, adorn'd with Domes; as are likewise the Caravanseras. On the other side the *Jews-Street*, to the Left-hand as you go to the Baths, is a Royal Mosque, in the Court whereof are the *Mausolea* of some of the Sultans, in certain Chappels strongly built, and separated from one another. We could meet with no body who was able to give us the Names of these Sultans. *Leunclave* may be consulted on this Point, who has written a very handsome Treatise concerning the Tombs of the Sultans.

[a] *Libitinarius Index Osmanidarum. Francofurti,* 1591.

[a] *Leuncl. Hist. Musulm. lib.* 5.

THE new [a] Seraglio is upon a steep Hill in the same Quarter: 'Tis the Work of *Mahomet* IV. for the old Seraglio was built in the Time of *Amurat*, or *Mourat* I. The Caravanseras of this City are fine and commodious. The *Bezestein* is a great House well built, wherein are many Warehouses and Shops, like those of the *Palais* at *Paris*; and there are all the Commodities of the *Levant* to be found, besides those which are work'd up in this City. They use here not only the Silk of the Country, which is reckon'd the best in *Turky*, but likewise that of *Persia*, which is not so dear, nor much esteem'd. The Silk of *Prusa* is worth fourteen or fifteen Piasters the Oque and half. All these Silks are well wrought; for it must be own'd that the best Workmen of all *Turky* are at *Prusa*; and that they imitate mighty well the Tapestries which are sent thither from *France* or *Italy*.

THE City is also very pleasant, well pav'd, neat, especially in the *Bazars* Quarter. They drink good Wine there at three Parats the Oque. Bread and Salt are very cheap. Butchers Meat is good. They have excellent Trouts, and good Barbel. The Carp are of a surprizing Beauty and

Large-

A View of Prusa from Mount Olympus

Largeneſs, but unſavoury and ſoft, which way ſoever they are dreſs'd. Lett. IX.
In coming from *Angora* to *Pruſa*, we paſs'd a fine River by a Bridge,
which was pretty well built: This River runs afterwards into the Valley of Oaks, on the South-ſide. I believe it is the *Zoufer*, which paſſes
towards *Montania*. There are in *Pruſa* ten or twelve thouſand Families
of *Turks*, which make above forty thouſand Souls, reckoning but four
Perſons to a Family. They reckon four hundred Houſes or Families
of *Jews*, five hundred of *Armenians*, and three hundred Families of
Greeks. And yet this City did not ſeem to us well peopled; and its
Circumference is not above three Miles about. The Walls are half
ruin'd, and were never good, tho they were fortified by ſquare Towers.
We found there neither old Marbles, nor Inſcriptions. Indeed we ſaw
but little Signs of Antiquity in the City, becauſe it has been rebuilt
many times. Its Situation is not ſo advantageous as it ſeems; for it is
commanded by ſome Hills towards the ſide of Mount *Olympus*. None
but Muſſulmans are permitted to dwell in the City. The Suburbs, which
are vaſtly larger, finer, and better peopled, are fill'd with *Jews*, *Armenians*, and *Greeks*. The Plane-Trees there are of a ſurprizing Beauty,
and make the Landſkip admirable, intermingled with Houſes, whoſe
Terraces have a charming View.

T H E Tombs of *Orcan*, his Wife, and Children, are in a *Greek* Church,
cover'd like a Moſque, which is neither large nor beautiful. At the Entrance
are two great Pillars of Marble, and at the farther end four ſmall ones, which
incloſe the Quire the *Turks* have not meddled with: So that their Baſes are
not in the place of their Capitals, and the Capitals in the place of their
Baſes, as Meſſieurs *Spon* and *Wheeler* have written. The Quire, tho cover'd with Marble, was never beautiful: The Stone is of a dirty white,
dull, and green in ſome places. The Sanctuary remains ſtill, with four
Steps into it. They ſhew Strangers, in the Porch of the Moſque, *Orcan*'s
pretended Drum, which is three times as big as the common Drums.
When it is jogg'd, it makes a great Noiſe, by means of certain Balls of
Wood, or ſome other Matter, which make it ſound, to the great aſto-
niſhment of the People of the Country. The Chapelet of this Sultan
is alſo in the ſame place; the Beads of it are of Jet, and as big as a Wall-
nut. There remains ſtill at the Door of the Moſque a piece of Marble,

on which was read formerly a *Greek* Infcription, but at prefent it can't be underftood. Befides the Mofques I have fpoken of, there are in *Prufa* many Colleges of Royal Inftitution, where the Scholars are maintain'd and taught *gratis* the *Arabick* Tongue, and the Knowledge of the *Alcoran.* They are diftinguifh'd by the white Seffe of their Turbants, which form a great Knot as big as the Fift, made up like Stars. In a *Turkifh* Chappel near the City they keep an old very large Sword, which they pretend was *Roland*'s Sword. The Chappel ftands upon an Eminence on the South-Weft fide.

THERE is a Baffa in *Prufa*; an Aga-Janizary, who commands about two hundred and fifty Janizaries; and a *Moula,* or great *Cadi,* who is the moft powerful Officer in the City. When we were there, it was the Son of the Mufti of *Conftantinople* who had this Poft; and at the fame time he had the Reverfion of the Charge of Mufti, which is a Thing without Example in *Turky.* A little time after he follow'd the Fortune of his Father: The Son was not only ftripp'd of all his Goods· and Honours, but was likewife put to death at the fame time when his Father was drawn upon a Hurdle at *Adrianople.*

THE *Armenians* have but one Church in *Prufa*: The *Greeks* have three. The *Jews* have four Synagogues. We were furpriz'd, as we were walking about the City, to hear them fpeak as good *Spanifh* there as at *Madrid.* The *Jews,* to whom I addreffed my felf, told me that they always preferv'd their natural Tongue ever fince their Fathers retir'd out of *Granada* into *Afia.* It's true, they have chofe the City which in all the World moft refembles *Granada* for Situation and Fountains, as I have faid before.

THE 21ft of *November* we fet out at Seven in the Morning, to go to fee Mount *Olympus,* the Afcent of which is eafy enough: But after three Hours riding, we faw nothing but Fir-Trees and Snow; fo that about Eleven of the clock we were oblig'd to ftop near a fmall Lake, in a very high Place. To go from thence to the top of the Mountain, which is one of the biggeft in *Afia,* and like the *Alps* or *Pyrenees,* the Snows muft be melted, and we muft travel a whole Day. The Seafon did not permit us to fee any of the more curious Plants. The Beeches, Yoke-Elms, Afps, Small-Nuts, are common enough here. The Firs don't

don't differ from ours; for we examin'd nicely their Leaves and Fruits. Lett. IX.
After all, we were not well fatisfied with our herborizing, tho we had
obferv'd fome fingular Plants among many others which are common in
the Mountains of *Europe*. 'Twas near this Mountain that our poor
Gauls were defeated by *Manlius*, who, under pretence that they had fallen in with *Antiochus*, was refolv'd to be reveng'd of them for the Mifchief their Fathers had brought upon *Italy*.

THE 23d of *November* we went to fee the new Baths of *Capliza*,
a Mile to the North-North-Weft of the City, to the Right of the Road
to *Montania*. The *Turks* call them *Jani-Capliza*, that is to fay, *New-Baths*. They are two Buildings near one another; the biggeft of which
is magnificent, and has four great Domes cover'd with Lead, bor'd like
a Skimmer, if I may ufe that Comparifon; and all the Holes of
thefe Domes are clos'd with Glafs-Bells, like thofe the Gardiners ufe to
cover Melons withal. All the Rooms of this Bath are paved with Marble: The firft is very large, and, as it were, divided into two by a
Gothick Arch. The Middle of this Room is taken up by a fine Fountain with many Pipes of cold Water; and round the Walls is a Bench
of two feet high, cover'd with Mats, upon which they undrefs themfelves. To the Right are the Rooms wherein they bathe, enlightned by
Domes pierc'd in the fame manner as the larger ones. In thefe Apartments they mix the Springs of hot Water with thofe of the cold.
The Referver, which is of Marble, wherein they bathe, and fwim if they
pleafe, is in the farthermoft Room. They fmoke in this Houfe, and
drink Coffee and Sherbet: This laft is only iced Water, wherein they
fteep a certain Confection of Grapes or Raifins. This Bath is only for
the Men. The Women bathe in the other; but it is not fo fine. The
Domes are fmall, and cover'd with that fort of hollow Tiles which at
Paris we call *Fequieres*.

THE Springs of hot Water run in the Road between the two Baths:
Their Heat is fo great, that Eggs will become like thofe that are foftboil'd in ten or twelve Minutes, and quite hard in lefs than twenty; fo
that one can't bear one's Finger in it. The Water, which is fweet, or rather infipid, fmells a little copperifh: It fmokes continually. The Sides
of the Canals are of a rufty Colour; and the Vapour of thefe Waters

fmells

fmells like addled Eggs. Thefe Baths are on a fmall Hill, which lofes it felf upon the large Plain of *Prufa*. Upon the Rifing between the Road to *Montania* and *Smyrna*, there are two other Baths; one of which is called *Cuchurtli*, becaufe its Waters fmell of Sulphur. ''Twas the Baffa *Ruftom*, Son-in-Law to *Solyman* II. who caufed it to be built.

[¹ *Leuncl. Ind. Libitin.*]

T W O Miles from *Prufa*, and one from the New Baths, in the Road from *Smyrna* to the City *Cechirge*, are the antient Baths of *Capliza*, which the *Turks* calls *Eski-Capliza*. Doctor *Mark Anthoey Cerci* accompanied us thither, and caufed us to obferve that there was in this Place a fine *Imaret*. 'Twas undoubtedly that which was founded by ² *Mourat* I. The Waters of Old *Capliza* are very hot. And tho this Building be much like that of the New Baths, and by confequence not old, it is very probable that thefe are the Royal Hot Waters us'd by the antient *Greeks* in the flourifhing Time of that Empire, which are mention'd by ³ *Conftantine* and ⁴ *Stephen* of *Byzantium*. *Mahomet* I. caufed them to be repair'd, and put into the Condition in which they now are. Befides this great Bath, there is a fmaller one in the fame Village, which the *Turks* frequent likewife; where they caufe themfelves to be pumped. The Waters of both the Old and New Baths make Oil of Tartar white; but make no Alteration upon blue Paper.

[² *Leuncl. Hift. Muful. lib. v. in* Murat Chan Gafi.]

[³ *De adminiftra. Imp. cap.* 50. Τα` δ' ε' εν Πϱου'ση βα- σιλιϰ͞α λεγό- μϑϱα.
⁴ *Stephan. ad vocem* Θέϱμα.]

W E were acquainted with two Botanifts at *Prufa*, one an *Emir*, the other an *Armenian*, who went for great Doctors. They furnifh'd us with the Root of the true *Black Hellebore* of the Antients, in what quantity we would, to make an Extract. 'Tis the fame Species with that of the *Anticyres*, and the Coafts of the *Black-Sea*. This Plant, which the *Turks* call *Zopléme*, and which is very common at the foot of Mount *Olympus*, has for its Root a Stump about the bignefs of the Thumb, lying along, three or four inches in length, hard, woody, divided into feveral Roots, fmaller and wreath'd. All thefe Parts put forth Shoots of two or three inches long, ending in reddifh Eyes, or Buds: But the Stump and the Subdivifions are blackifh without, and whitifh within. The Fibres which accompany them are bufhy, eight or ten inches long, from one to two lines thick, little or nothing hairy. The oldeft are black without, the others brown; the new ones white: One and t'other are of a brittle Flefh or Subftance, without Sharpnefs or Smell; and a reddifh Nerve

runs

runs through them.   They fmell like Bacon, when it's boiled in Water.

OUT of twenty five Pounds of the Root, we drew two Pounds and a half of an Extract, brown, very bitter, and refinous. It purges taken alone, from twenty Grains to half a *Gros*.   Three *Armenians*, to whom we gave it, all complained they were much troubled with *Naufeas*, Griping of the Guts, Heats, a Sharpnefs in the Stomach, along the *Oefophagus*, in the Throat and Fundament; of Cramps, Convulfive Motions, join'd with violent fhooting Pains in the Head, which alfo return'd again fome Days after. So that we abated one half of our Efteem for this great Remedy. As for the Roots, they muft be us'd as thofe of our *Hellebore*, boiling them to the quantity of a *Gros*, or a *Gros* and a half, in Milk, letting them infufe the whole Night, warming the Milk in the Morning the next Day, and ftraining it through a Cloth.

The *Turks* afcribe great Virtues to this Plant; but we could not learn them. M. *Anthony Cerci*, who has practifed Phyfick a great while at *Conftantinople*, *Cutaye*, and *Prufa*, told us he never us'd it, becaufe of the Accidents which it brings upon fick People. He inform'd us that they gather'd *Gum-Adragant* at *Caraiffar*, or *Black-Caftle*, four Days Journey from *Prufa*. Tho he be a Man of Parts, he has no Tafte for Antiquity. He laugh'd at us when we talk'd of beautiful *Greece*, and referr'd us to *Nice* and *Cutaye*. *Nice* is but one Day's Journey from *Prufa*, but on the other fide of a Mountain, which is infefted with Robbers to fuch a degree, that there is no paffing without a ftrong Guard. *Cutaye* is but three Days Journey from *Prufa*. The Baffa who commands there is accufed of having an Underftanding with the Robbers, and of having confiderable Fees of them. The Caravans are five Days going from *Cutaye* to *Prufa*: It is their Way from *Satalia*, or *Attalia*, an antient City of *Caramania*. They go from *Prufa* to *Montania* in four Hours, and from *Montania* to *Conftantinople* by Water in one Morning: So that there needs but one Day to go from *Prufa* to *Conftantinople*. On horfeback they are three Days going from *Prufa* to *Scutari*. Mount *Olympus* is called by the *Turks Anatolai-Dag :* The *Greeks* formerly call'd it the *Mountain of the Caloyers*, becaufe a great many had retir'd thither for Solitude.

THE

THE Name of *Prusa,* and the Situation at the foot of Mount O-
lympus, leave no room to doubt but this is the City they antiently call'd
Πϱουσα, built by *Hannibal,* according to *Pliny,* or rather by *Prusias,* King
of *Bithynia* ; who made War with *Crœsus* and *Cyrus,* according to *Strabo,*
and his Copyer *Stephen* of *Byzantium.* It muft be older ftill, if it be
true that *Ajax* ftabb'd himfelf here with his Sword, as is reprefented on
a Medal of *Caracalla.* 'Tis furprizing that *Livy,* who has fo well defcrib'd
the Neighbourhood of Mount *Olympus,* where the *Gauls* were defeated by *Manlius,* has not mention'd this Place. After *Lucullus* had
beaten *Mithridates* at *Cyziqua, Triarius* came to befiege *Prufa,* and took
it. The Medals of this City, ftamp'd with the Heads of the *Roman* Emperors, fhew that it was very faithful to them. The *Greek* Emperors
did not enjoy it fo quietly. The *Mahometans* plunder'd and ruin'd it
under *Alexis Comnenius.* The Emperor *Andronicus Comnenius,* as *Nicetas*
affirms, caus'd it to be fack'd, on occafion of a Revolt there begun. After the taking of *Conftantinople* by the Earl of *Flanders, Theodorus Lafcaris,* Defpot of *Romania,* got poffeffion of *Prufa,* by the help of the
Sultan of *Iconium,* under pretext of keeping the Places in *Afia* for his
Father-in-Law *Alexis Comnenius,* firnam'd *Andronicus.* *Prufa* was befieg'd by *Bem de Bracheux,* who had put to flight the Troops of *Theodorus Lafcaris.* The Citizens made a brave Refiftance, and the *Latins* were
oblig'd to raife the Siege, and the City remain'd to *Lafcaris* by the Peace
made in 1214, with *Henry* II. Emperor of *Conftantinople,* and Brother of
Baldwin.

   PRUSA was the fecond Seat of the *Ottoman* Empire in *Afia,* for
it muft be acknowledg'd that *Angora* was the firft place where the *Turks*
fix'd themfelves : they made themfelves Mafters of *Prufa* by Famine, and
the Negligence of the *Greek* Emperors. The illuftrious *Othoman,* who
may be compar'd to the greateft Heroes of Antiquity, block'd up the
City by two Forts, which hindred their receiving any Provifions. One
was at the old Baths of *Capliza,* with a ftrong Garifon of chofen Men,
under the Command of his Brother *Actemur,* a great Warriour. The
other, which was upon one of the Hills of Mount *Olympus,* which divided the City, was called the Fort of *Balabanfouc* : it was commanded by
a General Officer of great Reputation. As *Prufa* was continually more

*

and

and more preſs'd with the Scarcity of Proviſions, *Othoman*, who was Lett. IX.
kept in his Bed by the Gout, order'd his Son *Orcan* to carry on the Siege.
Others affirm that he was there in Perſon. Be that as it will, *Beroſes*,
the Governour of the Place, made as honourable a Capitulation as he
could, in the Year 1327. *Calviſius* places the Taking of *Pruſa* in the
Year 1326.

AFTER the Defeat of *Bajazet*, *Tamerlane* came to *Pruſa*, where
he found the Treaſures this Emperor had heap'd up, and which he had
wreſted from the other Princes his Neighbours. They meaſur'd, as *Ducas* ſays, the Precious Stones and Pearls by Buſhels. But when *Tamerlane* went down towards *Babylon*, Sultan *Mahomet*, Son of *Bajazet*, who
reign'd afterwards under the Name of *Mahomet* I. took poſſeſſion of *Pruſa*,
tho he had fix'd the Seat of his State at *Tocat*. *Iſa-beg*, one of his Brothers, came before the City; but the Inhabitants abandon'd it, and retir'd to the Caſtle, and there defended themſelves with a great deal of
Reſolution, inſomuch that *Iſa-beg*, not being able to take the Place, burn'd
and raz'd the City. It was rebuilt ſome time after by *Mahomet*, who
beat his Brother's Forces. It ſeems as if this Place was deſign'd to
hold the *Ottomans* in play. *Solyman*, who was one of the Sons of *Bajazet*, ſeiz'd the Caſtle of *Pruſa*, by means of a forg'd Letter, which he
cauſ'd to be deliver'd to the Governour, in the Name of his Brother *Mahomet*, wherein he orders him to deliver the Caſtle to *Solyman*; but *Mahomet* recover'd it again by means of the ſame Governour, who, thro
Remorſe of Conſcience that he ſhould be ſo deceiv'd, gave it up to its
former Maſter, when *Solyman* was oblig'd to go into *Europe* to defend
his Dominions, which another of his Brothers had invaded: and by a
very extraordinary Misfortune this Place, which did not expect to change
its Maſter, ſaw itſelf again expos'd to the Inſults of *Caraman*, Sultan of
Iconium, who had taken and plunder'd it in 1413. He took up the
Bones of *Bajazet*, and burned them, in revenge that this Emperor had
cauſ'd his Father's Head to be cut off. *Leunclave* adds, that *Caraman*
burnt *Pruſa* in 1415.

AFTER the Death of *Mahomet* I. his Son *Murat*, or *Amurat* II. who
reſided at *Amaſia*, came to *Pruſa*, to cauſe himſelf to be declar'd Emperor. We read in the *Annals of the Sultans*, that there was ſo great a

Fire at *Prusa* in 1490, that the twenty five Regions of it were confirmed; and by this we know that it was divided into many Regions or Quarters. *Zizime*, that illustrious *Ottoman* Prince, Son of *Mahomet* II. disputing the Empire with his Brother *Bajazet*, seiz'd on the City of *Prusa*, to secure *Anatolia*; but being beaten twice by *Achmet*, *Bajazet*'s General, he was forc'd to retire to the great Master of *Rhodes*. 'Twas the same *Zizime*, who came into *Italy* to Pope *Innocent* IV. and died at *Terracina*, as he accompanied *Charles* VIII. in his Voyage to *Naples*.

I am, MY LORD, &c.

LETTER X.

To Monseigneur the Count de Pontchartrain, Secretary of State, &c.

My Lord,

IN the Uncertainty under which we were, whether it was Journey to Smyrna and Ephesus. safer from Robbers to travel the great Road to *Constantinople*, or take the Route to *Smyrna*; we at last chose to go to *Smyrna*, in hopes not only of finding more rare Plants than we had met with upon the *Black Sea*, but likewise of approaching to *Syria*, whose Borders we intended to see.

WE set out therefore the 8th of *November* from *Prusa* for *Smyrna*, and lay at *Tartali*, a Village three hours and a half from *Prusa*. We pass'd by *Cechirge*, where are the antient Baths of *Capliza*, and from thence over the Bridge of the *Loufer* or *Merapli*, a small River which comes from Mount *Olympus*, and runs into the Sea near *Montania*. The Trouts of the *Loufer* are excellent, and all the Country is fine and well-cultivated. To the Left runs a Chain of Hills, on which stands *Phisidar*, a considerable Borough, inhabited by *Greeks*; who for the Pleasure of being alone, without any Mixture of *Turks*, pay a double Capitation, and see but once in a Year a Cadi-Itinerant.

THE 9th of *December*, after a Journey of nine Hours, we began to discover the Lake of *Abouillona*, which is five and twenty Miles about, and seven or eight Miles wide in some Places, sprinkled with several Isles and some Peninsulas; 'tis properly the great Sink of Mount *Olympus*. The biggest of the Islands is three Miles in circumference, and is called *A-*

bouillona.

bouillona, as well as the Village, which is doubtlefs the antient City of *Apollonia;* for 'tis from this Lake that the River *Rhyndacus* proceeds, which paffes to *Lopadi* or *Loubat.* *Caragas* is alfo a Village of *Greeks,* in another Ifland of the fame Lake, but there are fome *Turks* mingled with them. They both pafs in Caiques with Sails from one Ifland to another, to cultivate them. The Carps of this Lake weigh twelve or fifteen Pounds; but we did not find them to be better than thofe we had eaten at *Prufa.* This Lake was antiently called *Stagnam Artynia.* The *Rhyndacus* was call'd *Lycus;* and perhaps *Lopadi,* a fmall Town a League below, is the City of *Metellopolis* mention'd by *Pliny;* but it muft not be confounded with the *Metellopolis* of *Strabo.* According to this Author, the Lake of *Abouillona* was called *Apolloniatis;* and the City which was there, bore the Name of *Apollonia.* The Medal of *Septimius Severus,* the Reverfe of which reprefents a Ship failing, fhews that the Inhabitants gave themfelves much to Navigation, and that the City was confiderable. That of *M. Aurelius,* on the Reverfe of which is the *Rhyndacus* with a long Beard, lying along, and leaning upon his Urn, holding a Reed in his Left Hand, and with his Right fhoving a Boat, fhews that this River was navigable in that time.

M. *VAILLANT* affirms that he has feen the City of *Apollonia,* and places it upon a Hill, at the foot of which runs the *Rhyndacus,* fifteen Miles from the Sea; but no doubt this learned Man took *Lopadi* for *Apollonia,* which muft be the Village of *Abouillona.* *Apollo* was undoubtedly worfhip'd in this City; for befides that it bore his Name, this God is reprefented on a Medal of *M. Aurelius,* ftanding before a Tripos, round which a Serpent is twin'd. *Apollo* is there crown'd by *Diana* the Huntrefs. The Medal of *Lucius Verus* alfo reprefents *Apollo* ftanding, the Left Arm leaning on a Pillar, and holding a Branch of Laurel in his Right Hand. The fame Honour appears upon another Medal of *Caracalla,* where *Apollo* is ftanding among four Pillars of the Frontifpiece of his Temple. The fame Reprefentation is alfo upon the Medal of *Gordianus Pius.* The City of *Apollonia* continued to be very confiderable under the Emperor *Alexis Comnenus;* his Daughter *Ann* relates, that it was pillag'd by the *Turks* as well as *Prufa.*

W E

W E leave the Lake of *Abouillona* all the way on the Left to go to *Lopadi*, where we lay that Day, after having crofs'd a large Plain. The River comes out of the Lake about two Miles above the City ; but it is deep, and carries Boats, notwithſtanding no body has now a long time caus'd it to be clear'd. We pafs'd it at *Lopadi* upon a wooden Bridge, to the Left of which are the Ruins of an antient Stone-Bridge, which appears to have been well built. *Lopadi*, which the *Turks* call *Ulubat*, the *Franks Loubat*, and the *Greeks Lopadion*, contains but about two hundred Houfes of a very poor Appearance ; neverthelefs this Place was confiderable under the *Greek* Emperors. Its Walls, which are almoſt ruined, were defended by Towers, fome round, fome of five fides, and fome triangular ; the Circumference is almoſt fquare. There are Pieces of antique Marble, Pillars, Capitals, Bafs-Reliefs, and Architraves, but all broken and much abus'd. The Caravanfera where we lodg'd was very dirty and ill-built, tho there are fome old Capitals and Bafes of Marble.

T H E Emperor *John Comnenus*, who came to the Empire in 1118. built the Caftle of *Loubat*, when he was about to fight the *Perfians* : 'tis at prefent almoſt quite demolifh'd. *Nicetas* affirms that this Emperor built the City of *Lopadion*, when he went to retake *Caft ancone* upon the Coaft of the *Black Sea*. All this may be eafily reconciled, by faying that *John Comnenus* built the Caftle in one of his Journeys, and the Walls of the City in another. For it is certain that this City is antienter than that time, feeing it was plunder'd by the *Mahometans* under the Emperor *Andronicus Comnenus*, who reign'd in 1081. The Marble Remains which are found, fhew that it was older than the *Comneni*, unlefs they have been brought by Water from the Ruins of *Apollonia*. Indeed there is fome probability that the Inhabitants of this Place, for the convenience of their Commerce, did gradually remove to the Place where *Loubat* ftands, and that they call'd it *Apollonia*, after they had forfaken the antient *Apollonia*, which ftands upon the biggeft Ifle we before fpoke of : for *Ann Comnena* relates, that under *Alexis Comnenus*, *Helian* a famous *Mahometan* General, feizing *Cyziqua* and *Apollonia*, the Emperor fent thither *Euphorbenus Alexander*, to drive him thence. *Alexander* made himfelf Mafter of *Apollonia*, and *Helian* was forc'd to retire into the

Caftle ;

Caſtle; but the Succours appearing, the Chriſtians rais'd the Siege: and as they were about to retreat by the Sea, *Helian*, who was Maſter of the Bridge, hem'd them in by the River, and cut them to pieces. *Opus*, who commanded the Army after the Defeat of *Euphorbenus*, repair'd this Loſs; he not only took *Apollonia*, but oblig'd *Helian* to ſurrender himſelf, and ſent him to *Conſtantinople*, where he became a Chriſtian, with two of his moſt famous Generals. This ſeems to prove that *Lopadi* had taken the Name of *Apollonia* at that time.

A N D R O N I C U S C O M N E N U S ſent an Army to *Lopadi*, to reduce the Inhabitants to their Duty, who, after the Example of thoſe of *Nice* and *Pruſa*, had revolted from him. After the taking of *Conſtantinople* by the Earl of *Flanders*, *Peter de Bracheux* put to flight the Troops of *Theodorus Laſcaris*, who had *Lopadi* by the Peace made with *Henry*, Succeſſor of *Baldwin*, Earl of *Flanders*, and firſt *Latin* Emperor of the Eaſt.

A F T E R the great *Othoman* had defeated the Governor of *Pruſa*, and the neighbouring Princes, who had form'd themſelves into a League to ſtop the Progreſs of his Conqueſts, he purſu'd the Prince of *Teck* to the very Bridge of *Lopadi*, and ſent the Governor of the Place word, That if he did not ſend him his Enemy with his Throat cut, he would paſs the Bridge, and deſtroy all with Fire and Sword. The Governor anſwer'd, That he would ſatisfy him, provided he would ſwear that nei-ther he, nor any of his Succeſſors ſhould ever paſs that Bridge. Indeed, ſince that time the *Ottomans* always paſs that River by Boat. *Othoman* caus'd the Prince of *Teck* to be hew'd to pieces in ſight of the Citadel, and took poſſeſſion of the Place. *Lopadi* is as famous in the *Turkiſh* Hiſtory for the Defeat of *Muſtapha*, as the *Rhyndacus* is in the *Roman* Hiſtory for that of *Mithridates*.

T H E General, who was juſt beaten at *Cyziqua*, being inform'd that *Lucullus* beſieg'd a Caſtle in *Bithynia*, march'd thither with his Horſe and the remainder of his Foot, deſigning to ſurprize him. But *Lucullus* having Intelligence of his March, ſurpriz'd him, notwithſtanding the Snow and Rigour of the Seaſon. He beat him at the River *Rhyndacus*, and made ſo great a Slaughter among his Troops, that the Women of *Apollonia* came out of the City to plunder the Dead, and ſteal their Baggage. *Appian*, who agrees to this Victory, forgot the chief Circum- ftances, which *Plutarch* has related.

A S to the Battel which *Amurat* won over his Uncle *Muftapha*, Authors relate it differently. *Ducas* and *Leunclave* pretend that *Amurat* deftroy'd the Bridge at *Lopadi*, to hinder his Uncle from coming to him. We faw the Remains of it; and ever fince that time they have had a Bridge of Wood, over which they pafs to the City. *Muftapha* finding himfelf abandon'd by his Allies, thought only of paffing into *Europe*. *Calcondylas* affirms that *Amurat* caus'd a Bridge to be made over the River. *Leunclave* may be read concerning the other Particulars of the Action; for he pretends there was a bloody Combat, and that *Muftapha* was the Aggreffor.

M. *S P O N* had no reafon to take the Lake of *Lopadi* for the Lake *Afcanius*, no more than to affirm that the River of *Lopadi* throws it felf into the *Granicus*. The Lake *Afcanius* is the Lake of *Nice*, which the *Greeks* call *Nixaca*, and the *Turks Ifmich*. M. *Tavernier* fays, That this Lake is called *Chabangioul*, becaufe of the City *Chabangi*, which ftands upon the Borders of it, five or fix Miles from *Nice*. *Strabo* places the Lake *Afcanius* near this City. As for the *Granicus*, it is far enough off from *Lopadi*, as we fhall fee; and we obferv'd the Mouth of the *Rhyndacus* by an Ifland which the Antients call'd *Besbicos*.

W E ftaid at *Lopadi* the next Day, the 10th of *December*, becaufe five Jewifh Merchants of *Prufa*, who had the fame Carrier with us, had made their Bargain to reft the Sabbath-Day: So we quitted the great Caravan, and were but fix Perfons with Fufees, namely, us three, two Carriers, and the Jews, who all together had but one very indifferent Carabine with a Lock, very foul, and which we could not charge for want of a Gun-ftick. The good People were fo much afraid of the *Turks*, that they hid themfelves as foon as they faw any of them at a diftance. When they could not hide themfelves, they put off their Turbant with the white Seffe. We took white Turbants at *Angora*, that we might not be taken for *Franks* by the Robbers, who ufe fuch without Mercy. We met five arm'd with Lances between *Prufa* and *Lopadi*; but they pafs'd away very quietly.

THE

THE next Day, the 11th of *December*, we continued our Route in *Michalicia*, which is part of the *Myſia* of the Antients, and travell'd till Two of the clock in a great Plain, well cultivated, with ſome ſmall Hills on it, cover'd with Woods: But in our way we ſaw only *Squeticui*, a poor Village, to our Right. We had on our Left a Well with Buckets, for the Conveniency of Travellers. Afterwards we paſs'd a ſmall River, which throws it ſelf into the *Granicus*, and quickly found our ſelves upon the Banks of this River. The *Granicus*, whoſe Name we ſhall never forget ſo long as *Alexander* ſhall be remember'd, runs from South-Eaſt to North, and afterwards towards the North-Weſt, before it falls into the Sea. Its Banks are very high on the Weſt-ſide: ſo that the Forces of *Darius* had a conſiderable Advantage, had they known how to uſe it. This River, ſo famous for the firſt Battel the greateſt Captain of Antiquity gain'd upon its Banks, is at preſent call'd *Souſoughirli*, which is the Name of the Village by which it paſſes. We paſs'd the *Granicus* upon a wooden Bridge, which did not ſeem to us very ſafe. The Caravanſeras of *Souſoughirli* are vile Stables with Benches, which are but two feet high, and but juſt broad enough to lie down croſs-ways; ill pav'd, full of Filth, with very bad Chimneys, five or ſix feet from one another. There are however ſome Pillars, and antient Marbles in the Village, but without Inſcriptions. The *Agnus Caſtus* and *Yellow Daffodil* are common upon the Banks of the *Granicus*. M. *Wheeler* took this *Daffodil* for that with the fiſtulous Leaves: But I don't underſtand how he could ſuppoſe that *Alexander* met the Army of *Darius* upon the *Granicus* on this ſide of Mount *Taurus*, near the *Euphrates*.

THE 12th of *December* we ſet out at half an Hour paſt Four in the Morning, and arriv'd after twelve Hours Journey at *Mandragoia*, a ſorry Village; which we ſhould not have caſt our Eyes on, had there not been ſome old Marbles. The Pillars of the Caravanſera, where we lodg'd, as old as they are, are but rough form'd, and, according to appearance, will remain a great while in the ſame Condition.

THESE Remains of Antiquity have cauſed M. *Spon* to conjecture that *Mandragoia* may be the City of *Mandrapolis*, which *Pliny* ſpeaks of. To go from *Souſoughirli* to *Mandragoia* we croſs'd a Mountain, which M. *Wheeler* took for Mount *Timnus*: And we could not diſcover any

*

of

of the Ruins of that antient Citadel, which it's pretended *Alexander* caufed to be built after the Battel of the *Granicus*, becaufe we fet out before Day. Mount *Timnus* is not very high, but very wide; and its Sides are cover'd with fmall *Oaks*, *Spanifh Junipers*, and *Adrachnes*. The *Iron-Gate* is a very bad forfaken Caravanfera in one of its Valleys, upon a Brook, which runs towards the *Levant :* We happily pafs'd all thefe Defiles at a time when the Robbers could not keep the Field.

T H E 13th of *December*, after a Route of ten Hours, through Defiles fill'd with *Oaks*, *Pines*, and *Phillyrea*, which they often burn to encreafe the Pafturage, we lay at *Courougoulgi*, and found about half way from *Mandragoia* the Village of *Tchoumlekechi*. There are nothing but Storks Nefts upon the Caravanferas of this Route. Thefe Nefts are like great Baskets, hollow'd in form of a Bafin, made up of Branches of Trees laid confufedly together. The Storks come there every Year to hatch their Young; and the People of the Country, far from driving them away, have fo great a Veneration for them, that they don't dare touch their Nefts. A Stranger would be ill us'd if he fhould venture to fhoot at them.

A S to the Brook which runs a little way from *Mandragoia*, and which M. *Spon* took for the *Granicus*, 'tis the *Fourtiffar*, which falls from Mount *Timnus*, and which may be the *Caicus* of the Antients. We ate this day, the firft time, of the Fruit of the *Adrachne :* This Fruit is very thin upon Bunches, which are branch'd and purpurine, almoft oval, half an inch long, chagrin'd with flat Seeds, whereas thofe of the *Arbut-Tree* have pointed Seeds. That of the *Adrachne* ends in a fmall blackifh Nib, half a line long: The Flefh of it is reddifh, inclining to an orange, yellowifh within, more or lefs agreeable to the Tafte, according to the Condition of the Fruit. They feem to me rougher than thofe of the *Arbut-Tree :* Neverthelefs they are of the fame Make, divided into five Chambers, each fill'd with a flefhy *Placenta*, charg'd with Seeds one line long, brown, pointed at the Ends, a little crooked, and, as it were, triangular in their length: The Flefh of thefe Kernels is whitifh.

T H E *Origany*, which M. *Wheeler* obferv'd upon Mount *Sypilus*, is very common in all thefe Parts; as are likewife the *Sage of Candia*, mention'd by *Clufius*; the *Thyme of Crete*, fpoken of by the Antients; the

Turpentine, the *Echinophora* of *Columna,* the *After Tomentofus Verbafci Folio,* the *Valeriana Tuberofa Imp.* and many othe fine Plants.

THE 14th of *December* we travelled but about fix Hours, and pafs'd over a Mountain not fo high and rugged, extended, and divided by many little Dales, full of great and fmall Oaks, mix'd with fome *Pines* of *Tarara, Phillyreas, Adrachne, Turpentine-Trees.* We arriv'd at *Baskelambai,* a pretty handfome Borough, where we ate good Winter-Melons, as long as thofe of *Vera* in *Spain*; but their Flefh is white, not vinous, tho otherwife very pleafant. We pafs'd two Rivers before we came to *Baskelambai*; this Place is fituate on a well-cultivated Plain, and they drive a great Trade in Cotton.

THE 15th of *December* we continu'd our Journey in the Plain of *Baskelambai,* where runs a fmall River. We afterwards afcended a flat Mountain, and enter'd upon the great Plain of *Balamont,* where they cultivate a great deal of Cotton. *Balamont* was our Inn, after a Journey of eight hours. 'Tis a handfome Place, upon a Brook which runs to the South-Weft. There are feveral broken Pillars in this Plain; and the two Caravanferas of *Balamont,* which are feparated only by a large Court, are full of Pillars of Marble and Granate, which fupport its Beams. They have even heap'd together Pieces of Pillars mingled with Capitals and Bafes, which make but a very ill Performance. We obferv'd in the Village a Capital fo well made, that I could not forbear having it ingrav'd. The Hills, which are to the right and left, have between them very fine Plains fow'd with Cotton. *Ackiffar,* or the antient *Thyatira,* which is one of the feven Churches in the *Apocalypfe,* is to the left of the Road from *Balamont.* *Kircagan* is a great Mountain, an Hour and a half from *Baskelambai,* where there is another *Ackiffar.* The *Turks* much ufe the Names of *Ackiffar* or *Karaiffar,* that is to fay, *White Caftle* or *Black Caftle*; of *Eskiffar* or *Jeniffar, Old Caftle* or *New Caftle,* according as they fancy.

THE 16th of *December* we travel'd from Three in the Morning till Noon, in a pretty flat Country terminated by this great Plain of *Magnefia,* bounded on the South by Mount *Sypilus*; and this Mountain, tho very wide from the Eaft to the Weft, feem'd not by far fo high as Mount *Olympus:* the higheft Top of *Sypilus* is to the South-Eaft of *Magnefia,* and this City is not much more than half fo big as *Prufa.* Thefe two Cities

are

Capitals found at
Balamont

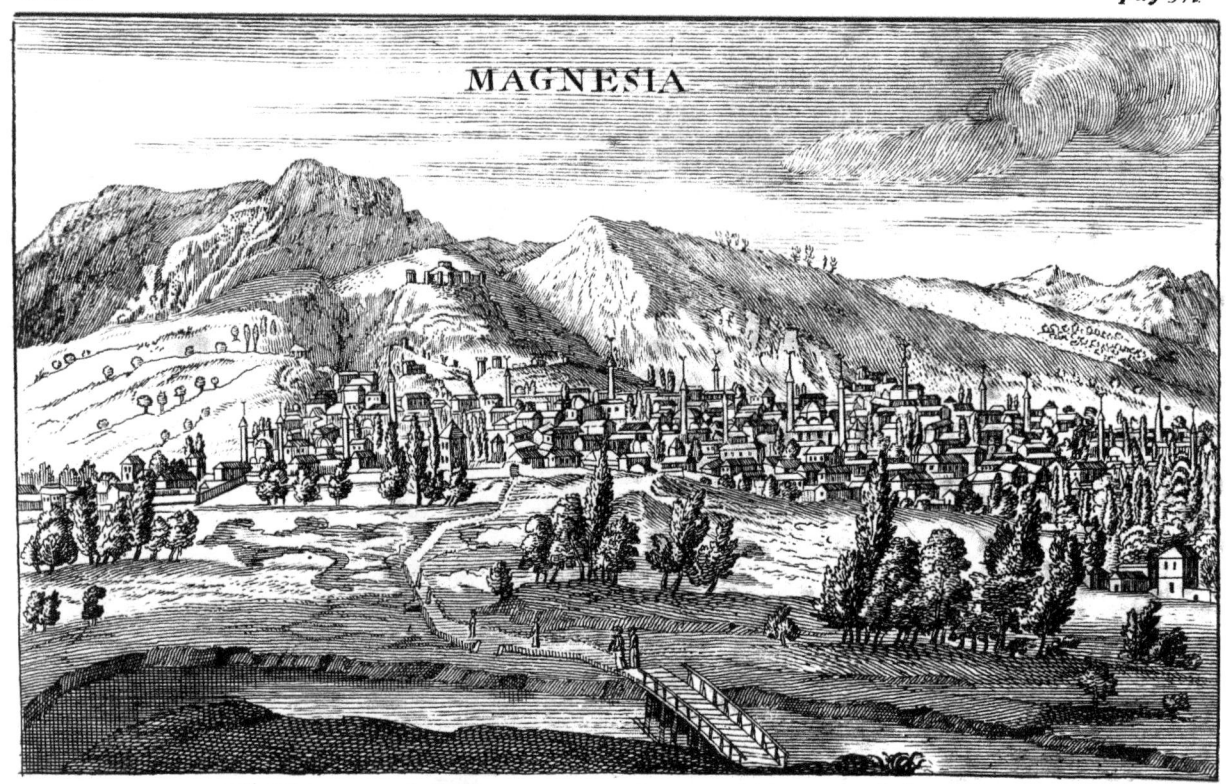

MAGNESIA

A View of Magnesia from Mount Sypili.   145.

are like one another only in Situation; for there are neither good Churches nor Caravanferas in *Magnefia*, and they trade only in Cotton. Moft
of the Inhabitants are *Mahometans*. The *Jews*, who are more numerous
than the *Greeks* or *Armenians*, have but three Synagogues. The Citadel
is fo much neglected, that it runs to ruin ; as does the Seraglio, whofe
Ornaments are nothing but fome old Cyprefs-Trees. The Verdure is
much finer in the Neighbourhood of *Prufa*, and Mount *Sypilus* is not to
be compared to Mount *Olympus* ; and alfo the River of *Hermus*, which
feem'd to us to be much bigger than the *Granicus*, is a great Ornament
to the whole Country. This River receives two others into it, whereof
one comes from the North, and the other from the Eaft. It runs half a
league diftant from *Magnefia* under a Bridge of Wood, fupported by Piles
of Stone. After having travers'd the Plain from the North-North-Eaft
towards the South, it makes a great Elbow before it comes to the Bridge ;
and running to the South, throws itfelf into the Sea between *Smyrna* and
Phocea, as *Strabo* has well obferv'd ; while all our Geographers make it
empty itfelf into the bottom of the Gulph of *Smyrna*, on this fide the
Plain of *Menimen*. This River forms at its mouth great Banks of Sand,
for which reafon, the Veffels which come into the Bay of *Smyrna* are
obliged to keep along the Coaft, and to pafs in view of the old Caftle
upon the Shore.

WE pafs'd the Morafs between *Hermus* and *Magnefia*, over a fine
Caufey of about a quarter of a league long, in which they have ufed a
great many antique Marbles and Jafpers ; there are fome in the Walls of
the City, but we found no Infcriptions. The Plain of *Magnefia*, tho of a
furprizing beauty, is almoft cover'd over with Tamarisks, and is not well
cultivated, except on the Eaft-fide : its Fruitfulnefs is exprefs'd by a Medal in the King's Cabinet ; on one fide is the Head of *Domitia*, Wife of
Domitian ; on the other, a River lying down, holding a Bow in his Right-
Hand and the Horn of Plenty in his Left. *Patin* has given us one of
the like figure : *Strabo* alfo obferves, that *Hermus* is one of thofe Rivers
which fatten the Earth with their Mud.

THEY burn nothing in this City but the Wood of *Adrachne*, with
which they are fupply'd from Mount *Sypilus*. The *Jewifh* Merchants of
our Caravan obliged us to lie by the 17th of *December* ; and to make up

B b b 2   for

for the Lofs of Time, provided us with good Wine among their Brethren there, at eight Parats for a thoufand Drachms, as they fpeak; thefe thoufand Drachms weigh two Oques, that is, five Pounds. It was very cold, and the North Wind blew very hard, but it did not freeze.

WE amufed ourfelves this day with herborizing upon Mount *Sypilus*, which is very fteep on the North-fide; and among the Plats of Laurel-Rofes and Adrachne, we found upon the Precipices feveral rare Plants which we had feen in *Candia*, efpecially the *Jacea*.

THE Goddefs *Sypilene* took her Name from this Mountain; or rather *Cybele*, the Mother of the Gods, was named *Sypilene*, becaufe fhe was worfhip'd in a particular manner upon Mount *Sypilus*: therefore 'tis not ftrange that we fee fo many Medals of *Magnefia*, on the Reverfe of which this Goddefs is reprefented, fometimes on the Frontifpiece of a Temple with four Pillars, and fometimes in a Chariot. They alfo in Affairs of Importance were ufed to fwear by the Goddefs of Mount *Sypilus*; as appears by that valuable Marble at *Oxford*, on which is cut the League of *Smyrna* and *Magnefia*, upon the *Meander*, in favour of King *Seleucus Callinicus*.

FROM the top of Mount *Sypilus* the Plain fhows admirable, and one fees with abundance of pleafure the Courfe of the River. Sometimes we thought on the great Armies of *Agefilaus* and *Tiffaphernes*, fometimes thofe of *Scipio* and *Antiochus*, who difputed the Empire of *Afia* upon thefe large Plains. *Panfanias* affirms, that *Agefilaus* beat the Army of the *Perfians* by the fide of the *Hermus*; and *Diodorus Siculus* relates, that the famous General of the *Lacedemonians*, defcending from Mount *Sypilus*, went and ravaged all the Neighbourhood of *Sardis*. *Xenophon* fays, the Battel was fought by the fide of the *Pactolus*, which throws itfelf into the *Hermus*.

AS to the Battel of *Scipio* and *Antiochus*, it was fought between *Magnefia* and the River *Hermus*, which *Titus Livy* and *Appian* call the *River of Phrygia*. This great Action, which gave the *Afiaticks* fo high an opinion of the *Roman* Valour, was perform'd in the Road from *Magnefia* to *Thyatira*, the Ruins whereof are at *Ackiffar* or *White Caftle*. *Scipio* had caufed his Troops to advance on this fide; but having intelligence that *Antiochus* was encamp'd advantageoufly about *Magnefia*, he pafs'd the River with his Army, and forced the Enemy to come out of their Trenches, and

give

give him Battel. There were, fays *Florus*, in this King's Army Elephants of a prodigious Bignefs, who fhin'd with Gold, Silver, Ivory, and Purple, with which they were cover'd. This Battel, which was the firft the *Romans* won in *Afia*, fecur'd them the Country till the Wars of *Mithridates*.

AFTER the Taking of *Conftantinople* by the Earl of *Flanders*, *John Ducas Vatatze*, Son-in-law and Succeffor of *Theodorus Lafcaris*, fix'd the Seat of his Empire at *Magnefia*, and reign'd there three and thirty Years. The *Turks* made themfelves Mafters of it under *Bajazet*; but *Tamerlane*, who took him Prifoner in the famous Battel of *Angora*, after having plunder'd *Prufa*, and the Places thereabout, came to *Magnefia*, and caus'd all the Riches of the Cities of *Lydia* to be carried thither.

THE *Sicilian* War being at an end between the Count *de Valois*, and *Frederick* King of *Sicily*, Son of *Peter* of *Arragon*, the *Catalans*, who had ferv'd under *Frederick*, enter'd themfelves among the Troops of *Andronicus*, Emperor of *Conftantinople*, who was at war with the *Turks*. *Roger de Flor*, Vice-Admiral of *Sicily*, came into *Afia*, at the head of the *Catalans*, and beat the *Mahometans* in 1304, and 1305 : but the Diforders and Violences committed by the *Catalans* againft the *Greeks*, having oblig'd thofe of *Magnefia*, fupported by *Ataliotes* their Governour, to rife againft the Garifon of the *Catalans*, and cut their Throats ; *Roger*, who had left his Treafures there, came and befieged the Place, which defended it felf fo well, that he was forc'd to retreat.

AMURAT II. chofe *Magnefia*, wherein to fpend the Remainder of his Days in quiet, after he had plac'd his Son *Mahomet* II. upon the *Ottoman* Throne; neverthelefs the Wars which the King of *Hungary*, and *John Hunniades*, rais'd againft him in *Europe*, forc'd him to quit his Retirement, for his Son was too young to bear the Burden. *Amurat* pafs'd the Canal of the *Black Sea* at *Neocaftron*, came to *Adrianople*, and march'd againft the Chriftian Princes : the King of *Hungary* was kill'd, and *Hunniades* put to flight.

AFTER this fignal Victory, the Vifiers, by their Inftances, prevail'd with the Sultan to take upon him the Adminiftration, and *Mahomet* retir'd to *Magnefia*. The *Turks* made a fmall Province of the Country about this Place, whereof *Magnefia* was the Capital, and where *Corcut* Son of *Bajazet*

zet II. reign'd. The great *Solyman* II. alfo refided at *Magnefia*, till the Death of his Father. Sultan *Selim* made himfelf Mafter of it, and drove out another *Corcut*, an *Ottoman* Prince. There is no Baffa at *Magnefia*, but one Mouffelin and one Sardar are there in Command. The *Greeks* there are very poor, and have but one Church.

THE 18th of *December* we again afcended Mount *Sypilus*, to go to *Smyrna*. The way is rough, and the Mountain is very fteep. *Plutarch* likewife fays it was call'd the *Thunder Mountain*, becaufe it thundred there more frequently than in other Places thereabout; and it is probably for this reafon, that at *Magnefia* they have ftamp'd Medals of *M. Aurelius, Philip* the elder, *Herennia* and *Etrufcilla*, whofe Reverfe reprefents *Jupiter* arm'd with Thunder-bolts. After eight Hours Journey, we arriv'd at *Smyrna*. There is nothing commoner in this Route, than the *Adrachne*; with it they heat Ovens, and cover the tops of Garden-Walls and Vineyards, to fecure them from the Rains.

SMYRNA is the fineft Port at which one can enter into the *Levant*, built at the bottom of a Bay, capable of holding the biggeft Navy in the World. Of the feven Churches in the *Apocalyfe*, 'tis the only one which remains in any Reputation: It owes this Advantage to St. *Polycarp*, to whom St. *John*, who had rais'd it into a Bifhoprick, writ by Command of our Lord, *Be thou faithful unto Death, and I will give thee a Crown of Life*. The other Cities St. *John* counfel'd by our Lord's Command, are either miferable Villages, or utterly ruin'd. The illuftrious City of *Sardis*, fo renowned for the Wars of the *Perfians* and *Greeks*; *Pergamus*, the Capital of a fine Kingdom; *Ephefus*, which gloried in being the Metropolis of all *Afia*; thefe three famous Cities are fmall Boroughs built with Clay and old Marbles. *Thyatira, Philadelphia, Laodicea*, are not known but by fome remaining Infcriptions, wherein we find their Names mention'd.

SMYRNA is one of the largeft and richeft Cities of the *Levant*. The Goodnefs of Port, fo neceffary for Trade, has preferv'd it, and caus'd it to be rebuilt feveral times, after it had been deftroy'd by Earth-quakes. 'Tis as it were the Rendevous of Merchants from the four Parts of the World, and the Magazine of the Merchandize they produce. They reckon fifteen thoufand *Turks* in this City, ten thoufand *Greeks*, eighteen

MAGNESIA

hundred *Jews*, two hundred *Armenians*, as many *Franks*. The *Turks* have nineteen Mofques, the *Greeks* two Churches, the *Jews* eight Synagogues, the *Armenians* one Church, and the *Latins* three Convents of Religious. The *Latin* Bifhop has but an hundred *Roman* Crowns Income; the *Greek* Bifhop has one thoufand five hundred Piafters. Tho the *Armenian* Bifhop fubfifts barely on the Alms of thofe of his Nation, he is better provided for than all the Chriftian Prelates. They gather thefe Alms on Feftivals and Sundays, and they fay it amounts to fix or feven Purfes a year.

THE Situation of *Smyrna* is admirable. The City extends itfelf all along the Shore, at the foot of a Hill which commands the Port. The Streets are there better enlightned, better pav'd, and the Houfes better built than in other Cities upon the Continent. The *Franks* Street, which is the fineft in *Smyrna*, runs all along the Port. It may be faid it is one of the richeft Magazines in the World: the City is plac'd in the Center of the Trade of the *Levant*, eight days Journey from *Conftantinople* by Land, and four hundred Miles by Water; five and twenty days Journey from *Aleppo*, by the Caravans; fix days Journey from *Cogna*, feven from *Cutaya*, and fix from *Satalia*.

THERE is no Baffa in *Smyrna*, but only one Sadar, who commands two thoufand Janizaries, lodg'd in and about the City. Juftice is adminiftred there by a Cadi. The *French* in 1702 had about thirty Merchants there well fettled, without reckoning many other *Frenchmen*, who drive a lefs confiderable Trade. The *Englifh* were as numerous, and their Trade flourifhing.

AT the time when we were at *Smyrna*, the *Dutch* were not above eighteen or twenty Merchants, well fettled, and much efteem'd. There were but two *Genoefe*, who traded under the Protection of *France*. There was a Conful from *Venice*, tho there was not one Merchant of that Nation. It was Signior *Lupazzolo*, a venerable old Man, of one hundred and eighteen Years of Age, who boafted he was in the third Century of his Life, for he was born about the End of 1500, and we look'd upon him as the Head or the oldeft of all Mankind. He was of a middling Stature, and fquare; he died a little after. They faid he had had near fixty Children of five Wives he had married, without reckoning;

koning his Miftreffes and Slaves, for the good Man was of an amo-
rous Difpofition. It is very certain that his eldeft Son died before him
at the Age of eighty five, and the youngeft of his Daughters was but
fix Years old at that time.

THE Caravans of *Perfia* are continually arriving at *Smyrna* from *All*
Saints to *May* and *June*. They bring thither fometimes near two thoufand
Bales of Silk a Year, without reckoning the Drugs and Cloths. Our
French bring from thence Cochineel, Indigo, Sarfaparilla, Brafil, Campe-
chy, Verdigreafe, Almonds, Tartar-Powder, Cinnamon, Cloves, Ginger,
Nutmegs. Cloths of *Languedoc*, Serges of *Beauvais*, Serge *de Nifmes*,
Pinchinats, the Satins of *Florence*, Paper, fine Tin, good Steel and Enamels
of *Nevers*, go off very well there. Before our Trade was fettled thither,
the Merchants of other Nations call'd us *Mercanti di Barretti*, becaufe
we then, as now, furnifh'd them with almoft all their woollen Bonnets
and Caps. We alfo carried thither Earthen Ware, but the greateft Quantity comes thither from *Ancona*. The *French* Foines are much in efteem
there, efpecially thofe of *Dauphine*, which are us'd for Furs. A Fur for
a Veft is fold from fifty to eighty Crowns: they mix thofe of the deep-
eft Colours with the Samour, which is the Sable or Foine of *Mufcovy*.
They ufe more of thefe Foine-Skins, which are brought from *Sicily*, than
of thofe which come from *France*; but they are cheaper, becaufe
thofe from *France* are upon the foot with the Foines of *Armenia* and
Georgia.

BESIDES the Silks of *Perfia*, and the Thred made of the Goats Hair
at *Angora* and *Beibazar*, which are the richeft Commodities of the *Le-*
vant, our Merchants bring from *Smyrna* Cotton fpun, or *Caragack*, Cotton rough in Bags, fine Woollens, Baftard-Woollens, and thofe of *Metelin*,
Nut-Galls, Wax, Scammony, Rhubarb, Opium, Aloes, Tutty, Galbanum,
Gum-Arabick, Gum-Adragant, Gum-Ammoniack, Semen-contra, Frankin-
cenfe, Zedoaria, large and ordinary Carpets.

THE whole Trade is carried on by the Interpofition of *Jews*, one can
buy or fell nothing but what muft pafs thro their Hands. We may call
them *Chifous*, and miferable, but 'tis they put all into motion. We muft
do them juftice, and own they have better Capacities than other Merchants; befides, they live at *Smyrna* well enough, and make a very

*

handfome

handſome Appearance, which is very extraordinary among a People who ſtudy nothing but how to ſave. Foreign Merchants live together very genteelly, and don't fail in any Viſits of Ceremony or Decency. The *Turks* are ſeldom ſeen in the *Franks* Street, which is the whole Length of the City. When we are in this Street, we ſeem to be in *Chriſtendom*; they ſpeak nothing but *Italian, French, Engliſh* or *Dutch* there. Every body takes off his Hat, when he pays his reſpects to another. There one ſees Capuchins, Jeſuits, Recolets. The Speech of *Provence* ſhines there above all others, becauſe there are more from *Provence* than any other Parts. They ſing publickly in the Churches; they ſing Pſalms, preach, and per-form Divine Service there without any trouble; but then they have not ſufficient Regard to the *Mahometans,* for the Taverns are open all Hours, Day and Night. There they play, make Good-Cheer, dance after the *French,* the *Greek,* and the *Turkiſh* Manner. This Quarter would be very fine, if there was a Key at the Port; but the Sea beats up to the very Sides of the Houſes, and the Boats enter, as I may ſay, into the very Warehouſes.

M. *ROTER,* our Conſul, maintains the Honour of our Nation there very worthily; he dwells in a ſmall Palace, where Men of Faſhion are receiv'd very agreeably: he is withal very well made, wiſe, of good Parts, generous, and applies himſelf very much to every thing which regards the Honour or Intereſt of the *French.* As he had the Complaiſance to lodge us in his Houſe, we were there when the *Engliſh* and *Dutch* Mer-chants came to wiſh him a merry Chriſtmas. His Bufet was well furniſh'd; for beſides the Wines of the Country, there was plenty of *French, Italian,* and *Spaniſh* Wines; there was no want of Liquors, or the different Fruits, according to the Seaſon: thus they ſpent the Feaſt, to which our chief Merchants were invited for the Honour of our Na-tion. After the ordinary Compliments were over, they gave every bo-dy to drink; and you muſt pledge, or ſeem to do ſo by putting the Glaſs to the Mouth. The Conſul was oblig'd to drink above a hundred times of all ſorts of Wine. When the *Engliſh* and *Dutch* were retir'd, came the *Greeks* and *Armenians* in their turn. Our Merchants go likewiſe to make their Compliments to the *Engliſh* and *Dutch* Conſuls, by whom they are receiv'd much in the ſame manner, that is to ſay, with Bottle and Flag-gons; but by good Luck not on the ſame Day, for they reckon accoring to the Old Stile. The Confuls don't vifit one another upon thefe occafions, but fatisfy themfelves with fending their mutual Compliments by their Interpreters.

AFTER we had refted ourfelves fome days at M. *Royer's*, where we found every thing we could wifh for, to make amends for what we had undergone in fuch long Journeys ; that is to fay, abundance of Good-Cheer, charming Converfation, all the Gazettes, and a Library : we went to take a walk by the fide of the Caftle which ftands on the fhore, with the Chancellor of the Nation and fome of his Friends well arm'd, as were likewife their Servants. This Precaution is necefary when there are any *Barbary* Veffels near *Smyrna* ; for the Soldiers and Seamen, who ramble about upon the fhore, feize on Perfons as foon as they perceive they have difcharg'd their Fuzees at any fort of Game.

THE Caftle, of which I have the honour to fend you a Plan, is a fquare Fort, whofe Sides are about a hundred paces long, flank'd with four mean Baftions, and defended by a fquare Tower which ftands in the middle : the Inclofure of it is low, with Battlements ; the Cannon, which are without Carriages, are as big as at the Caftles of the *Dardanelles*. This Place is furrounded with Marfhes, which are paffable, and full of Snipes. After having pafs'd a fmall Foreft of Olives, we found at the foot of one of the Hills which face the Road where the Ships ride, fome hot Baths almoft abandon'd. Perhaps thefe are the fame *Strabo* fpeaks of, in his Defcription of the Places which lie in the way between *Clazomene* and *Smyrna* : this Author affures us, that he there found a Temple of *Apollo*, and hot Water. Of the antient Building of thefe Baths, which were very find if we may judge by the Ruins, there remains nothing at prefent but one little Cellar, in which is the Refervoir into which two Pipes empty themfelves, one of hot Water, the other of cold. Thefe Baths are to the South-Eaft of *Smyrna*, but the Water feem'd not fo hot as that at *Milo*. As for the Temple of *Apollo*, it can't be far off, and the *Englifh* Conful's Chaplain affured me he had difcover'd the Ruins of it. He is a pretty Gentleman, and a good Antiquary ; I communicated to him the Infcriptions I had copy'd at *Angora*. We were at my Return from *Ephefus* to have had fome Converfations upon our Difcoveries, but during my abfence he went to *Conftantinople* to my Lord *Paget*, and then

*

into

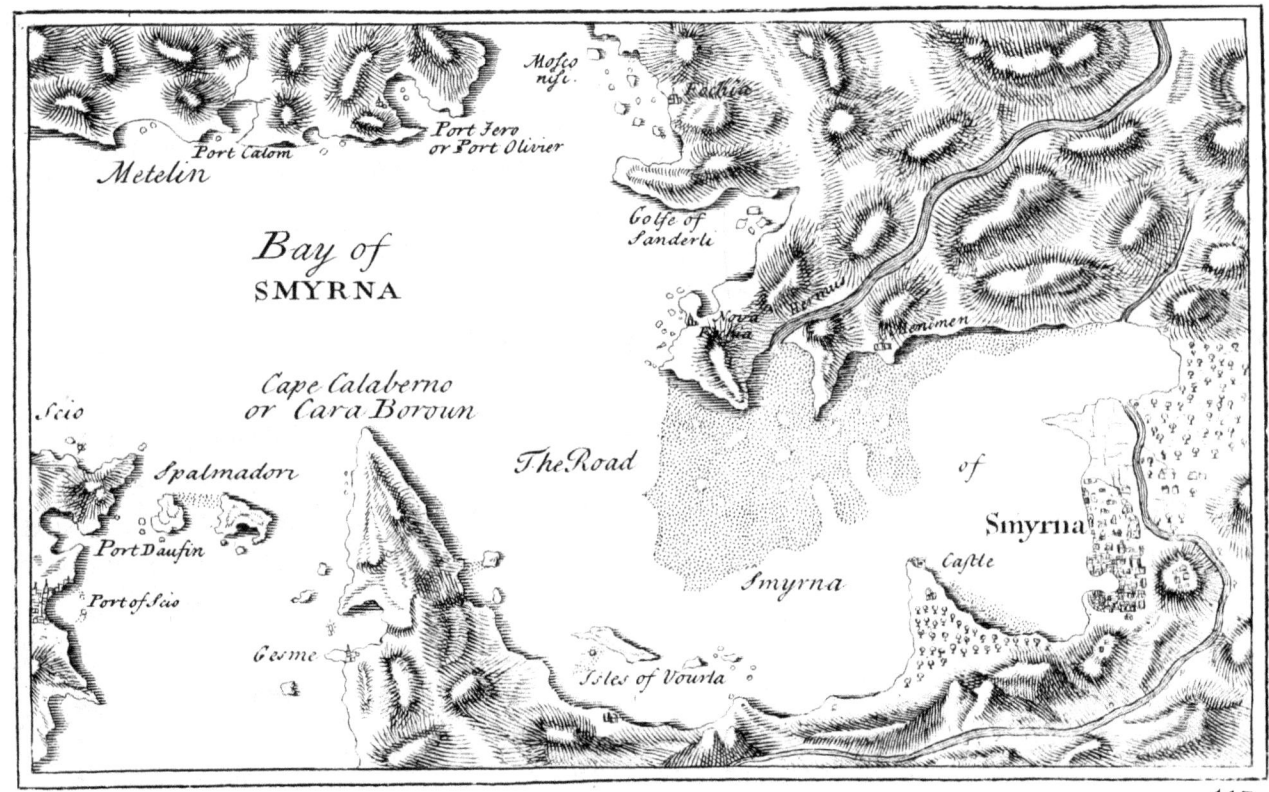

into *England*; fo that I learnt nothing more of the Temple of *Apollo*. I hope Mr. *Sherrard*, who is at prefent Conful of that Nation, will inform us of all the Antiquities of *Smyrna*, and the Places adjacent; for he is a very learned Man, and full of Zeal for the Perfection of the Sciences : he has given me fome light into the Situation of *Clazomene*, and its Iflands.

CLAZOMENE, which they take to be the Village of *Vourla*, was an illuftrious City in the flourifhing time of *Greece*, and had great part in the *Peloponnefian* War. The *Perfians* thought it fo neceffary for heir Defigns, that they not only feiz'd on it, but likewife took care to referve it to themfelves in that famous Peace of *Antalcidas*. *Auguftus* is call'd the Founder of this City, upon a Medal in the Elector of *Brandenburgh's* Cabinet; but this Emperor was only the Reftorer of that Place. *Clazomene* formerly kept *Smyrna* and all the Country about the Bay in fo much awe, that *Tzachas*, a famous *Mahometan* Corfair, was obliged to get poffeffion of it, when he fix'd himfelf at *Smyrna* under *Alexis Comnenus*.

ONE can't better fet out the Situation of *Clazomene*, than by the Iflands which are at the Entrance of the Bay of *Smyrna*, after doubling the Cape of *Carabouron*. *Strabo* reckons eight in number : *Pliny* fpeaks but of four : they are near the fhore, on this fide the Caftle. The *Turks* know them by the name of the Ifles of *Vourla*.

PAUSANIAS fays that *Clazomene* was firm Land, and that the *Ionians* fortify'd it, to put a ftop to the Conquefts of the *Perfians*; however, they were fo terrify'd with their Progrefs, after the taking of *Sardis*, that they pafs'd into one of the Ifles over againft the City, thinking themfelves much fafer there, becaufe the *Perfians* had yet no Fleet. Afterwards *Alexander* the Great made it a Peninfula, by a Jettee of two hundred and fifty paces long, on which they went from the Ifland to the Continent. To avoid the great and dangerous Tour of *Carabouron*, this great Prince open'd a Plain crofs Mount *Mimas*, which led to *Erythrea*, a famous City and Sea-Port over againft *Scio*; fo that difembarking at *Erythrea*, they pafs'd by this new Road to *Clazomene*, in the fame manner as now difembarking at *Seagi* they go to *Smyrna* by Land, without entring the Bay. Perhaps *Seagi* is a Corruption of *Teus*, for the *Greeks* for the moft part pronounce the *T* like *S*; of *Teus* they make *Seus*, and fo *Seagi*.

Tis

Tis a Country of good Wine : we had a Medal of *Augustus*, with a Legend of this City, and a Reverse representing *Bacchus* standing, clothed like a Woman, holding a Pitcher in his Right-hand, and a Thyrsus in his Left : by Flattery they have set round the Head of *Augustus*, that he was the Founder of this City.

THE Antients call that Chain of Mountains *Mineas*, which occupy the Peninsula which they named *Myonnese*, or the *Isle of Field-Mice*, wherewith all the Coast of *Asia* is infested. The two principal Summits of this Mountain are call'd *the Brothers*, because they seem equal, and stand one by the other like Twins. The Country Folk call them *Poussos*, that is *Breasts*, according to the Fancies of the antient *Greeks*, who thought the Points of Mountains resembled Breasts. M. *Morel*, who surpass'd the greatest Antiquaries of his Time, by the wonderful Correctness of his Designs, thought *Clazomene* was the antient City of *Gryniam*, which gave the Name of *Grynæus* to *Apollo*. *Cybele*, the Mother of the Gods, was much worship'd at *Clazomene*, and bore the Name of the City, as one may see upon the Medals of *Valerian*. They also there worship'd *Diana with white Eye-brows*, as we learn from some Medals of *Gallienus*. It would be very pleasant to go and rake among the Ruins of *Vourla*.

SOME days after, we went to the old Castle of *Smyrna*, situate on a Hill which commands the City. The *Turks* have quite demolish'd one of the finest marble Theatres in *Asia*, which stood upon the Brow of this Mountain, on the side which looks to the Road where the Ships lie. They have used all these Marbles in building a fine Bezestein and a great Caravansera. The antient Castle, built by *John Ducas*, is upon the top of this Hill ; its Circumference is irregular, and favours of the Times of the later *Greek* Emperors, under whom they used the finest Marbles in the building of the Walls of Cities. One sees before the Gate of this Castle, a famous Tree, which the *Greeks* pretend to be a Shoot of St. *Polycarp*'s Staff. As far as I can judge of it, at the beginning of *January*, by a Branch I cut off from it, which began to lose its Leaves, it is the *Micocoulier* which we observ'd in our Route of *Tocat*. To the right, and by the side of the Gate, is mortiz'd into the Wall the Bust of the pretended *Amazon Smyrna*, about three feet high ; but it does not seem to

have

The Castle upon the Point at Smyrna

An Amazons Head at Smyrna.

ha ve been ever very handfom, and the *Turks* have ufed it ill, by ftriking their Fuzees againft it to break the Nofe off. It is certain, this Buft has none of the Attributes of an *Amazon* : whereas on the Medals which are ftamp'd with the Legend of this City, the *Amazon* who founded it is diftinguifh'd by an Ax with a double Edge, and a Shield. In the firft Times the Figure of this Heroine was as the Symbol of the City, as appears by the Reverfe of the Medals which were ftamp'd in token of the Alliances made between the *Smyrneans* and their Neighbours.

THERE is nothing in the Caftle which is worth feeing; the *Turks* have built an ordinary Mofque there. Upon the North Gate there are two Eagles, very ill defign'd, and an Infcription fo high, that we could not read it. The Place where the Caftle now ftands, was taken up, in the flourifhing time of *Greece*, by a Citadel under the protection of *Jupiter Acræus*, or who prefided over lofty Places. *Paufanias* affures us, that the top of the Mountain of *Smyrna*, call'd *Coryphus*, gave the Name of *Coryphæus* to *Jupiter*, who had a Temple there. M. *de Camps* has a fine Medaillon, whereon this God *Acræus* is reprefented fitting, as he is likewife on another Medal of *Vefpafian*, where the fame God fits, holding a Victory in his Right-hand, and a Spear in the other.

MANY other Medals of *Smyrna* help us to know the Rank it held among the Cities of *Afia*. The Citizens boaft, fays *Tacitus*, to be the firft in all *Afia*, who rais'd a Temple to *Rome* under the Name of *Rome the Goddefs*, in the very time while *Carthage* ftood, and that there were powerful Kings in *Afia*, who as yet knew nothing of the *Roman* Valour. *Smyrna* was made *Neocore* under *Tiberius* with a great deal of diftinction; and the moft famous Cities of *Afia* having ask'd permiffion of that Emperor to dedicate a Temple to him, *Smyrna* was prefer'd to them. It became *Neocore* of the *Cæfars*, whereas *Ephefus* was only fo of *Diana*; and at that time the Emperors were much more fear'd, and confequently more honour'd than the Goddeffes. *Smyrna* was declared *Neocore* the fecond time under *Adrian*, as the *Oxford* Marbles fhew. Again it had the fame Honour, and took the title of *First City of Afia* under *Caracalla*, which it retain'd under *Julia Mæfa*, *Alexander Severus*, *Julia Mammæa*, *Gordianus Pius*, *Otacilla*, *Gallienus*, and *Saloninus*.

GOING

GOING out of the Caftle, we went to fee the Remains of the Circus, which are on the left. We pafs'd before a Chappel half ruin'd, where they fhew us the Fragments of the Tomb of St. *Polycarp*, who was the firft Bifhop of *Smyrna*; who not only had the happinefs to be a Difciple of St. *John*, but was made a Bifhop by the Apoftles themfelves. After having govern'd his Church a long time, he was burnt alive at the Age of Ninety Five or Six, under *Aurelius* or *Antoninus Pius*. The Acts of his Life fay this holy Tragedy was acted in the Amphitheatre of *Smyrna*; fo that it is more probable it was done in the Theatre which we have been fpeaking of, than in the Circus we are going into.

THIS Circus is fo much deftroy'd, that no more of it remains, as I may fay, but the Mould; they have carry'd away all the Marbles, but the Pit retains its antient Figure. It is a kind of Dale of four hundred fixty five feet long, and one hundred and twenty wide; the Top is terminated in a Semicircle, and the Bottom opens in a Square. This Place is made very pleafant by the Moufe-Ear, for the Waters don't ftand there. We muft not judge of the true Bignefs of the Circus or Stadium by the Meafures we have given; we know that this fort of Places were ordinarily but one hundred and twenty five paces long, and that they were call'd *Diauli*, when they were twice as long. From this Hill we difcover all the Champain of *Smyrna*, which is perfectly fine; the Wines whereof were much efteem'd in the Times of *Strabo* and *Athenæus*.

NOTHING can give a finer Idea of the Magnificence of the antient *Smyrna*, than the Defcription *Strabo* has given us of it. *When the* Lydians, fays that Author, *had deftroy'd* Smyrna, *all that part, for about four hundred Years, was inhabited only in fmall Villages*; but Antigonus *rebuilt it, and afterwards* Lyfimachus. *'Tis at prefent the fineft City in* Afia. *One part is built upon the Mountain, but the greateft part ftands in the Plain upon the Port, over againft the Temple and Gymnafium of* Cybele. *The Streets are the moft beautiful that can be, running at right Angles, and paved with fine Stones. There are large and fine Porticos, a publick Library, and a fquare Portico, where ftands the Statue of* Homer; *for the Inhabitants of* Smyrna *are very fond of having* Homer *to have been born there, and they have ftamp'd a Copper Medaillon, which they call* Homerion. *The River* Meles

SUCH was *Smyrna* in the time of *Auguftus,* and it feems as if they
had not then built either the Theatre or the Circus, for *Strabo* would
not have forgot them.   So that M. *Spon* very well conjectures, that the
Theatre was built under *Claudius,* for one finds the Name of that Emperor upon a Pedeftal.   *Strabo* informs us, that the *Lydians* had deftroy'd
a City more antient than that which he defcribed, and 'tis of this that
Herodotus fpeaks, when he fays that *Giges* King of *Lydia* declared War
with the *Smyrneans,* and that *Halyattes,* his Grandfon, took it.   It was
afterwards ill ufed by the *Ionians,* furprized by the *Colophonians*; afterwards reftored to its own Citizens, but difmember'd from *Æolia.*   M. *Spon*
writes, that this antient *Smyrna* was between the Caftle on the fhore and
the prefent City; there remain ftill fome of its Ruins upon the Water-
fide.

THE *Romans,* to preferve to themfelves the fineft Port in *Afia,* always treated the *Smyrneans* very kindly; and they, not to expofe themfelves to the *Roman* Arms, carry'd it very fair with them, and were very
faithful to them.   They put themfelves under their protection during
the War with *Antiochus*; only *Craffus,* the *Roman* Proconful, was ever
unfortunate near this City.   He was not only overcome by *Ariftonicus,*
but taken and put to death; his Head was prefented to his Enemy, and
his Body bury'd at *Smyrna.*   *Perpenna* foon avenged the *Romans,* and
took *Ariftonicus* captive.   In the Wars of *Cæfar* and *Pompey, Smyrna* declared for the latter, and furnifh'd him with Ships.   After the death of
Cæfar, Smyrna, which inclined to the fide of the Confpirators, refus'd entrance to *Dolabella,* and receiv'd the Conful *Trebonius,* one of the principal Authors of the Dictator's death; but *Dolabella* impofed upon him fo
well, that entring the City by night, he feiz'd him, and martyr'd him in
two days.   *Dolabella* however could not keep the Place; *Caffius* and *Bru-
tus* came thither to take their meafures.

ALL that was pafs'd was forgotten when *Auguftus* was become peaceable Poffeffor of the Empire.   *Tiberius* honour'd *Smyrna* with his good
Will, and regulated the Rights and Privileges of the City.   M. *Aurelius*
rebuilt it after a great Earthquake.   The *Greek* Emperors, who poffefs'd

it

it after the *Romans*, loſt it under *Alexis Comnenus*. *Tzachas*, a famous *Mahometan* Corſair, ſeeing the Affairs of the Empire very much embaraſs'd, ſeiz'd *Clazomene*, *Smyrna*, and *Phocea*. The Emperor ſent thither his Brother-in-law *John Ducas*, with an Army by Land, nd *Caſpax* with a Fleet. *Smyrna* ſurrender'd without ſtriking a Blow : that Government was given to *Caſpax*, who returning to the City, after he had been to accompany *Ducas*, was ſtabb'd with a Sword by one *Sarraſin* : this Wretch had robb'd one of the Citizens of a large Sum of Money, and ſeeing his Condemnation unavoidable, vented his Fury upon the Governour.

THE *Mahometans*, in the time of *Michael Paleologus*, who drove the *Latins* from *Conſtantinople*, ſeiz'd on almoſt all *Anatolia*. *Atin*, one of their chief Generals, took *Smyrna*, under *Andronicus* the elder. *Homur* his Son ſucceeded him ; and as he was taken up in ravaging the Coaſts of the *Propontis*, the Knights of *Rhodes* took poſſeſſion of the Country about *Smyrna*, and built the Fort *St. Peter*. *Homur* return'd to *Smyrna*, and viewing the Fort, which was not yet finiſh'd, receiv'd a Wound with an Arrow, of which he died. During the Life of *Homur*, who was call'd the *Prince of* Smyrna, the *Latins* burnt his Fleet, and took the City. The Patriarch of *Conſtantinople*, who had been made by the Election of the Pope, judging it proper to ſay Maſs in the principal Church, was there ſurpriz'd by *Homur*'s Troops, who having put the *Latins* to flight, beheaded him in his Pontifical Habit, and maſſacred the Nobility who were about him. Some *Genoeſe* Hiſtorians refer an Expedition the *Genoeſe* made upon theſe Coaſts, under the Doge *Vignoſi*, to the Year 1346, wherein they added to their former Domains, *Scio*, *Smyrna*, and *Phocea*. It ſeems as if they did not keep *Smyrna* long, becauſe *Morbaſſan* beſieg'd it by Order of *Orcan* II. Emperor of the *Turks*, who had married one of the Daughters of the Emperor *Cantacuzenus*.

AFTER the Battel of *Angora*, *Tamerlane* beſieg'd *Smyrna*, and encamp'd very near to Fort *St. Peter*, which the Knights of *Rhodes* had built, and whither the greateſt part of the Chriſtians of *Epheſus* had retir'd. *Ducas*, who has given an account of this Siege, relates two Circumſtances of it which are very ſingular. 1. That *Tamerlane* caus'd the Entrance of the Port to be fill'd up, by ordering every Soldier to caſt in a Stone. 2. That he had built there a Tower, after a new Order of Architecture,

compos'd

compos'd in part of Stone, and in part of dead Mens Skulls, rang'd in order like inlaid Work, fometimes full-fac'd, and fometimes fideways. After the Retreat of the *Tartars*, *Smyrna* remain'd in the power of *Cineites*, Son of *Carafupafi* Commandant of *Ephefus*, who had been Governour of *Smyrna* under *Bajazet*. Neverthelefs, *Mufulman*, one of the Sons of *Bajazet*, jealous of the Greatnefs of *Cineites*, pafs'd into *Afia* in the Year 1404, with defign to humble him. *Cineites* made a ftrong League with *Caraman*, Sultan of *Iconium*, and *Carmian* another *Mahometan* Prince, but they made Peace without coming to an Engagement. *Cineites* had not fuch good Succefs with *Mahomet* I. another Son of *Bajazet*. *Mahomet* came to befiege *Smyrna*, which they had well fortified, and ftor'd with Ammunition. *Cineites* retir'd to *Ephefus*, and the Great Mafter of *Rhodes* endeavour'd with all poffible Expedition to repair Fort *St. Peter*, which *Tamerlane* had raz'd; the City furrender'd after ten Days Siege. *Mahomet* caus'd the Walls to be demolifh'd, and beat down a Tower the Great Mafter of *Rhodes* had caus'd to be built at the Entrance of the Port. Since that time the *Turks* have remain'd peaceable Mafters of *Smyrna*, and have rebuilt the Tower, or to fpeak more properly, have built a kind of a Caftle on the Left of the Entrance into the Galley-Port, which is the antient Port of the City.

W E walk'd out at the other end of *Smyrna*, at the end of the *Franks* Street, toward the Gardens which are water'd by the River *Meles*. 'Tis the nobleft Stream in the World, in the Republick of Letters. The greateft Poet was born upon its Banks, and as the Name of his Father was unknown, he bore the Name of this River. A fair Adventurer nam'd *Critheis*, driven from the City of *Cuma*, by the Shame of finding herfelf with Child, and being deftitute of Lodging, came to lie in here. Her Child afterwards loft his Sight, and was therefore called *Homer*, that is to fay, *Blind*. It is not neceffary to fay his Mother married *Phanius*, a Schoolmafter and Mufician of the City. An ingenious Woman never wanted a Husband. *Smyrna*, illuftrious for the Birth of fo great a Poet, did not only erect a Statue and Temple to him, but likewife ftamp'd Medals with his Name. *Amaftris* and *Nice*, its Allies, did the like, one with the Head of *M. Aurelius*, and another with the Head of *Commodus*. As for the River *Meles*, tho it hardly turns two Mills, I leave you to guefs

Melefigene, born on the Banks of the Meles.

whe-

whether it was forgot upon thefe Medals. It is become a very poor one fince the time of *Paufanias*, who calls it the *fine River*. This Stream, at the Head of which *Homer* employed himfelf in a Cavern, is reprefented upon a Medal of *Sabin*, under the Figure of an old Man, leaning with his Left-Hand upon an Urn, holding a Horn of Plenty in his Right. It is alfo reprefented upon a Medal of *Nero*, with the fimple Legend of the City, as likewife upon thofe of *Titus* and *Domitian*.

A M I L E or thereabouts on the other fide the *Meles*, in the Road to *Magnefia*, to the Left in the middle of a Field, they ftill fhew the Ruins of a Building they call the *Temple of* Janus, and which M. *Spon* fuppos'd to be that of *Homer*; but fince the Departure of that Traveller, they have utterly demolifh'd it, and that Quarter is fill'd with fine antient Marbles. Some Paces thence runs an admirable Spring, which turns conftantly feven Mill-ftones in one Mill. What pity it was that *Homer's* Mother did not come to be deliver'd near fo fine a Fountain. One fees there the Fragments of a great Marble Edifice, call'd the *Baths of* Diana; thefe Fragments are very magnificent, but there are no Infcriptions.

I F we go from the Baths of *Diana* into the Fields of *Meneme*, befides that they are very fruitful in Melons, Wines, and all forts of Fruits, we find the Earth there full of a natural fix'd Salt, which they ufe inftead of Saltweed to make Soap.

T H E 25th of *January* we went from *Smyrna* for *Ephefus*, about nine in the Morning. At going out of the City, we enter'd upon a *Military Way*, which is ftill pav'd with large Pieces of Stone, cut almoft like Lozenges. Three Hours from *Smyrna* we pafs a pretty handfome Stream, which runs into the Sea; but we met another near four Hours from thence, which may pafs for a little River. The Country is flat, uncultivated, cover'd in fome Places with fmall Wood like Underwood, mix'd with Pines. We drank good Coffee on the Road, in a Meadow where a *Turk* had a Stall, or fmall moveable wooden Houfe. We arriv'd about half an hour after four at *Tcherpicui*, a poor Village in a great uncultivated Plain, where we faw the Remains of a great old Wall of Stone, which has been an Aqueduct, according to the People of the Country, to carry Water to *Smyrna*.

F R O M

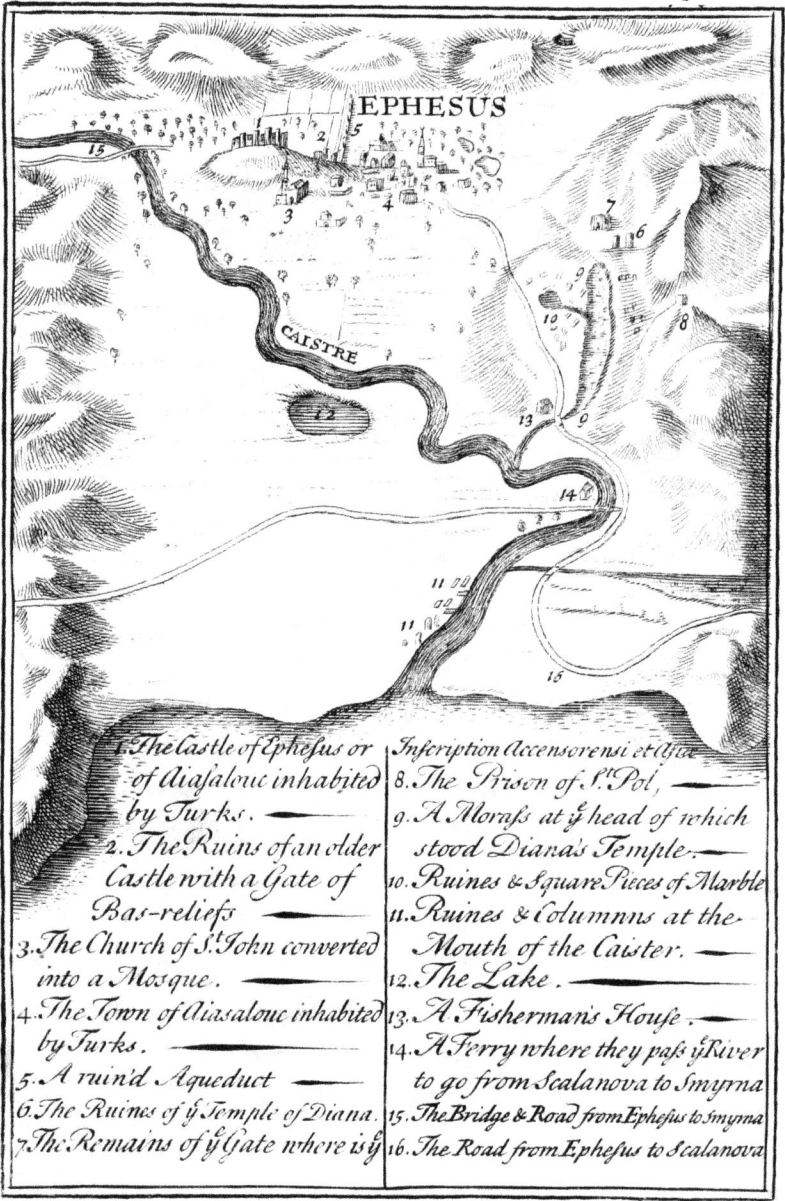

EPHESUS

CAISTRE

1. The Castle of Ephesus or of Aiasalouc inhabited by Turks.
2. The Ruins of an older Castle with a Gate of Bas-reliefs
3. The Church of S.ᵗ Iohn converted into a Mosque.
4. The Town of Aiasalouc inhabited by Turks.
5. A ruin'd Aqueduct
6. The Ruines of y̆ Temple of Diana.
7. The Remains of y̆ Gate where is y̆ Inscription Accensorensi et Afiæ
8. The Prison of S.ᵗ Pol.
9. A Morafs at y̆ head of which stood Diana's Temple.
10. Ruines & Square Pieces of Marble
11. Ruines & Columns at the Mouth of the Caister.
12. The Lake.
13. A Fisherman's House.
14. A Ferry where they pafs y̆ River to go from Scalanova to Smyrna
15. The Bridge & Road from Ephesus to Smyrna
16. The Road from Ephesus to Scalanova

FROM the Plain of *Tcherpicui* to *Ephefus* is a continued Range of Mountains, whofe Woods and Defiles are full of Robbers in the fine Seafon. We met with nothing but Stags and Wild-Boars; but we were agreeably furpriz'd to fee the Hills naturally cover'd with fine Olives, which without Culture bring excellent Fruit, which is all loft, for want of People to gather it. As we drew nigh to *Ephefus* on the Right, the Mountains are frightful, fteep, and perpendicular, and make but a hideous Sight. We pafs the *Cayftre*, half a League on this fide *Ephefus*. This River, which is very fwift, runs under a Bridge built with antique Marbles, and turns fome Mills. We enter afterwards upon the Plain of *Ephefus*, that is to fay, into a great Bafon fhut up on all fides, except towards the Sea, with Mountains; the *Cayftre* winds in this Plain, but it does not make fo many Turnings as M. *Spon* reprefents by much: and thofe of the *Meander*, which are much more twifted, don't come near thofe of the *Seine* below *Paris*; I am furpriz'd that our Poets have never defcrib'd them. The *Cayftre* has been reprefented on Medals; there are fome with the Heads of the Emperors *Commodus, Septimius Severus, Valerian, Gallienus.*

WE in vain fought for another River, which the Antients fpeak of, which water'd the Country about *Ephefus*; without doubt it throws itfelf into the *Cayftre* above the Bridge. In reality they told us at *Ephefus*, that the *Cayftre* receives a confiderable River beyond the Mountains which lie to the North-Eaft, which agrees very well with the Medal of *Septimius Severus*, on which the *Cayftre* is reprefented under the Form of a Man, as being a River which difcharges itfelf into the Sea; and the *Kenchrios*, which is the River we are fpeaking of, under the Figure of a Woman, to fignify that it runs into another. Befides thefe Figures, a *Diana with feveral Breafts* is reprefented on one fide upon the fame Reverfe, and on the other a Horn of Plenty. All this fignifies the Fruitfulnefs which thefe Rivers procure to the Lands of *Ephefus*. The *Seine* and the *Marne*, which bring fo great Riches to *Paris*, would, in my opinion, well deferve a Medal.

'TIS a melancholy thing to fee *Ephefus*, a City formerly fo famous, that *Stephanus Byzantinus* calls it *Epiphaneftate*, at prefent reduced to a miferable Village, inhabited by thirty or forty *Greek* Families, which certainly, as M. *Spon* obferves, *are not capable of underftanding the Epiftle*

St.

St. Paul *writ to them.* The Threatning of our Lord is fulfill'd upon it, *I will remove thy Candleſtick out of its place, except thou repent.* Theſe poor *Greeks* are among old Marbles, and near a fine Aqueduct built of the ſame Stones. The Citadel, where the *Turks* are retired, ſtands upon a little Hill, which ſtretching from North to South, commands the whole Plain; this is perhaps the Mount *Pion* of *Pliny.* The Incloſure of this Citadel, which is fortify'd with many Towers, has nothing magnificent; but ſome paces thence, on the South ſide, one ſees the Remains of another Citadel more antient, much finer, and whoſe Works were cover'd with the fineſt Marbles of antient *Epheſus.*

THERE remains ſtill a Gate of a very good Taſte, built of the ſame Fragments. I don't know for what reaſon it is call'd the *Gate of the Perſecution.* It is remarkable for three Bas-Reliefs upon the Mould; that on the left was the fineſt of all, but it is moſt abuſed. It is about five feet long, and two and a half high, and repreſents a Bacchanal of Children, who roll upon Vine-Branches. That in the middle is one foot higher than the other, and twice as long. The laſt is almoſt as high, but not above four feet long. The *Gate of the Perſecution* turns from the South to the South-South-Eaſt; this Gate was defended by Works which were pretty irregular, which were enlarg'd as there was occaſion, as may be ſeen by the Ruins; for as they tumble down, one ſees other Marble Works which had been cover'd over.

TO the South, and at the foot of the Hill whereon the Caſtle is built, ſtands the Church of *St. John,* converted into a Moſque. I don't know whether it be the ſame which *Juſtinian* cauſed to be built there; but it is certain, that from this great Evangeliſt comes the name of *Aiaſaloue,* under which *Epheſus* is known by the *Greeks* and *Turks.* The *Greeks* call St. *John Aios Scologos,* inſtead of *Agios Theologos,* the *Holy Divine,* becauſe they pronounce the *Theta* as a *Sigma:* from *Aios Scologos* they have made *Aiaſaloue.* The Outſide of this Church has nothing extraordinary. They ſay there are fine Pillars within: but beſides that the fineſt Pieces of the Ruins of *Epheſus* were carry'd to *Conſtantinople* for the Royal Moſques, the *Turk* who keeps the Key was abſent when we were there. 'Tis believed, that after the death of Jeſus Chriſt, St. *John* choſe *Epheſus* for the Place of his Reſidence, and that the Holy Virgin retired thither alſo.

*

St.

The material originally positioned here is too large for reproduction in this reissue. A PDF can be downloaded from the web address given on page iv of this book, by clicking on 'Resources Available'.

St. *John,* after the death of *Domitian,* came to take the Cafe of the Church of *Ephefus,* and found that St. *Timothy,* its firft Bifhop, had been martyr'd there.

THE Aqueduct, which ftill remains to this day, tho half ruin'd, is to the Eaft; it was the Work of the *Greek* Emperors, as alfo the ruin'd Citadel. The Pillars which fupport the Arches are built of very fine Pieces of Marble, intermingled with Pieces of Architecture; and there are Infcriptions which fpeak of the firft *Cæfars.* Thefe Pillars are fquare, higher or lower according as the Level of the Water required, but the Moulds of the Arch are all of Brick. This Aqueduct ferv'd to bring Water to the Citadel and to the City, from the Spring of *Halitee* which *Paufanias* fpeaks of. It was fpread over the City by Brick Pipes or Gutters, made in fmall fquare fhape, and faften'd upon fome one of the Pillars. This City extended itfelf principally to the South, and all this part is full of Ruins; but *Ephefus* has been demolifh'd fo many times, that one can know nothing.

AS for the Infcriptions, we copy'd fome; for befides that we could read but a few, the others are fo high, that it is impoffible to explain them: we can get neither Ladders nor Treffels among the *Greeks.*

THE next day we travers'd the Plain to go and view the Ruins of the famous Temple of *Diana,* which pafs'd for one of the Wonders of the World. This great Edifice was fituate at the foot of a Mountain, and at the head of a Morafs. *Pliny* thinks they chofe that marfhy place, as lefs expos'd to Earthquakes; but at the fame time they enter'd into a vaft Expence, for they muft make Drains to carry off the Water which came down the Hill, and throw it into the Morafs and the *Cayftre.* Thefe Drains or Vaults are what they now unreafonably take for a Labyrinth; by looking into them, one may be fully convinced, that they never were of any other ufe but to carry off the Water. My Opinion is confirm'd by *Philo Byzantinus,* who agrees that they were obliged to make very deep Ditches and Paffages, wherein they ufed fuch a quantity of Stone, that they almoft empty'd all the Quarries in the Country. For the fecuring the Foundation of thefe Conduits or Sewers, which were to bear a Building of fo prodigious a weight, *Pliny* fays they laid Beds of Charcoal well ramm'd, and upon that other Beds of Wool. This wonderful Temple,

ple, built at the charge of the moſt powerful Cities of *Aſia*; two hundred Years before *Pliny* ſpoke of it, was four hundred and twenty five feet long, and two hundred and twenty feet wide.   There were one hundred and twenty ſeven Pillars, at the charge of the Kings of *Aſia*, and theſe Pillars were each ſixty feet high.   Six and thirty of them were cover'd with Bas-Reliefs; and among theſe, one was done by *Scopas* the famous Sculptor.   *Cherſiphron* was the Architect of this Building.   There remains little of it at preſent, but ſome large Pieces, which have nothing extraordinary, except their Thickneſs: the moſt part are of Brick cover'd with Marble, all pierc'd with holes for the Cramps of thoſe Plates of Braſs with which it is believ'd it was adorn'd.   One ſees now among the Ruins only four or five broken Pillars.

THIS was not the firſt Temple the *Epheſians* built in honour of *Diana*. *Dionyſius* the Geographer informs us, that the firſt Temple was a kind of Nich of a ſingular beauty, which the *Amazons*, Miſtreſſes of *Epheſus*, had caus'd to be made in the Trunk of an Elm, where probably the Image of the Goddeſs was placed.   'Twas not doubtleſs of this Temple of the *Amazons* that *Pindar* ſpeaks, when he ſays they caus'd a Temple to be built at *Epheſus*, at the time that they made war with *Theſeus*. *Pauſanias* maintains, that it was the Work of *Crœſus*, and *Epheſus* the Son of *Cayſtre*, and that it was famous before *Nileus* Son of *Codrus* his paſſing into *Aſia*.   This being ſo, the Temple muſt be older than the City; for *Strabo* thinks that *Androclus*, Son of *Codrus*, built it; and *Pauſanias* ſpeaks of the ſame *Androclus*, who drove the *Carians* thence.

THE Temple which that Fool *Heroſtratus* burnt on *Alexander*'s Birthday, was not the ſame with that which was in being in *Pliny*'s time; for *Alexander* would have caus'd it to be rebuilt when he went to *Epheſus*. This great Prince propos'd to the *Epheſians*, that he would freely be at the expence, provided they would put his Name upon the Front of it; but they anſwer'd with a great deal of Politeneſs, *That it was not fit that one God ſhould build Temples to other Deities*.   *Strabo*, who relates this Paſſage, affirms that *Cherſiphron* was indeed the firſt Architect of the Temple of *Diana*, but that another Architect enlarg'd it.   After it was burnt by *Heroſtratus*, the *Epheſians* not only ſold the Pillars which had been uſed in the former Temple, but likewiſe all the Jewels of the Ladies of the

†

City were turn'd into Money, and this Money employ'd in building an Edifice much finer than that which had been burnt. *Cheiromocrates* was the Architect; 'twas he that built the City of *Alexandria*, and who would have made Mount *Athos* into a Statue of *Alexander*. In this Temple were to be feen Performances of the moft famous Sculptors of *Greece*. The Altar was almoft wholly the Work of *Praxiteles*. *Strabo* fpeaks of it, as having feen it in *Auguftus*'s time; and its Privilege of Afylum reach'd to one hundred and twenty five feet about it. *Mithridates* enlarged it to a Bow-fhot. *M. Anthony* doubled this diftance, and took in part of the City; but *Tiberius*, to prevent the Abufes committed on account of thefe fort of Privileges, abolifh'd them at *Ephefus*. They don't exprefs the Afylum upon the Medals of this City, till after the Emperor *Philip* the Elder had been there, and then only upon that of *Otacilla*; the Reverfe reprefented *Diana* of *Ephefus* with her Attributes, the Sun on one fide, and the Moon on the other. We have a Medal of *Philip* the Younger with the fame Reprefentation, but the Legend is different. That which was ftamp'd with the Head of *Etrufcilla*, reprefents *Diana* with her Attributes and Stags; the Legend is the fame with that upon the Medal of *Otacilla*. As for the coming of *Philip* to *Ephefus*, it is mark'd upon a Medal of that Emperor, the Reverfe whereof is charg'd with a Ship which is carry'd along with Oars and Sails.

IN the time of *Herodotus*, the City of *Ephefus* was at a diftance from the Temple of *Diana*; but this Author fays nothing of the Statue of Gold which was fet up there, according to *Xenophon*. *Strabo* affirms that the *Ephefians*, in acknowledgment, had made in their Temple a Statue of Gold to *Artemidorus*. *Syncellus*, who fays this Temple was burn'd, probably fpeaks of a burning which did no more damage than what might be repair'd without altering the whole; and fo the Temple *Pliny* defcribes, was the fame which *Strabo* faw. The fame Temple was rifled and burnt by the *Scythians* in the Year 263. The *Goths* plunder'd it under the Emperor *Gallienus*. We have feveral Medals, on the Reverfe of which the Temple is reprefented with a Frontifpiece fometimes of two Pillars, of four, of fix, and even of eight, with the Heads of the Emperors *Domitian, Adrian, Antoninus Pius, M. Aurelius, Lucius Verus, Septimius Severus, Caracalla, Macrinus, Heliogabalus, Alexander Severus, Maximinus*.

BESIDES

BESIDES the Bas-Reliefs and the Statues, this Temple muſt have been adorn'd with wonderful Paintings; for *Apelles* and *Parrhaſius*, the two moſt famous Painters of Antiquity, were of *Epheſus*. About the Ruins of this Temple, are to be ſeen the Fragments of divers Houſes built of Brick, in which perhaps dwelt the Prieſts of *Diana*, who often came from far to be honour'd with this Dignity. To them was committed the Care of the Virgin Prieſteſſes, but not till they were made Eunuchs. There are few Cities, of which there remain ſo many Medals. Some inform us, that it was three times *Neocore* of the *Cæſars*, and once of *Diana*. Others, that it was built on occaſion of a Wild-Boar. Some prove that the·Citizens call'd themſelves *the firſt People of Aſia*. Moſt of theſe Pieces repreſent *Diana*, or a Huntreſs, either with ſeveral Breaſts, or ſet out with her Attributes.

ONE ſees now no more fine Ruins at *Epheſus*, thoſe which remain are very ſcarce. The Fragments of ſome Caſtles built with Marble, ſhew nothing worthy of the antient City. I have cauſ'd to be grav'd a Port which is to the left of the Road of *Scalanova*. The Mould of the Arch, which is good, is not proportion'd to the Shafts which ſupport it, for it makes more than a Semicircle; the Frizes are cut very handſomly, and upon the Remainder of this Building we read within and without the part of an Inſcription which I here give you: it is in *Roman* Characters, but we don't comprehend what they can mean.

A C C E N S O
R E N S I E T A S I Æ.

THE Daffodils with yellow Flowers, a ſtrait Stalk, and without Indentings, ſhine among ſeveral other rare Plants.

THE Caſtle, which they call *the Priſon of St.* Paul, is not antient, and was never fine. The Grotto of the *Seven Sleepers* might deſerve to be view'd, if one could be aſſured of the Truth of the Story. As we go out of the Ruins of the Temple, we enter upon an ugly Moraſs, full of Ruſhes and Reeds, which empties itſelf into the *Cayſtre*. On the other ſide that River is a very muddy Lake; perhaps it ſeem'd ſo to us, becauſe of the great Rains which had fallen: this muſt be the Lake of

Seli-

The Ruines of an antient Building of Marble at Ephesus.

The Gate of the Persecution at Ephesus.

Selinufia, mention'd by *Strabo*.   As we go to the Port, we fee upon the
Banks of the River a great many antient Ruins and old Marbles.   This
was properly that part of *Ephefus* which *Lyfimachus* built, and where the
Arfenals were, which *Strabo* fpeaks of.   They pafs the *Cayftre* fome paces
beyond, in a Ferry-boat with a Rope, to go from *Scalanova* to *Smyrna*,
without coming over the Bridge.   'Tis the antient way from *Ephefus* to
Smyrna, for it is the fhorteft, and *Strabo* fays they went in a direct Line
from one City to the other; it is at prefent the moft hazardous way.

NOTWITHSTANDING the Plain of *Ephefus* be fine, the
Situation of *Smyrna* has fomething in it more grand; and the Hill,
which is at the bottom of the Gulph, is like an Amphitheatre defign'd to fhew a fine City, whereas *Ephefus* lies in a hollow.   Moreover,
tho this City has been the Seat of the *Roman* Conful, and the Rendevouz of Strangers who went into *Afia*, its Port was never comparable to
that of *Smyrna*.   This of *Ephefus*, on account of which they have ftruck
fo many Medals, is nothing but an open Road expos'd to Dangers; at
prefent 'tis not much frequented.   Formerly the Veffels ran up into the
very River, but the Mouth of it has been fince fill'd up with Sand.

NOTHING is more tirefome, than to fearch in the antient Books
for the Founders of *Ephefus*.   What is it to us to know how it was call'd
in the time of the *Trojan* War? or whether it took its Name from *Ephefus*, Son of *Cayftre*, and the *Amazon Ephefe*?   'Tis hardly of any more
confequence, to know whether it be the Work of the *Amazons*, or of
Androclus, or of one of the Sons of *Codrus* King of *Athens*: this can
only ferve to clear up a Paffage in *Syncellus*, where he fays, that it was
Andronicus, inftead of *Androclus*, who built *Ephefus*.   Who will trouble
himfelf to know whether there was one Quarter in *Ephefus* call'd *Smyrna*?
this fort of Learning is of no ufe to us.   But it is pleafant to remember,
that during the Wars of the *Athenians* and *Lacedemonians*, *Ephefus* was fo
politick as to keep a good Underftanding with the ftrongeft fide: That on
Alexander's Birth-day, the Soothfayers of this City began all to cry out,
that the Deftroyer of *Afia* was come into the World: That *Alexander* the
Great, on whom the Prophecy fell, came to *Ephefus* after the Battel of
Granicus, and there eftablifh'd a Democracy: That the Place was taken

by *Lyſimachus*, one of his Succeſſors: That, in fine, *Antigonus* in his turn had poſſeſſion of it, and there ſeiz'd the Treaſures of *Polyſper-chon*.

CAN one be ignorant that *Hannibal* had an Interview with *Antiochus* at *Epheſus*, to concert Meaſures againſt the *Romans?* That the Procon-ful *Manlius* ſpent the Winter there, after the Defeat of the *Galatians?* All theſe Events renew the great Ideas we have of the antient Hiſtory. Nothing is more terrible than the Maſſacre of the *Romans* in this City, by the order of *Mithridates*. *Lucullus* made great Feaſts at *Epheſus*. *Pompey* and *Cicero* did not fail to ſee this famous City. *Cicero* made no ſtep in *Greece*, without finding new Subjects of Admiration. *Scipio*, the Father-in-Law of *Pompey*, had leſs reſpect for *Epheſus*, for he ſeiz'd the Treaſures of the Temple; but nothing is ſo comfortable to Chriſtians, as to follow St. *Paul* to *Epheſus*. *Auguſtus* honour'd this Place with one of his Viſits, and they built there Temples to *Julius Cæſar* and the City of *Rome*. *Epheſus* was rebuilt by the Care of *Tiberius*. On the other ſide, the *Perſians* plunder'd it in the third Century, and the *Scythians* did not ſpare it ſome time after. There is a great deal of probability that the famous Temple of *Diana* was deſtroy'd under *Conſtantine*, in conſequence of the Edict by which that Emperor commanded to demoliſh all the Tem-ples of the Heathens.

EPHESUS was a Place too conſiderable not to be expoſed in its turn to the Ravages of the *Mahometans*. *Anna Comnena* relates, that the Infidels having render'd themſelves maſters of *Epheſus* under the Reign of her Father *Alexis*, he ſent thither *John Ducas* his Father-in-Law, who de-feated *Tangriperme* and *Marace* the *Mahometan* Generals. The Battel was fought in the Plain below the Citadel; by which it appears that the fineſt part of the City was deſtroy'd for that time. The Chriſtians had the advantage; they took two thouſand Priſoners, and the Government of the Place was given to *Petzeas*. The Citadel of which *Comnena* ſpeaks, was probably the antient abandon'd Marble Caſtle. *Theodorus Laſcaris* made himſelf maſter of *Epheſus* in 1206. The *Mahometans* return'd thi-ther under *Andronicus Paleologus*, who began to reign in 1283. *Manta-chias*, one of their Princes, conquer'd all *Caria*; and *Homur*, Son of *Atin*

Prince

*

Prince of *Smyrna*, ſucceeded him. *Tamerlane*, after the Battel of *Angora*,
commanded all the leſſer Princes of *Anatolia* to come and join him at E-
pheſus, and employ'd a whole Month in plundering the City and its Neighbourhood. *Ducas* ſays that all was drain'd away, Gold, Silver and Jewels; they took even their very Clothes. After the Departure of the Conqueror, *Cineites* a great *Turkiſh* Captain, Son of *Caraſupaſi*, who had been
Governour of *Smyrna* under *Bajazet*, declar'd War againſt the Children of
Atin, who had ſettled at *Epheſus*. He immediately ravag'd the Country,
at the head of five hundred Men: afterward he came before the Citadel with a greater Number of other Troops, and eaſily gain'd it; but ſome
time after, another Son of *Atin*, who was called *Homur*, (the Name of his
Brother who was juſt dead) join'd himſelf to *Mantachias* Prince of *Caria*,
who accompanied him to *Epheſus* with an Army of ſix thouſand Men.
Caraſupaſi, Father of *Cineites*, commanded in the City where this ſame
Cineites, who was at *Smyrna*, had left but three thouſand Men. Notwithſtanding the vigorous Defence made by the *Epheſians*, the Beſiegers ſet
fire to the City, and in two days time, all that had eſcaped the Fury of
the *Tartars*, was reduc'd to Aſhes. *Caraſupaſi* being retir'd to the Citadel, bore the Siege till Autumn; but his Son not being able to ſuccour
him, he ſurrender'd to *Mantachias*, who return'd the Country of *Epheſus*
to *Homur*, and ſhut up *Caraſupaſi* and his principal Officers in the Caſtle
of *Mamalus*, on the Borders of *Caria*. Then *Cineites* went from *Smyrna*
with a Galley, and gave his Father notice of his Arrival at *Mamalus*.
The Priſoners made the Guards drink ſo much till they were drunk;
and then taking the advantage of this Device, they let themſelves down
by Ropes, and eſcap'd to *Smyrna*. At the beginning of the Winter they
undertook the Siege of *Epheſus*. *Homur* in his turn retir'd to the Citadel. The City was deliver'd to the Soldiers; they committed there all
manner of Wickedneſs and Cruelty. In the midſt of ſo many Misfortunes, *Cineites* reconcil'd himſelf with *Homur*, and gave him his Daughter in Marriage. *Epheſus* afterwards fell into the hands of *Mahomet* I.
who having overcome not only all his Brothers, but alſo all the *Mahometan*
Princes who embaraſs'd him, remain'd peaceable Poſſeſſor of the Empire.

pire. From that time *Ephefus* has remain'd to the *Turks* ; but its Trade has been carried to *Smyrna* and *Scalanova*.

WE departed from *Ephefus* the 27th of *January* to go to this laſt Place, which the *Turks* call *Coufada*, and the *Greeks Scalanova*, an *Italian* Name, which the *Franks* gave it perhaps after the Deſtruction of *Ephefus*. What is obſervable in the Change of the Name is, that it anſwers to the antient Name of this City, which is the *Neapolis* of the *Milefians*. Notwithſtanding a very great Rain, we arriv'd in three Hours. When we are near the Ruins of the Temple of *Ephefus*, we muſt go directly to the South, then to the South-Eaſt, to gain the Sea. Thence we take to the Left at the foot of ſome Hills, where ſtands the Priſon of *St. Paul*, leaving to the Right the Morafs, which empties itſelf into the *Cayſtre*. This way is very narrow in many Places, by means of the River which winds, and comes beating againſt the foot of the Mountains ; after which it runs directly into the Sea. One can hardly diſcern the Way becauſe of the great Quantity of *Tamarisk* and *Agnus Caſtus*. The Road of *Ephefus* is terminated in this Place, which is to the South-Weſt, by a Cape which muſt be left on the Right, and upon which one muſt go to take the way to *Scalanova*. At length we come to the Shore, from whence we diſcover'd the Cape of *Scalanova*, which advances much farther into the Sea. Two Miles on this ſide this City we paſs thro the Breach of a great Wall, which, as they pretend, ſerv'd for an Aqueduct to carry the Water to *Ephefus* ; but there are no Arches. One ſees however the Continuation of the Wall, which approaches to the City, round the Compaſs of the Hills. The Avenues to *Scalanova* are made very pleaſant by the Vineyards. They drive there a conſiderable Trade in Red and White Wines, and dried Raiſins ; they likewiſe prepare there a great many Goats Skins, or what we call *Spaniſh* Leather.

SCALANOVA is a very handſome City, well built, well pav'd, and cover'd with hollow Tiles like the Roofs in our Cities in *Provence*. Its Circumference is almoſt ſquare, and ſuch as the Chriſtians built it. There live only *Turks* and *Jews*. The *Greeks* and *Armenians* inhabit the Suburbs only. You ſee a great many old Marbles in this City.

THE

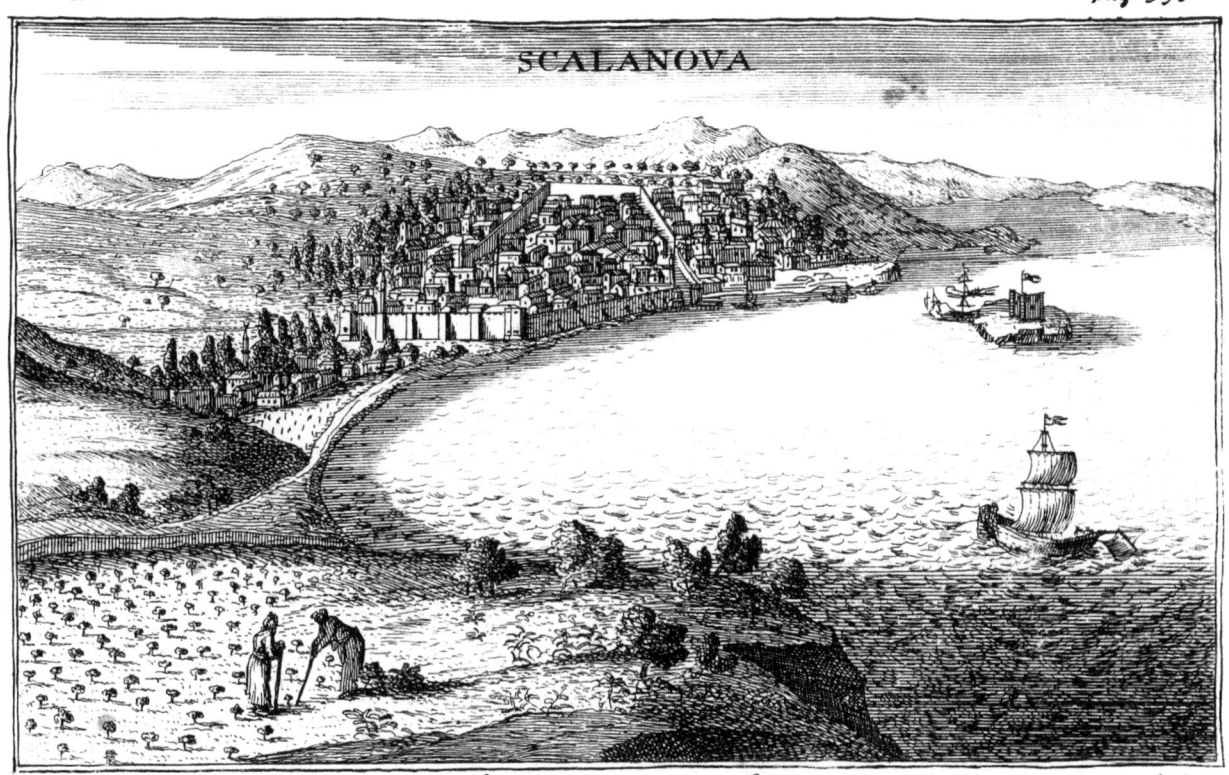

SCALANOVA

A View of Scalanova near Smyrna

THE Church of St. *George of the Greeks* is in the Suburbs, upon the Brow of a Hill which encompaffes the Port; over-againft it is Shelf on which they have built a fquare Caftle, where they keep a Garifon of twenty Soldiers. The Port of *Scalanova* is a Station for the Navy, and looks towards the Weft and North-Weft. There are about a thoufand Families of *Turks* in this City, fix hundred Families of *Greeks*, ten Families of *Jews*, and fixty of *Armenians*. The *Greeks* have there the Church of St. *George*, the *Jews* a Synagogue, the *Armenians* have no Church there. The Mofques there are fmall. They maintain in and about the City not above one hundred Janizaries. Their Trade is not confiderable, becaufe they are prohibited loading any Goods for *Smyrna*; fo that they only load Corn and Kidney-Beans. There is in this Place a Cadi, a Difdar, and a Sardar. They reckon it but one Day's Journey to *Tyre*, as much to *Guzetliffar*, or *Fine Caftle*, which is the famous *Magnefia*, upon the *Meander*, one Day's Journey and a half from the Ruins of *Miletum*.

THE 25th of *March*, in returning from *Samos*, we went from *Scala-nova* to *Ephefus*. The next Day we departed to return to *Smyrna*, and we lay that Day at *Tourbale*, which is fix Hours from *Smyrna*. *Tour-bale* is a poor Village, in which we fee feveral old Marbles, which pleafe Strangers, for otherwife the *Turks* who inhabit it are not very civil. One fees alfo in the Caravanfera Pillars of Granat or white Marble. Three Miles from *Tourbale*, at the foot of the Mountain, near a Burying-place, are the Fragments of an antient City, but we met with nothing whence we might learn its Name. All this Part is full of *Leon-topetalon*, and *Anemonies* of a bright fhining Fire-Colour. We found nothing to eat at *Tourbale* but *Dora* Bread, which is very heavy without being very unpleafant. The 27th we arriv'd at *Smyrna*, where we ftaid waiting an Opportunity to embark.

MAUNDY-THURSDAY, the 13th of *April* 1702, we fet fail with the Wind at South-Eaft, in the Ship call'd the *Golden Sun*, commanded by Captain *Laurent Guerin* of *la Cioutad*, carrying fix Pieces of Iron Cannon, and eight Patereroes: It was laden with Silk, Cotton, Goat's Hair, and Wax for *Leghorn*. The Veffel was of about 6000 Quintals. After

forty

forty Days Sail, in which time we had endured great Storms and contrary Winds, which oblig'd us to take in Refreshments at *Malta*, we arrived at *Leghorn* the 23d of *May*, and went into the *Lazaret*. The 27th we came out of the *Lazaret*, and embark'd on a Felucca, which brought us to *Marseilles* the 3d of *June*, being the Vigil of Pentecost, where we return'd Thanks to God, that he had preserv'd us thro the Course of our Journey.

I am, My LORD, &c.

F I N I S.

An ALPHABETICAL TABLE of the Principal Matters in both Volumes.

[*N. B.* The Letters ſhew the Volume, the Figures the Pages of each.]

An Alphabetical Table.

An Alphabetical Table.

An Alphabetical Table.

Chap-

An Alphabetical Table.

Corn,

An Alphabetical Table.

An Alphabetical Table.

Gala-

An Alphabetical Table.

An Alphabetical Table.

rini

An Alphabetical Table.

An Alphabetical Table.

An Alphabetical Table.

An Alphabetical Table.

St.

An Alphabetical Table.

*

Its

An Alphabetical Table.

An Alphabetical Table.

† *At*

An Alphabetical Table.

Their

An Alphabetical Table.

The End of the Table.

For EU product safety concerns, contact us at Calle de José Abascal, 56–1°,
28003 Madrid, Spain or eugpsr@cambridge.org.

www.ingramcontent.com/pod-product-compliance
Ingram Content Group UK Ltd.
Pitfield, Milton Keynes, MK11 3LW, UK
UKHW051425240426
470322UK00020B/623